Enzymes and Immobilized Cells in Biotechnology

BIOTECHNOLOGY

JULIAN E. DAVIES, *Editor*
Biogen, S.A.
Geneva, Switzerland

BIOTECHNOLOGY SERIES

Enzymes and Immobilized Cells in Biotechnology

Edited by

Allen I. Laskin
New Jersey Center for
Advanced Biotechnology and Medicine
Piscataway, New Jersey

1985
THE BENJAMIN/CUMMINGS PUBLISHING COMPANY, INC.
Advanced Book Program
Menlo Park, California

LONDON • AMSTERDAM • DON MILLS, ONTARIO • SYDNEY • TOKYO

7269-1955

CHEMISTRY

Library of Congress Cataloging in Publication Data

Main entry under title:

Enzymes and immobilized cells in biotechnology.

    Bibliography: p.
    Includes index.
    1. Immobilized enzymes — Industrial applications.
I. Laskin, Allen I.
TP248.E5E584   1985   660'.63   84-21457
ISBN 0-8053-6360-2

abcdefghij-MA-8987654

CONTRIBUTORS

Harold E. Swaisgood
Department of Food Science
North Carolina State University
School of Agriculture and Life Sciences
Raleigh, NC 27650

Pekka Linko
Helsinki University of Technology
Laboratory of Biotechnology and Food Engineering
SF-02150 Espoo, Finland

Ichiro Chibata
Research & Development Headquarters
Tanabe Seiyaku Co., Ltd.
16–89, Kashima-3-chome,
Yodogawa-ku, Osaka, Japan

Tetsuya Tosa and Tadashi Sato
Research Laboratory of Applied Biochemistry
Tanabe Seiyaku Co., Ltd.
16–89, Kashima-3-chome
Yodogawa-ku, Osaka, Japan

Joachim Klein
Institut für Technische Chemie
Technische Universität Braunschweig
Hans-Sommer-Strasse 10; D-3300
Braunschweig, West Germany

Sidney A. Barker and Graham S. Petch
Department of Chemistry
The University of Birmingham
P.O. Box 363
Birmingham B15 2TT England

Peter Brodelius
Institut für Biotechnologie
Eldg. Technische Hochschule Zürich
ETH-Hönggerberg
CH-8093 Zürich, Switzerland

Atsuo Tanaka and Saburo Fukui
Laboratory of Industrial Biochemistry
Department of Industrial Chemistry
Faculty of Engineering
Kyoto University
Yoshida, Sakyo-ku, Kyoto 606, Japan

vii

Christian Wandrey and R. Wichmann
Institut für Biotechnologie der Kernforschungsanlage
Postfach 1913
D-5170 Jülich 1 West Germany

Isao Karube and Shuichi Suzuki
Research Laboratory of Resources Utilization
Tokyo Institute of Technology
Nagatsuta Campus
4259 Nagatsuta, Midori-ku, Yokohama, 227 Japan

George G. Guibault and Graciliano de Olivera Neto
Instituto de Quimica
Cidade Universitaria
Universidade de São Paulo, Brazil

Thomas M. S. Chang
Artificial Cells and Organs Research Centre
McGill University
3655 Drummond Street
Montreal, PQ, Canada H3G 1Y6

John Geigert and Saul L. Neidleman
Cetus Corporation
1400-53rd Street
Emeryville, CA 94608

Garfield P. Royer
Standard Oil Company (Indiana)
Amoco Research Center
P.O. Box 400
Naperville, IL 60566

CONTENTS

In the late 1960's and early 1970's a great deal of enthusiasm developed for the potential of enzyme technology, especially centered around extensive developments in the immobilization and stabilization of enzymes. A few early successes fueled the enthusiasm and stimulated much additional research activity. There followed a period of relative quiet. More recently, however, the rapid and widespread rise in interest in biotechnology, resulting from the revolutionary developments in molecular genetics and related disciplines, has rekindled new and expanded interests in the subdiscipline of enzyme technology.

During this period a number of volumes on methods of enzyme immobilization, enzyme technology, enzyme engineering, and related subjects have appeared. This present volume therefore makes no attempt to be comprehensive but presents a spectrum of new contributions from many of the most active researchers in the field, covering most of the areas of current interest. Thus there are chapters that deal with both immobilized enzymes and cells, including plant cells; chapters on some of the more established enzyme technologies (e.g., high fructose corn syrup, 6-APA production); chapters on specific applications (e.g., foods, analytical, diagnostic, clinical), and still others on newer enzymes and on enzyme-like molecules ("synzymes"). It is hoped that this selection of topics will provide an appreciation of the status, scope, and potential of the use of immobilized enzymes and cells for biotechnological applications.

Somerset, New Jersey   **Allen I. Laskin**

1

Immobilization of Enzymes and Some Applications in the Food Industry

Harold E. Swaisgood

I. INTRODUCTION

Although the immobilization of an enzyme was first reported nearly 70 years ago (Nelson and Griffin, 1916), most of the development in this area of biotechnology has occurred in the last two decades. It is fitting that invertase, a food-related enzyme, was the first to be immobilized, although its commercial use is still not anticipated. Nevertheless, since food is a biological material whose nutrition, flavor, aroma, and texture must be closely guarded during processing, its chemical manipulation by biological rather than chemical catalysts is quite naturally preferred. Thus the specificity afforded by enzyme catalysts permits selective changes that enhance a food's flavor, texture, and the like, without sacrificing its nutritional value or causing other undesirable changes resulting from "side reactions".

In view of such considerations it is not surprising that most of the present commercial applications of immobilized enzymes occur in the food industry. This is true even though the industry is rather conservative and operates with a relatively small profit margin. However, the commercialization of new technology is slow, and few new commercial processes have been implemented since the last reviews of this field (Olson and Cooney, 1974; Brodelius, 1978;

1

Pitcher, 1980; Chibata, 1978; Weetall, 1977; Sweigart, 1979). Accordingly, this discussion will focus on a brief consideration of the principal factors affecting the use of immobilized enzymes prior to listing their current and some potential applications in the food industry.

II. FACTORS INFLUENCING UTILIZATION OF IMMOBILIZED ENZYMES

A. Economic Factors

An economic incentive currently exists for conversion of labor-intensive processes into more capital-intensive operations that permit continuous, more automated production with better quality control and reduced plant size (Roland, 1980). As population growth in Western nations slows, industrial growth based on volume production will be limited. Thus growth potential will arise principally from technical advancement in production methods and development of new products. These considerations account for the incentives for development and use of immobilized enzyme technology.

Some of the primary factors affecting the cost of using an immobilized enzyme process are (1) enzyme cost, (2) immobilization cost, (3) system performance, (4) capital investment, and (5) cleanup costs. Enzyme cost is related to both the availability of an enzyme and its cost for isolation. The cost for immobilization is governed by labor and chemical costs and thus is related to the complexity of the method used. However, these factors must all be weighed against parameters pertaining to a system's performance. As illustrated in Fig. 1.1, system performance parameters can be classified into three main areas. This analysis emphasizes the obvious trade-off between degree of purification (enzyme cost) and the catalyst's specific activity, which determines reactor size. Likewise, the interrelationship between immobilization costs and catalyst specific activity and stability must be optimized. Enzymes immobilized covalently are less subject to leaching from the matrix, but the immobilization costs are usually greater than that for adsorption methods. No matter what method or synthetic matrix is used for immobilization, it is almost essential economically

FACTORS EFFECTING THE COST OF
AN IMMOBILIZED ENZYME PROCESS
- ● CATALYST SPECIFIC ACTIVITY
  - EFFECTIVENESS OF THE IMMOBILIZATION PROCEDURE
  - LOADING
- ● CATALYST STABILITY
  - STORAGE STABILITY
  - OPERATIONAL STABILITY
- ● REGENERATIVE CAPACITY OF THE CATALYST
  - CAPABILITY OF ACTIVITY REGENERATION
  - CAPABILITY OF SUPPORT REUSE

FIGURE 1.1   Items that must be optimized to minimize the cost of an immobilized enzyme process.

that the support can be reclaimed and reused following decay of catalytic activity. The obvious exception is, of course, when the living cells themselves provide the matrix in which the enzyme is immobilized.

Capital investments include the reactor vessel for containment of the immobilized enzyme, equipment for automated control of the process, and the enzyme support. These costs usually represent a small fraction of the long-term total cost per unit volume of product. However, it is a significant factor, especially where existing technology successfully accomplishes a similar result. For example, this factor may account for the lack of development of immobilized glucoamylase for industrial use, since a low-cost soluble enzyme process is already established (Reilly, 1980). It appears that a major improvement in the stability of immobilized forms of glucoamylase permitting long-term operation at 55–60°C will be necessary before its commercial use becomes attractive.

The cleanup cost is usually related to the stability of the immobilized enzyme. For example, if the feed stream being processed contains inhibitors and/or microbial organisms, these may have to be removed prior to contact with the enzyme. The half-life of immobilized lactase is substantially improved by deproteinization and demineralization of the feed stream (Coughlin and Charles, 1980).

Another, perhaps less obvious, economic factor pertains to how well the process accomplished with the immobilized enzyme integrates into the total operations and products of a processing plant. This may be illustrated best by the integration of immobilized lactase for whey processing in a cheese plant (Roland, 1980). Thus a complete whey recovery and utilization program (1) reduces costs for waste water treatment, (2) allows production of whey proteins that have high nutritional and functional qualities, and (3) by treatment with immobilized lactase, permits production of a concentrated syrup from whey permeate that is comparable to corn syrups. While any one of these processes alone would not be economically feasible, taken together and with large volumes the recovery becomes profitable. As an example of the significance of various cost factors for an immobilized enzyme process, the relative costs estimated for use of immobilized lactase with a half-life of 100 days are shown in Fig. 1.2.

B. Type of Support Matrix

Some of the support matrices that have been used for immobilization of enzymes are listed in Table 1.1. Key factors that must be considered in choosing a support are (1) matrix stability, (2) matrix compatibility with the enzyme, and (3) matrix compatibility with the substrate and the material being processed. The support matrix must be stable during processing to prevent leaching of that enzyme and, most importantly in food operations, to prevent adulteration of the product. Even the most stable matrix may exhibit some leaching, so it is unlikely that any support having suspected carcinogenic properties would be

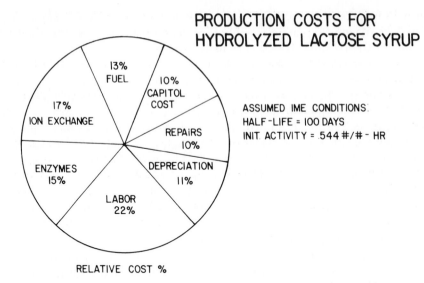

PRODUCTION COSTS FOR HYDROLYZED LACTOSE SYRUP

ASSUMED IME CONDITIONS:
HALF-LIFE = 100 DAYS
INIT. ACTIVITY = .544 #/# - HR

RELATIVE COST %

FIGURE 1.2   Relative costs for the various factors in the cost of production of hydrolyzed-lactose syrup by an immobilized enzyme process. (From Roland, 1980.)

acceptable. For this reason, biopolymer or inorganic supports are more attractive for food processing uses than some of the synthetic organic matrices.

Porous rather than nonporous supports have been used almost exclusively, since these materials possess a much greater surface area per unit volume and thus allow preparation of an immobilized catalyst with a much higher specific activity. To use the matrix effectively, however, it is imperative that the en-

TABLE 1.1   Types of Support Matrices Used for Immobilization of Enzymes

1. Organic supports
   - Polystyrene
   - Nylon
   - Phenol-formaldehyde resins
   - Acrylic copolymers (e.g., polyacrylamide)
2. Biopolymer supports
   - Cellulose
   - Polydextrans (Sephadex)
   - Agarose
   - Collagen
   - Chitin
3. Inorganic supports
   - Glass beads (porous and nonporous)
   - Stainless steel
   - Metal oxides (porous ceramics, e.g., ZrO_2, TiO_2, Al_2O_3, NiO)
   - Sand

zyme be capable of penetrating all of the pore volume. This principle is illustrated by results for immobilization of sulfhydryl oxidase on porous glass (Janolino and Swaisgood, 1978). As shown by the data in Fig. 1.3, the loading for this preparation of membrane-associated enzyme actually increases with increasing pore size, although the surface area is decreasing.

The size of the substrate molecule is an important consideration in selecting the matrix. For example, proteinases entrapped in gels or coupled to small pore supports may be highly active toward synthetic small substrates and yet have only limited or no activity toward proteins. The net charge on the substrate may also be a factor if the enzyme is attached to an ionic support, especially when operated at low ionic strength. In general, the diffusivity of the substrate in the matrix, the possible partitioning of substrate or product molecules due to attractive or repulsive interactions with the matrix, and the potential establishment of pH gradients in ionic matrices must all be taken into account in choosing a matrix.

C. Method of Immobilization

There are three principal methods for containment of an enzyme in a reactor: entrapment, adsorption to a solid, and covalent attachment to a solid. These

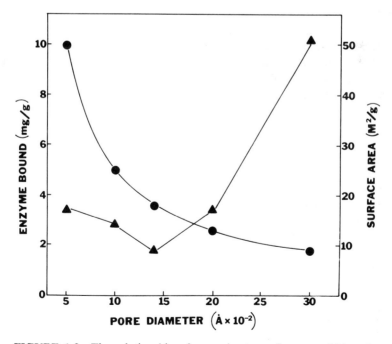

FIGURE 1.3   The relationship of pore size to surface area (●) and to amount of enzyme bound (▲) per gram of glass beads. (From Janolino and Swaisgood, 1978.)

methods are summarized in Table 1.2 and have been extensively reviewed (Mosbach, 1976; Goldstein and Manecke, 1976). Enzymes immobilized by entrapment are more subject to diffusion inhibition, since by definition the matrix or membrane pore size must be sufficiently small to prevent enzyme loss by diffusion. The key distinction is that the substrate and product must be able to diffuse through the barrier, whereas the enzyme must not. Enzymes retained by microencapsulation (Chang, 1976) or with a semipermeable membrane (Chambers et al., 1976) are, in effect, soluble enzymes; hence these methods do not permit potential structural stabilization by matrix–enzyme interactions. Solution polymerization has been used to entrap enzymes or whole cells in polyacrylamide gels (Larsson and Mosbach, 1976), fibers (Dinelli et al., 1976), and calcium alginate (Kierstan and Bucke, 1977). Of more commercial significance, enzymes can be immobilized by cross-linking the whole cells containing the enzyme, for example, with glutaraldehyde (Poulsen and Zittan, 1976).

Although an enzyme immobilized by adsorption to a solid is more subject to leaching than enzyme covalently immobilized, adsorptive methods are commercially attractive because initial loading and reloading following activity decay is so simple. In general, enzymes immobilized by adsorption are more intimately associated with the matrix, as required by the need for tenacious binding to prevent loss, than are covalently attached enzymes, which can be separated from the matrix by a "tethering" chain. Thus the physical and chemical characteristics of the adsorptive matrix can profoundly affect enzyme activity. Use of ion exchange materials as an adsorbent is an obvious choice, and since most enzymes have a net negative charge in the pH range of 7–9, anion exchangers such as DEAE-cellulose or DEAE-Sephadex are often used (Pitcher, 1980). Many enzymes, especially membrane-associated enzymes, have "hydrophobic patches" on their surface and consequently will tenaciously bind to

TABLE 1.2   Methods Most Commonly Used for Immobilization of Enzymes

1. Entrapment
   - Gels (organic and biological polymers)
   - Vesicles – microencapsulation
   - Semipermeable membranes
2. Adsorption
   - Ionic adsorption
   - Hydrophobic adsorption
3. Covalent bonding (reactive functional group)
   - Enzyme amino groups; activating reagents: water-soluble carbodiimides, glutaraldehyde, cyanogen bromide, N-hydroxysuccinimide esters, triazines, cyanuric chloride, carbonyl diimidazole.
   - Enzyme carboxyl groups; activating reagents: water-soluble carbodiimides, N-ethyl-5-phenylisoxazolium-3-sulfonate.
   - Enzyme tyrosyl residues; activating reagents: diazonium compounds.
   - Enzyme sulfhydryl groups; activating reagents: dithiobis-5,5′-(2-nitrobenzoic acid), glutathione-2-pyridyl disulfide.

polymers with long alkyl side chains, for example, hexyl- or octyl-derivatives of agarose (Porath and Caldwell, 1977; Cashion et al., 1982). Enzymes are also adsorbed, although usually less strongly than to ion exchangers, to ceramics, or to glass (Pitcher, 1980). Recently, a spherical porous glass bead (Spherosil) developed for this purpose has shown some useful characteristics.

Covalent attachment to the solid provides the most secure method for immobilization. If activity is to be maintained, attachment must occur through a nonessential functional group on the surface of the protein molecule. The more abundant the reactive group on the surface, the less likely that reaction will occur with an essential residue. Consequently, carboxyl or amino groups of enzymes often have been the targeted functional groups for attachment (Srere and Uyeda, 1976). As is indicated by a partial list in Table 1.2, a multitude of reagents have been used to activate groups on the matrix or, in some instances, groups on the enzyme's surface. Water-soluble carbodiimides can be used to attach through either the enzyme's carboxyl or amino groups and, in the latter case, in the presence or absence of the carboxyl-activating reagent during incubation of the soluble enzyme with the matrix (Janolino and Swaisgood, 1982). Health-hazard criteria for use as a food processing aid severely limits the type of reagents that can be considered for covalent immobilization. It appears that, at present, glutaraldehyde is the only reagent employed commercially for preparation of covalently immobilized enzymes.

III. REACTOR DESIGN

A. Types of Reactors

Numerous types of reactors for containment of the immobilized enzyme during processing of a liquid have been designed (Fig. 1.4). The type of reactor chosen for an application must be compatible with the characteristics of the fluid being processed and the kinetic properties of the enzyme. The simplest design is a batch reactor; however, this type is not the most efficient, and it does not accommodate automation, which is usually considered as one of the advantages of using an immobilized enzyme process. The other reactors shown in Fig. 1.4 are all variants of continuous systems, the characteristics of which have been reviewed in some detail (see, for example, Lilly and Dunnill, 1976; Vieth et al., 1976; Pitcher, 1980). The continuous systems can be generally classified as three types: (1) stirred tank reactors, (2) fluidized- or expanded-bed reactors, or (3) fixed-bed reactors. A number of factors should be considered in the choice of a reactor type. The major factors are discussed briefly below.

Effect of back-mixing on reactor efficiency.   Kinetically, there are two extremes: that characterized by complete back-mixing, as exemplified by an ideal continuous stirred tank reactor (CSTR), and that characterized by zero back-mixing, as exemplified by an ideal fixed-bed plug-flow reactor (PFR). In an

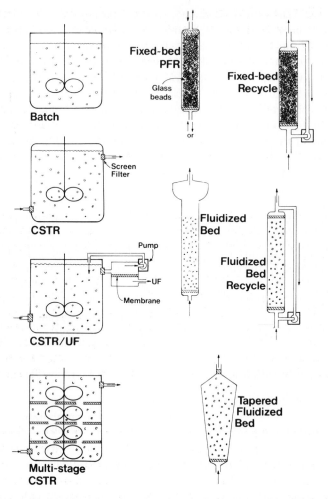

FIGURE 1.4   Types of reactors for containment and use of immobilized enzymes.

ideal CSTR, all of the enzyme is operating at the exit stream substrate concentration. However, the fluid velocity profile in an ideal PFR is constant through the cross-section (that is, fluid flows as a plug through the reactor); consequently, the enzyme is exposed to substrate concentrations ranging from that in the inlet stream to that in the outlet stream. The relative efficiencies for these two extremes depend on the feed stream substrate concentration, more specifically $[S]/K_m$, and the desired fractional conversion of substrate. In kinetic terms, relative efficiency depends on the substrate concentration dependence of the rate, which, for most enzyme-catalyzed reactions, varies between zero and first order. For operation approaching zero-order conditions, that is, large values of $[S]/K_m$ and fractional conversions less than 0.9, the effect of back-mixing does not greatly reduce the efficiency. However, for operating conditions that result in kinetics approaching first-order, back-mixing greatly reduces the efficiency. For example, an ideal CSTR operating at $[So]/K_m = 0.1$ and 0.99 fractional conversion would require 21-fold more immobilized enzyme than an ideal PFR (Lilly and Dunnill, 1976).

The efficiency of fluidized-bed reactors under optimum conditions approaches that of ideal PFR reactors. Of course, recycling converts fixed-bed and fluidized-bed reactors into back-mix reactors. On the other hand, the design of multistage CST reactors confers some plug-flow character on the CSTR.

Effect of diffusion limitations on reactor efficiency.   Immobilized catalysts are subject to external diffusion inhibition and, if they are on porous matrices, to internal diffusion limitations. External diffusion, reflecting substrate transport from the bulk solution to the catalyst surface, is a function of the stirring or linear flow rate and is relatively easy to minimize with most reactor types. Enzyme loading and particle size are the major factors influencing internal diffusion inhibition, which reflects the diffusion of substrate within the matrix of the catalyst particle. Most porous catalysts exhibit some degree of internal diffusion limitation, which results in an increase in the apparent Michaelis constant and thus leads to more first-order character for the reactor kinetics. Consequently, the effect of diffusion inhibition is to decrease the efficiency of back-mix reactors (Lilly and Dunnill, 1976).

Effect of substrate or product inhibition on reactor efficiency.   Since an ideal CSTR functions at the exit substrate concentration, it is the least subject to inhibition by substrate. Likewise, however, this type of reactor also operates at the exit product concentration, so that its efficiency is most affected by product inhibition. Since most enzymes are inhibited by products rather than substrates, the complete back-mix reactors, again, are often the least efficient.

Reactor size.   The fixed-bed reactor exhibits the greatest activity per unit volume of reactor; hence the containment vessel would be smallest for this type of reactor. Similar considerations show that CSTR systems would be the largest, with fluidized-bed reactors falling between these two extremes and depending on the desired degree of bed expansion. Expressing the percentage of substrate conversion as a function of "normalized residence time" (Pitcher, 1975) provides a convenient method for visualizing the kinetic behavior of a reactor. As illustrated by the data in Fig. 1.5, the results coincide for batch, PFR, and recycle reactors when there are no external diffusion limitations. Such representation readily permits scaling the reactor size according to the desired conversion and throughput.

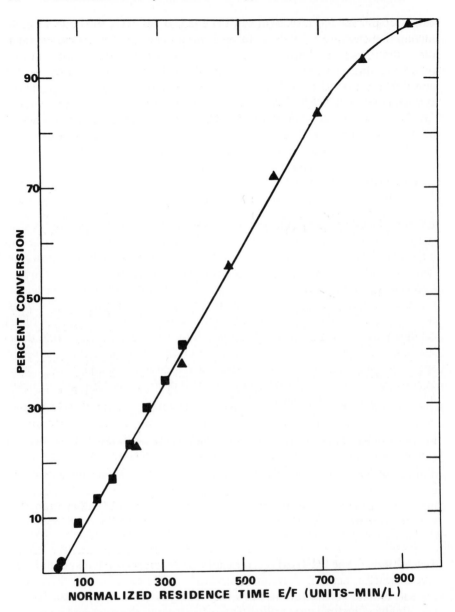

FIGURE 1.5 Conversion of substrate (percent oxidation of thiol groups) as a function of normalized residence time. ●, fixed-bed reactor assays; ■, recirculation fluidized-bed reactor assays; ▲, batch reactor assays. Immobilized sulfhydryl oxidase reactors were operated with 0.8 mM GSH as substrate dissolved in ultra-high-temperature processed milk. (From Swaisgood et al., 1980.)

Matrix properties.   Characteristics of the matrix material must also be considered in the choice of reactor. Stirred reactors are the most abrasive, and some abrasion occurs in fluidized beds as well, which can result in loss of catalyst. Accordingly, materials that are most susceptible to such losses would function best in fixed beds. However, some materials—for example, agaroses and Sephadex—are subject to particle deformation with increasing flow rates and hence have limited function in a fixed bed. Decreasing particle size decreases diffusion inhibition, but it greatly increases back-pressure in fixed beds.

Characteristics of the fluid.   Inlet streams that contain colloidal or fine particulate matter tend to plug fixed-bed reactors. However, fluidized-bed reactors or CSTR systems can operate with fluids containing particulates. The fluid viscosity and density are important factors to be considered in fluidized-bed systems.

B. Methods for Incorporation

The method of integration of an enzyme reactor into the processing stream of a product is dictated by the position of the enzymic step in the overall process and the constraints placed on its operation by the total process. Continual decline in activity is a problem that must be addressed for all enzyme-catalyzed processes. Economically, it is usually mandatory that a reactor be used through several half-lives. For single reactors the activity decline is usually compensated by increasing the residence time in the reactor to achieve the same percent conversion. However, should a constraint of the total process be constant reactor throughput, relatively constant activity can be achieved with a multireactor system (Havewala and Pitcher, 1974). Thus if the acceptable fluctuation is represented by the ratio of the maximum to the minimum production rate, R_p, then the number of reactors required in a system is $N = -(0.693H/\ln R_p)$, where H is the number of half-lives for which each reactor is used. Reactors are replaced in this system on a schedule at a frequency of $Ht_{1/2}/N$, the oldest being the one replaced.

In addition to allowing for maintenance of production rate, the design must permit proper sanitation, possible activity regeneration, and easy replacement of the reactor.

IV. OPERATIONAL FACTORS

A. Factors Affecting Stability

As was indicated previously (Fig. 1.1), stability of the immobilized catalyst is often the primary factor limiting its commercial utilization. Loss of catalyst activity results from three main causes: (1) loss of enzyme from the reactor, (2) adsorption of components that limit access to the enzyme, and (3) chemical inactivation of the enzyme. Active enzyme can be lost from the reactor owing to desorption, cleavage of unstable chemical bonds, solubilization of matrix particles, and attrition of the matrix, yielding "fines" that are not retained by the reactor. Desorption of adsorbed enzymes always occurs, the rate depending upon the method of adsorption and conditions of the feed stream. For example, enzymes ionically adsorbed may not be suitable for use under conditions of high ionic strength or extremes of pH. Most covalent bonds remain stable under normal operating conditions; however, certain bonds such as thioester and disulfide bonds or linkages to cyanogen bromide-activated matrices can undergo cleavage under some conditions. Solubilization of the catalyst particle can be minimized by making the proper derivative (Weetall, 1976; Pitcher, 1980), and fixed-bed reactors cause very little attrition.

Components of the feed stream, such as proteins or microorganisms, can become attached to the surface, blocking access to pores or actually covering the immobilized enzyme. For example, operation of immobilized pepsin with skim milk resulted in rapid inactivation concomitant with the adsorption of a visible film (Richardson and Olson, 1974). Likewise, inactivation of immobilized sulfhydryl oxidase during treatment of milk was attributed to adsorption of milk components (Swaisgood et al., 1982).

The principal causes of chemical inactivation are the reaction with irreversible inhibitors, the induction of irreversible conformational changes, the loss of essential cofactors, and the occurrence of covalent changes. Although the feed stream may contain only low concentrations of an irreversible inhibitor, such as a heavy metal ion or an oxidizing agent, continuous exposure during operation can lead to gradual inactivation. Irreversible conformational changes resulting from interaction with or binding of components in the substrate feed stream can also lead to eventual inactivation. In many enzymes an essential cofactor (e.g., a metal ion), although very tightly bound, is not covalently attached; consequently, significant activity can be lost over long periods of operation. However, covalent modification of critical functional groups or of the protein's integrity is probably the most serious cause of inactivation, and such modification leads to permanent loss of activity. Proteolysis of the immobilized enzyme by proteinases in the substrate feed stream or by microbially produced proteinases resulting from reactor contamination is often the chief cause of reactor inactivation. Of course, the presence of chemical reagents in the substrate solution that are reactive with functional groups of proteins can also inactivate the enzyme; however, such reagents are unlikely to be present in a food.

In many cases, immobilization leads to an apparent stabilization of enzymic activity (see, for example, Richardson and Olson, 1974). In some cases the apparent stabilization of activity may actually result from an increase in structural stability due to the introduction of cross-links in the enzyme's structure or to multipoint attachment to the matrix. Often, however, the apparent stabilization may be a result of diffusion inhibition of the activity of the porous catalyst (Ollis, 1972). As the number of active enzyme molecules immobilized decreases, the effectiveness factor increases, so the change in activity per unit volume of catalyst does not reflect the actual amount of enzyme inactivation. In fact, if the catalyst is severely diffusion-limited even when, for example, one third of the original enzyme has been inactivated, the observed activity change would be small. In a practical sense the net result is that the reactor will maintain its productivity through a half-life longer than the true half-life of enzyme stability.

B. Regeneration of Reactor Activity

Theoretically, permanent loss of reactor activity should occur only when covalent modification is the cause of inactivation. Regeneration of activity may be possible only with covalently immobilized enzymes, since substantial unfolding of the existing structure using very good protein solvents (e.g., urea or guanidinium chloride) may be required to overcome the energy barriers between the inactive and the native conformations. It has been demonstrated (Horton and Swaisgood, 1974, 1976; Swaisgood et al., 1978; Janolino et al., 1978) that high percentages of native structure can be generated from the random coil state of immobilized enzymes. Thus thermally inactivated enzymes have been regenerated by forcing their structure through a random coil state (Klibanov and Mozhaev, 1978). In principle, it should be possible to remove tenaciously bound ligands or metal ions from the enzyme and microorganisms or other components adsorbed to the matrix using such an approach. Regeneration by refolding from the random coil state apparently has not been examined in a practical application. However, the activity of a sulfhydryl oxidase reactor was repeatedly regenerated over a long period of time by washing with 4 M urea following operation with ultra-high-temperature processed milk (Swaisgood et al., 1982). The commercial feasibility of regeneration processes will depend upon the comparative costs of the process versus that for reactor or enzyme replacement.

C. Sanitation of Reactors

In addition to the undesirable potential for microbial contamination of a food product, growth of microorganisms in a reactor reduces its productivity and shortens its half-life. Consequently, it is essential to control the microbial population either by the choice of operating conditions or by using a periodic sanitation procedure. If the immobilized enzyme is sufficiently stable, it may be desirable to operate above 55°C, both to increase the reaction rate and to control microbial growth (Havewala and Pitcher, 1974; Harper et al., 1974). A combination of low pH (e.g., pH 3.5) and high temperature of operation is also very effective (Harper et al., 1974).

A number of compounds have been tested as potential sanitizing agents for reactors, for example, H_2O_2, quaternary ammonium compounds, chlorinated sanitizers, iodophor, and glutaraldehyde (Roland, 1980). Concentrations of H_2O_2 ranging from 0.01% to 0.24% have been used to sanitize reactors containing pepsin (Ferrier et al., 1972), β-galactosidase (Barndt et al., 1975), and sulfhydryl oxidase (Swaisgood and Horton, 1976) without adverse affects on the activity. Quaternary ammonium chloride also proved to be a very effective sanitizer with immobilized β-galactosidase (Barndt et al., 1975); however, it is not effective against gram-negative organisms (Harju, 1977). Lactoperoxidase and low concentrations of H_2O_2 are effective against gram-negative organisms (Harju, 1977); thus it may be possible to control microorganisms within a reactor by coimmobilization with this enzyme.

D. Effect of Inhibitors
The effect of irreversible inactivators was discussed previously, so here we will be concerned only with the effect of reversible inhibitors that may be present in the substrate feed. Dependence of the activity of enzyme reactors on inhibitor concentration often does not follow that predicted from studies of the soluble enzyme owing to effects of diffusion inhibition. Since the reversible inhibitor simply reduces the activity per unit surface area of the catalyst particle (analogous to reducing the enzyme loading), increasing the inhibitor concentration results in an increase of the effectiveness factor for catalyst particles showing significant diffusion limitations (Engasser and Horvath, 1976). Consequently, these antagonistic effects, the effect due to chemical inhibition and that due to diffusion inhibition, tend to cancel each other out, so the resulting effect of reversible chemical inhibitors on reactors displaying severe diffusion limitations may not be very significant. Although this means that the activity per unit weight of enzyme may be low owing to diffusion inhibition, it may actually be advantageous in a practical sense, since fluctuations in reactor productivity will not occur because of possible daily variation in substrate quality.

V. CURRENT INDUSTRIAL USES (FOOD INDUSTRY)

A. Production of High Fructose Corn Syrup (HFCS)
Isomerization of glucose in high-glucose corn syrup produces a sweetener that is competitive with sucrose. This product was based on the development of immobilized glucose isomerase, and the process remains the largest commercial use of an immobilized enzyme. The process also represented the first industrial use of an immobilized enzyme in the United States, having been introduced by Clinton Corn Products in 1972. A company using an immobilized enzyme process has the options of (1) producing and immobilizing the enzyme, (2) purchasing the enzyme and carrying out the immobilization, or (3) purchasing the immobilized enzyme reactor. Immobilized enzymes currently used, companies marketing the immobilized form, the methods used for immobilization, and the products for which it is used as a processing aid are listed in Table 1.3.

TABLE 1.3 Commercial Sources, Immobilized Forms, and the Resulting Products of Immobilized Enzymes Used Currently on an Industrial Scale

Enzyme	Company Marketing the Immobilized Enzyme	Immobilization Method	Product
Glucose isomerase	Novo Enzyme Corp.	Glutaraldehyde cross-linked, pelletized cells	HFCS
	Clinton Corn Processing	Adsorption on DEAE-cellulose	HFCS
	ICI	Chemically flocculated, pelletized cells	HFCS
	Corning Glass Works	Adsorption on porous alumina	HFCS
	Gist-Brocades	Mycelium-fixed enzyme entrapped in gelatin and cross-linked with glutaraldehyde	HFCS
	Snamprogetti	Entrapment in cellulose triacetate fibers	HFCS
	Car-Mi	Glutaraldehyde cross-linked cells	HFCS
Aminoacylase	Tanabe Seiyaku	Adsorption on DEAE-Sephadex	L-amino acids
Aspartase	Tanabe Seiyaku	Entrapment of whole cells in polyacrylamide followed by cell lysis	L-aspartic acid
α-Galactosidase	Hokkaido Sugar Co.	Entrapment in mycelial pellets	Sucrose
	Great Western Sugar Co.	Entrapment in mycelial pellets	Sucrose
β-Galactosidase (lactase)	Snamprogetti	Enzyme from *S. lactis* entrapped in cellulose triacetate fibers	Lactose-hydrolyzed milk
	Corning Glass Works	Enzyme from *A. niger* bound to porous ceramic	Lactose-hydrolyzed whey
Glucoamylase	Tate and Lyle	Entrapment in a gel which coats an inert particle	High-glucose syrups
	Corning Glass Works	Covalently bound to alkylamine silica	High-glucose syrups

Glucose isomerase is obtained from a variety of microorganisms, for example, *Streptomyces wedmorensis* (Clinton), *Bacillus coagulans* (Novo), *Arthrobacter* (ICI), *Streptomyces olivaceus* (Corning), and *Actinoplanes missouriensis* (Gist-Brocades). Standardized HFCS is 71% solids composed of 42% fructose, 52% glucose, and 6% other saccharides. Roughly 1.5 million tons (on a dry weight basis) were produced in 1977, and nearly 4.5 million tons are forecast for 1985 (MacAllister, 1980).

Typically, a feed solution of high glucose syrup containing roughly 50% solids, of which 95% is glucose, produced by enzyme-catalyzed starch hydrolysis using amylases and glucoamylase, is treated with the immobilized glucose isomerase at pH 7.0–7.5 at 60°C. Productivity is generally about 2000 pounds of HFCS per pound of catalyst per 1000 hours of operation (Sweigart, 1979; Venkatasubramanian and Harrow, 1979). The half-life of a reactor operating under the above conditions is typically about 500 hours, and generally the immobilized enzyme is used for two or three half-lives in a multireactor system.

B. Production of L-Amino Acids by Resolution of Racemic Mixtures

Amino acid supplementation of foods is used to improve the nutritive quality of proteins that are deficient in certain essential amino acids. Although amino acids are readily and economically obtained by chemical synthesis, a racemic mixture is obtained; but with rare exceptions, only the L-form is biologically useful. Chibata and co-workers at Tanabe-Seiyku Co. in Japan (for a review, see Chibata (1978)) developed immobilized forms of aminoacylase from *Aspergillus oryzae* that display a general specificity for asymmetric hydrolysis of acyl-D,L-amino acids. Tanabe-Seiyku has been using this enzyme immobilized on DEAE-Sephadex for production of L-methionine, L-alanine, L-phenylalanine, L-tryptophan, and L-valine since 1969; thus it represents the first commercial use of an immobilized enzyme in the world. This company had previously used the soluble enzyme, but conversion to use of the immobilized form greatly improved the ease of product isolation and its yield, lowered enzyme costs by an order of magnitude, and substantially reduced labor costs as a result of automation. Thus the overall cost of L-amino acid production using the immobilized enzyme was 60% of that for the soluble enzyme process.

The production process includes treatment of the racemic mixture of the acylamino acid in a fixed-bed reactor, resulting in specific hydrolysis of the acyl-L-amino acid, yielding the L-amino acid, which is readily separated from the acyl-D-amino acid owing to solubility differences. After removal of the crystalline L-amino acid the acyl-D-amino acid is again racemized for another pass through the reactor. If a 1000-liter reactor is used, the productivity ranges from about 200 kg to 700 kg of the L-amino acid, depending on the amino acid being produced, in a 24-hour period. If the reactor is operated at 50°C, the half-life is roughly 65 days, after which complete restoration of activity can be achieved by loading with fresh enzyme. The matrix itself has been used for seven years without detectable deterioration.

More recently, other methods for immobilization of aminoacylase have been tested for production of L-amino acids on a pilot plant scale. The enzyme covalently immobilized on porous glass exhibited a productivity comparable to that for the enzyme adsorbed on DEAE-Sephadex and, as might be expected, had a much greater half-life (Yokote et al., 1976). Of course, reloading the matrix would be more difficult. Also Rhone Poulenc has obtained a patent for adsorption of the enzyme to polymer-coated TiO_2, silica, or alumina matrices (Pilato and Reichle, 1978). A fiber-entrapped enzyme using cellulose triacetate has also been tested by Snamprogetti in Italy (Bartoli et al., 1978).

C. Production of L-Aspartic Acid

Synthesis of this amino acid on an industrial scale is accomplished by the aspartase-catalyzed addition of ammonia to the unsaturated bond in fumaric acid. Various immobilized forms of the enzyme have been investigated; however, as an industrial catalyst, Tanabe-Seiyaku has chosen the entrapment of *Escherichia coli* cells in polyacrylamide (Chibata, 1978). Since it is an intracellular enzyme, the activity is increased tenfold by cell lysis induced by incubation in substrate solution for 48 hours at 37°C following immobilization. Divalent cations such as Mn^{2+}, Mg^{2+}, or Ca^{2+} enhance the stability of the enzyme; hence one of these ions is included in the substrate feed stream. A half-life of more than 120 days is realized when operating with 1 M ammonium fumarate adjusted to pH 8.5 and 1 mM Mg^{2+} at 37°C. Contaminants, including microbial cells and proteins, are not present in the effluent from the reactor, so L-aspartate of high purity is obtained in high yield, thus reducing the production costs by 40% in comparison to batch processes using intact cells. Tanabe-Seiyaku has been using this process for commercial production of L-aspartic acid since 1973.

D. Production of Sucrose from Sugar Beets

Raffinose present in sugar beets retards crystallization of sucrose; consequently, hydrolysis of raffinose increases the sucrose yield because of both increased crystallization and the formation of sucrose from the hydrolysis. α-Galactosidase is specific for α-D-galactopyranosyl residues and thus hydrolyzes raffinose to give sucrose and galactose. The enzyme from *Morteriella vinacea* is immobilized in mycelial pellets for commercial use (Reilly, 1980; Brodelius, 1978; Kobayashi and Suzuki, 1972). Hydrolysis is performed in U-shaped reactors operating at pH 5.0–5.2 and 48–52°C. The advantages of elimination of the effects of product inhibition due to diffusional limitations for the immobilized system were noted. As a result of implementation of immobilized enzyme-catalyzed raffinose hydrolysis, the capacity for beet processing was increased 11–12%, the sucrose extraction was increased from 87.8% to 90.7%, and the amount of molasses that had to be discarded was completely eliminated.

E. Lactose Hydrolysis

Hydrolysis of the milk sugar lactose has application in two segments of the industry. One is the hydrolysis of lactose in milk, which requires an enzyme with a pH optimum near 7; the other is hydrolysis of lactose in acid whey, which requires an enzyme with a pH optimum between 4 and 5. The potential market for milk whey may be sufficiently increased by lactose hydrolysis, which eliminates the milk intolerance suffered by those individuals, mostly non-Caucasians, who lack sufficient β-galactosidase to metabolize the lactose. Lactose-hydrolyzed milk also functions better in the production of yogurt and cheeses owing to more rapid fermentation of the monosaccharides by the microorganisms (for a review, see Coughlin and Charles (1980)). β-Galactosidase (lactase) from *Saccharomyces lactis*, with a pH optimum of 6.8–7.0, is best suited for treatment of milk. This enzyme entrapped in cellulose triacetate fibers (Morisi et al., 1973; Brodelius, 1978) was investigated for treatment of milk by workers at Snamprogetti. These studies led to an industrial-scale plant constructed by Snamprogetti in cooperation with "Centrale del Latte" of Milan, Italy, in 1977, which is capable of processing 10 tons of milk per day. By using 0.5 kg of enzyme-fiber in a batch reactor, 10,000 liters of milk could be processed with less than 10% loss of activity.

Immobilized forms of β-galactosidase from *Aspergillus niger* with a pH optimum of 4.0–4.5, which is used for hydrolysis of lactose in acid-whey, has been the subject of much investigation (Coughlin and Charles, 1980). The increased sweetness, solubility, and fermentability of the monosaccharides glucose and galactose greatly expands the potential for utilization of whey in beverages, as syrups, in frozen desserts, and as a fermentation medium. Enzyme covalently immobilized on porous glass and titania has been developed and investigated by scientists at Corning. By using demineralized, deproteinized whey a half-life of 60 days at 50°C was obtained. Pilot plant studies indicated that a typical catalyst having 300 IU/g activity at 35°C and operated continuously at 35–50°C, the temperature increasing to compensate for activity loss, would have a lifetime of 559 days. The Corning process went into commercial operation in the fall of 1983 at the Nutrisearch facility in Winchester, Kentucky (Corning Glass works – Kroger joint venture). This plant is processing 100,000 gal of cottage cheese whey per day by ultrafiltration to remove the protein which is sold in the forms of 60% and 80% protein concentrates for use as a food ingredient. The deproteinized stream is then fed to the immobilized enzyme reactor and the resulting lactose hydrolysate is used as a growth replacement. As was noted previously, the combination of reducing the costs of waste treatment and recovery of nutritious whey proteins make the process economically viable.

Whey treatment in fluidized-beds containing β-galactosidase adsorbed on alumina and cross-linked with glutaraldehyde has been extensively investigated by Coughlin and co-workers (Coughlin and Charles, 1980). The stability of this preparation was similar to that immobilized by the Corning process. These authors, however, reported fewer problems with microbial contamination and no reactor plugging since a fluidized-bed was put into use.

F. Production of High Glucose Syrups by Hydrolysis of Dextrin

Conversion of starch to glucose requires both α-amylase to form dextrins and glucoamylase to hydrolyze the dextrins to glucose. Both enzymes have been traditionally used in the soluble form. However, a larger number of studies have been undertaken of immobilized forms of glucoamylase from *Aspergillus niger* (see Reilly (1980), and Brodelius (1978) for reviews). Perhaps the most extensive investigations are those that resulted in pilot plant studies at Iowa State University (D. D. Lee et al., 1976, 1980; Weetall, 1975). Glucoamylase is an exohydrolase that catalyzes hydrolysis of α(1-4) bonds, and much more slowly α(1-6) bonds, from the nonreducing end of maltooligosaccharides. Thus it works more slowly on maltose than on longer chains; moreover, it also exhibits some transferase activity, resulting in the formation of isomaltose or nigerose. The latter compounds are known as reversion products. Examination of the enzyme covalently immobilized on alkylamine porous silica, using glutaraldehyde for activation, showed that pore diffusion limitations lowered the maximum glucose yield, owing to an increased rate of formation of reversion products including isomaltose. However, very small immobilized enzyme particles gave glucose yields comparable to the yield for the soluble enzyme (D. D. Lee et al., 1976, 1980). Pilot plant operation at pH 4.4 with a 30% dextrin feed solution yielded a reactor half-life of 7.5 h at 70°C and 581 h at 55°C.

Although the required reaction time can be reduced from 75 hours for the soluble enzyme process to less than an hour for the immobilized enzyme (Weetall, 1975; Brodelius, 1978), it appears that two main obstacles must be overcome before the immobilized enzyme process is adopted commercially. These limitations are the lowered glucose yield due to pore diffusion inhibition and possible microbial contamination due to required operation at lower temperatures.

Recently, some of these objections may have been alleviated by an immobilization process patented by Tate and Lyle, Ltd., London. In this process the enzyme is immobilized in a gel surrounding an inert support such as bone char (Daniels and Farmer, 1981). The gel is formed by first coating the matrix with acetone and concentrated enzyme solution, followed by cross-linking with glutaraldehyde. The resulting immobilized glucoamylase exhibited a half-life of 40 days at 60°C while operating with a 42 DE syrup. This process is apparently being used or considered for use in some North American corn-processing plants.

VI. EXAMPLES OF POTENTIAL INDUSTRIAL USES

This discussion would not be complete without considering some examples of potential applications of immobilized enzymes that have been or currently are being investigated. Although applications of enzymes requiring coenzymes for catalysis certainly exist, these will not be considered here, since it appears that economic means for recycling these compounds have not yet been developed.

A. Glucose Oxidase/Catalase

These two enzymes are usually paired, since H_2O_2, a product of glucose oxidase catalysis, is inhibitory toward glucose oxidase. Glucose oxidase catalyzes the oxidation of glucose to gluconolactone, which is hydrolyzed to gluconic acid, with molecular oxygen being reduced to H_2O_2, while catalase dismutes H_2O_2, giving molecular oxygen and water. Thus the net reaction is the formation of gluconic acid and the consumption of glucose and oxygen. Hence the system has many uses where removal of glucose or oxygen or the formation of gluconic acid is beneficial. For example, removal of oxygen improves the preservation of many foods by decreasing oxidation, "desugaring" of eggs prior to drying prevents Maillard browning, and commercial production of both gluconic acid and crystalline fructose from HFCS is possible. Soluble enzymes are currently used for many of these applications; but if immobilized forms that are sufficiently stable can be prepared, there are advantages to the adoption of such systems (Bouin et al., 1976; Greenfield and Laurence, 1975).

Catalase alone may have application in the "cold" pasteurization of milk for cheese making with H_2O_2. The 0.05% H_2O_2 used for this purpose is readily removed by treatment with catalase.

B. Sulfhydryl Oxidase

Because of their extremely low organoleptic threshold concentrations, thiols are often responsible for undesirable off-flavors in foods. For example, the undesirable "cooked" flavor in ultra-high-temperature (UHT) processed milk is related to formation of these compounds. When such processed milk is aseptically packaged, it is shelf-stable at room temperature for long periods, for example, 6–12 months if a holding time of 7–10 s at 150°C is used. However, the more severe heat treatment causes a more intense cooked flavor. By treatment of such milk with immobilized sulfhydryl oxidase, it is possible to completely eliminate the cooked flavor (Swaisgood, 1980; Swaisgood et al., 1982). The half-life of enzyme immobilized on porous silica was about one week for treatment of UHT milk at 30°C; however, activity could be regenerated by washing with a sterile 4 M urea solution.

C. Treatment of Milk with Trypsin for Protection Against Oxidized Flavor Development

It has been known for a long time that treatment of milk with trypsin protects against development of oxidized flavor during storage. Such treatment is ideally suited for use of immobilized trypsin, since the extent of reaction can then be readily controlled. The enzyme immobilized on porous silica was investigated for this purpose and found to be effective in prevention of oxidized flavor development (E. C. Lee et al., 1974). The enzyme used in either a fixed-bed or a fluidized-bed reactor appeared to be stable over a 14-day period.

D. Applications of Limited Proteolysis

Use of immobilized proteinases or peptidases for limited proteolysis is particularly advantageous owing to the inherent control achieved. Continuous production of curd for cheese making using immobilized chymosin, pepsin, or microbial rennets has been investigated by a number of laboratories; however, unless the stability of such preparations under conditions required for the process is increased by several orders of magnitude, it is not likely to be commercialized. Nevertheless, the advantages of continuous operation and potential automation and production of a whey protein solution that is not contaminated with clotting enzymes make further research worthwhile.

The functionality of soy protein is improved by limited hydrolysis with several proteinases. Since the degree of hydrolysis is critical to the property desired, an immobilized system is obviously advantageous.

A haze commonly develops in beer during cold storage as a result of complexes formed containing protein. Thus beers are "chillproofed" by treatment with a proteinase, usually papain; however, extensive hydrolysis impairs foaming and thus beer quality. Several types of immobilized papain have been examined for treatment of beer with moderate success; the chief limitation apparently being the enzyme's stability. Papain was chosen as the soluble enzyme for treatment because of its more limited proteolytic action in comparison to other proteinases and the need to inactivate the enzyme by pasteurization when the desired degree of hydrolysis is achieved. Bearing this in mind, perhaps more stable proteinases should be examined for an immobilized enzyme beer treatment process. Similar haze formation problems occur with certain fruit juices, wine, and sake; consequently, the development of a generalized process could have wide application.

Another area in which limited proteolysis with immobilized proteinases could have application is the inactivation of soluble enzymes added as processing aids. In some cases a more mild alternative to heating, blanching, or addition of denaturants is desirable.

E. Production of Maltose Syrups Using β-Amylase and Pullulanase

β-Amylase catalyzes hydrolysis of $\alpha(1\text{-}4)$ glycosidic bonds in maltodextrins, while pullulanase hydrolyzes $\alpha(1\text{-}6)$ bonds. Therefore pure maltose can be produced from dextrins when these two enzymes are used together. More efficient hydrolysis is obtained when both enzymes are coimmobilized on the same support (for reviews, see Reilly (1980) and Brodelius (1978)).

F. Production of Isomaltulose

The sugar isomaltulose, although less sweet than sucrose, may have utility as a noncariogenic bulking agent in foods. Recently, an efficient process for its production from concentrated sucrose solutions using immobilized nongrowing, intact *Erwinia rhapontici* cells has been developed (Cheetham et al., 1982). These cells immobilized in calcium alginate were 350-fold more stable than free cells with a half-life of 8600 hours.

G. Debittering Citrus Juices

Certain citrus juices — such as those from naval oranges, Shamouti and Murcott oranges, Ponkan, Iyokan, and Wase mandarin oranges, and grapefruit — develop bitterness during storage due to formation of limonin (Hasegawa and Maier, 1983). Several microorganisms such as *Arthrobacter globiformis*, *Pseudomonas* 321-18, and *Corynebacterium fascians* produce enzymes, particularly limonoate dehydrogenase, which metabolize this compound to nonbitter substances. Hence these juices can be debittered by treatment with immobilized forms of such microorganisms or possibly enzyme systems derived from them.

VII. SUMMARY

During the past 20 years a great deal of research has been devoted to the characterization of immobilized enzymes and development of methods for their immobilization and industrial use. These studies have taught us a great deal, not only about the use of enzymes, but also about the structure and function of enzymes. Although the industrial development of immobilized enzyme processes has not proceeded as rapidly as some initially predicted, there has been a gradual adoption of this technology. Much of the rapid surveying, "try-it-and-see-what-happens" type of research has been done. What remains is the less rapid advances due to insight and understanding of enzyme structure and function. For example, a significant breakthrough could result from the discovery of general methods for greatly increasing enzyme structural stability or the ability to regenerate activity of immobilized forms, thus removing the limitation of short half-lives now experienced in many cases. Also, developments that lead to more effective use of multienzyme systems and enzymes that require coenzymes would certainly mark an expansion in the utility of immobilized forms.

Since the technology for fabrication of organized structures similar to cellular structures is not yet available, the immobilization of whole cells or organelles will continue to be used for catalysis of the more complex pathways. Marriage of the potential fruits of genetic engineering with the developments in enzyme technology has the tremendous possibility of tailoring catalysts for specific purposes. Such an approach, for example, could lead to the synthesis of enzymes having greatly enhanced stability under desired processing conditions. Finally, it should be noted that analytical applications of immobilized enzymes were not considered in this discussion. Obviously, such applications are numerous in the food industry.

REFERENCES

Barndt, R. L., Leeder, J. G., Giacin, J. R., and Kleyn, D. H. (1975) *J. Food Sci. 40*, 291–296.

Bartoli, F., Bianchi, G. E., and Zaccardelli, D. (1978) in *Enzyme Engineering*, Vol. IV (Broun, G. B., Manecke, G., and Wingard, L. B., Jr., eds.), p. 279, Plenum Press, New York.

Bouin, J. C., Dudgeon, P. H., and Hultin, H. O. (1976) *J. Food Sci. 41*, 886–890.

Brodelius, P. (1978) *Adv. Biochem. Eng. 10*, 75–129.

Cashion, P., Javed, A., Harrison, D., Seeley, J., Lentini, V., and Sathe, G. (1982) *Biotechnol. Bioeng. 24*, 403–423.

Chambers, R. P., Cohen, W., and Baricos, W. H. (1976) *Methods Enzymol. 44*, 291–317.

Chang, T. M. S. (1976) *Methods Enzymol. 44*, 201–218.

Cheetham, P. S. J., Imber, C. E., and Isherwood, J. (1982) *Nature 299*, 628–631.

Chibata, I. (1978) *Immobilized Enzymes*, 284 pp., Halsted Press, Division of John Wiley and Sons, New York.

Coughlin, R. W., and Charles, M. (1980) in *Immobilized Enzymes for Food Processing* (Pitcher, W. H., Jr., ed.), pp. 153–173, CRC Press, Boca Raton, Fla.

Daniels, M. J., and Farmer, D. M. (1981) UK Patent Application GB 2 070 022 A.

Dinelli, D., Marconi, W., and Morisi, F. (1976) *Methods Enzymol. 44*, 227–243.

Engasser, J.-M., and Horvath, C. (1976) *Appl. Biochem. Bioengr. 1*, 127–220.

Ferrier, L. K., Richardson, T., Olson, N. F., and Hicks, C. L. (1972) *J. Dairy Sci. 55*, 726–734.

Goldstein, L., and Manecke, G. (1976) *Appl. Biochem. Bioengr. 1*, 23–126.

Greenfield, P. F., and Laurence, R. L. (1975) *J. Food Sci. 40*, 906–910.

Harju, M. (1977) *N. Eur. Dairy J. 43*, 155–159.

Harper, W. J., Okos, E., and Blaisdell, J. L. (1974) in *Enzyme Engineering*, Vol. II (Pye, E. K., and Wingard, L. B., Jr., eds.), pp. 287–292, Plenum Press, New York.

Hasegawa, S., and Maier, V. P. (1983) *Food Technol. 37*, 73–77.

Havelawa, N. B., and Pitcher, W. H. (1974) in *Enzyme Engineering*, Vol. II (Pye, E. K., and Wingard, L. B., Jr., eds.), pp. 315–328, Plenum Press, New York.

Horton, H. R., and Swaisgood, H. E. (1974) in *Immobilized Biochemicals and Affinity Chromatography* (Dunlap, R. B., ed.), pp. 329–338, Plenum Press, New York.

Horton, H. R., and Swaisgood, H. E. (1976) *Methods Enzymol. 44*, 516–526.

Janolino, V. G., and Swaisgood, H. E. (1978) *J. Dairy Sci. 61*, 393–399.

Janolino, V. G., and Swaisgood, H. E. (1982) *Biotechnol. Bioeng. 24*, 1069–1080.

Janolino, V. G., Sliwkowski, M. X., Swaisgood, H. E., and Horton, H. R. (1978) *Arch. Biochem. Biophys. 191*, 269–277.

Kierstan, M., and Bucke, C. (1977) *Biotechnol. Bioeng. 19*, 387–397.

Klibanov, A. M., and Mozhaev, V. V. (1978) *Biochem. Biophys. Res. Comm. 83*, 1012–1017.

Kobayashi, H., and Suzuki, H. (1972) *J. Ferment. Technol. 50*, 625–630.

Larsson, P.-O., and Mosbach, K. (1976) *Methods Enzymol. 44*, 183–190.

Lee, D. D., Lee, Y. Y., Reilly, P. J., Collins, E. V., Jr., and Tsao, G. T. (1976) *Biotechnol. Bioeng. 18*, 253–267.

Lee, D. D., Lee, G. K., Reilly, P. J., and Lee, Y. Y. (1980) *Biotechnol. Bioeng. 22*, 1–17.

Lee, E. C., Senyk, G. F., and Shipe, W. F. (1975) *J. Dairy Sci. 58*, 473–476.

Lilly, M. D., and Dunnill, P. (1976) *Methods Enzymol. 44*, 717–738.

MacAllister, R. V. (1980) in *Immobilized Enzymes for Food Processing* (Pitcher, W. H., Jr., ed.), pp. 81–111, CRC Press, Boca Raton, Fla.

Morisi, F., Pastore, M., and Vigilia, A. (1973) *J. Dairy Sci. 56*, 1123–1127.

Mosbach, K. (1976) *Methods in Enzymology*, Vol. XLIV, *Immobilized Enzymes*, 999 pp., Academic Press, New York.

Nelson, J. M., and Griffin, E. G. (1916) *J. Amer. Chem. Soc. 38*, 1109–1115.

Ollis, D. F. (1972) *Biotechnol. Bioeng. 14*, 871–884.

Olson, A. C., and Cooney, C. L. (1974) *Immobilized Enzymes in Food and Microbial Processes*, 268 pp., Plenum Press, New York.

Pilato, L. A., and Reichle, W. T. (1978) Rhone Poulenc Patent BE 855 051, *Chem. Technol. 8*, 309.

Pitcher, W. H., Jr. (1975) in *Immobilized Enzymes for Industrial Reactors* (Messing, R. A., ed.), pp. 151–199, Academic Press, New York.

Pitcher, W. H., Jr. (ed.) (1980) *Immobilized Enzymes for Food Processing*, 219 pp., CRC Press, Boca Raton, Fla.

Porath, J., and Caldwell, K. D. (1977) in *Biotechnological Applications of Proteins and Enzymes* (Bohak, Z., and Sharon, N., eds.), pp. 83–102, Academic Press, New York.

Poulsen, P. B., and Zittan, L. (1976) *Methods Enzymol. 44*, 809–821.

Reilly, P. J. (1980) in *Immobilized Enzymes for Food Processing* (Pitcher, W. H., Jr., ed.), pp. 113–151, CRC Press, Boca Raton, Fla.

Richardson, T., and Olson, N. F. (1974) in *Immobilized Enzymes in Food and Microbial Processes* (Olson, A. C., and Cooney, C. L., eds.), pp. 19–40, Plenum Press, New York.

Roland, J. F. (1980) in *Immobilized Enzymes for Food Processing* (Pitcher, W. H., Jr., ed.), pp. 55–80, CRC Press, Boca Raton, Fla.

Srere, P. A., and Uyeda, K. (1976) *Methods Enzymol. 44*, 11–19.

Swaisgood, H. E. (1980) *Enzyme Microb. Technol. 2*, 265–272.

Swaisgood, H. E., and Horton, H. R. (1976) NSF/RANN Publication: *Enzyme Technology*, University of Pennsylvania Press, State College, PA, pp. 94–99.

Swaisgood, H. E., Janolino, V. G., and Horton, H. R. (1978) *Arch. Biochem. Biophys. 191*, 259–268.

Swaisgood, H. E., Janolino, V. G., and Sliwkowski, M. X. (1980) in *Proceedings of the International Conference on UHT Processing and Aseptic Packaging of Milk and Milk Products*, Nov. 27–29, 1979, Raleigh, NC, pp. 67–76.

Swaisgood, H. E., Sliwkowski, M. X., Skudder, P. J., and Janolino, V. G. (1982) in *Use of Enzymes in Food Technology* (Dupuy, P., ed.), pp. 229–235, Technique et Documentation Lavoisier, Paris.

Sweigart, R. D. (1979) *Appl. Biochem. Bioengr. 2*, 209–218.

Venkatasubramanian, K., and Harrow, L. S. (1979) *Ann. N.Y. Acad. Sci. 326*, 141–153.

Vieth, W.R., Venkatasubramanian, K., Constantinides, A., and Davidson, B. (1976) *Appl. Biochem. Bioengr. 1*, 221–327.

Weetall, H. H. (1975) *Process Biochem. 10*, 3–12, 22, 30.

Weetall, H. H. (1976) *Methods Enzymol. 44*, 134–148.

Weetall, H. H. (1977) in *Biotechnological Applications of Proteins and Enzymes* (Bohak, Z., and Sharon, N., eds.), pp. 103–126, Academic Press, New York.

Yokote, Y., Fujita, M., Shimura, G., Noguchi, S., Kimura, K., and Samejima, H. (1976) *J. Solid-Phase Biochem. 1*, 1–13.

2

Immobilized Lactic Acid Bacteria

Pekka Linko

I. INTRODUCTION

Lactic acid bacteria have been utilized by humans since ancient times for the production of sour milk products, cheese, fermented vegetables, and sour dough bread, although the first pure culture was isolated only about 100 years ago. Lactic acid was also the first biotechnically produced industrial chemical; the commercialization was pioneered by Avery in Littleton, Massachusetts, in 1881 (Lockwood, 1979). In 1980 a total of about 40,000 tons of lactic acid were produced in the United States and in Europe both by batch fermentation and by chemical synthesis (Eveleigh, 1981). Whereas the chemical route for synthesis of lactic acid results in a racemic mixture of the L- and D-acids, the ratio of L- and D-lactic acid may be controlled by microbial selection. Most lactic acid is used today by the food industry as an acidulant and preservative, but Lipinsky (1981) has recently emphasized the potential importance of biotechnically produced lactic acid as a chemical feedstock via lactonitrile and lactides. For example, lactides can form both homopolymers and, with a number of hydroxyacids, heteropolymers, which can be processed to biodegradable translucent films and other plastics. Inasmuch as homofermentative lactic acid bacteria form no carbon dioxide from sugars, lactic acid fermentation is an interesting alternative to biotechnical ethanol production. Theoretically, two moles of lactic acid are obtained from one mole of glucose, but typical yields in practice are of the order of 93–95% of theoretical. Nevertheless, although many lactic acid bacteria convert sugars efficiently to lactic acid, the separation and purification steps form a large part of the total production costs (Eveleigh, 1981). Different carbohydrates may be employed as substrate in lactic acid fermentations (Casida, 1964; Villet, 1979; Zeikus, 1980). Thus several lactic acid bacteria ferment glucose, maltose, and sucrose; *Lactobacillus bulgaricus* is typical of the lactose fermenting bacteria; *Lactobacillus pentosus* and *Lactobacillus pentoaceticus* ferment xylose and other pentoses to produce a mixture of lactic and acetic acids; *Lactobacillus amylophillus* can ferment starch directly; and the heterofermentative *Thermoanaerobicum brockii* can utilize hexoses, cellobiose, and starch, producing lactic acid as the main product on a yeast extract–rich medium and ethanol in the absence of sufficient yeast extract. Also, fungi have been employed in commercial lactic acid production, and Showa Chemical Industries, Japan, produces about 360 tons of lactic acid annually from glucose using *Rhizopus* sp. (Yamada, 1977).

Batch fermentation technology is still used in industrial biotechnical lactic acid production because residual sugar in the fermentation broth interferes with lactic acid separation. For the same reason, substrate sugar content should normally not exceed 12%. Nevertheless, continuous culture technology with *Streptococcus lactis* was reported by Whittier and Rogers (1931) as early as 1931. They claimed up to 90% yield of lactic acid both from lactose (6%) and from cheese whey under continuous flow conditions at about a 24-hour residence time, employing a mixed culture of *S. lactis* and a mycoderm. The process was successfully tested in pilot scale with a fermentor of about 2500-kg capacity, using slow agitation, a temperature of $43° \pm 1°C$, and a pH of 5.0–5.8 controlled by feeding dry hydrated lime. Residual lactose was maintained below 0.5%. In 1945, batch processes employing ammonia as a neutralizing agent in lactic acid fermentation of whey for ruminant feed were patented (Jansen, 1945; Perquin, 1945). It was later shown that automatic pH control can significantly reduce fermentation time (Czarnetzky, 1958), and successful pilot-scale experiments were also reported (Henderson et al., 1973). More recently, Keller and Gerhardt (1975) developed a mathematical model for computer simulation of continuous lactic acid fermentation of whey employing ammonia as the neutralizing agent. Steady state conditions could be maintained for 42 days with less than 0.2% of compounds other than lactic acid detected (Stieber and Gerhardt, 1979). Furthermore, if two fermentors were used in series at pH 5.5, residual lactose concentration was less than 0.1%.

Milko et al. (1966) obtained 57% lactic acid yield and 6 g/Ld biomass production at a 3-hour residence time, and 90% yield with 1.5 g/Ld biomass production at a 20-hour residence time in continuous fermentation with *Lactobacillus delbrueckii* on sucrose; they concluded that a two-stage process should be preferred. The first fermentation stage would have a short residence time for rapid cell growth, and the second stage would have a relatively long residence time to maximize lactic acid production and to minimize biomass formation. This agrees well with Marshal's (1970) later observation. Childs and Welshby (1966) have also reported on continuous lactic acid fermentation with *L. delbrueckii*. They obtained the highest volumetric productivity of 3.7 g/Lh with 97 g/L glucose feed containing some ammonium sulfate and sodium sulfite. The optimum pH, with calcium carbonate as neutralizing agent, was 4.9–5.2, and that with sodium carbonate was 5.5–5.8. Linklater and Griffin (1971a) maintained the pH at 5.4–6.0 in continuous culture of *S. lactis* C 10 employing a pH-stat with feedback control. Cox and MacBean (1977) reported that when lactic acid was produced from casein whey (11 g/L protein), an increase in yeast extract or cornsteep liquor from 2 to 12 g/L resulted in an increase in volumetric productivity from about 3 to more than 10 g/Lh.

Clearly, attempts to establish a continuous lactic acid fermentation process have met with some success, although a number of difficulties still remain to be overcome, and no full-scale commercial operation was known in 1978 (Miall, 1978). However, recent developments in membrane and hollow fiber technology and in the utilization of immobilized living microbial cells have resulted in a renewed interest in continuous biotechnical lactic acid production.

II. MEMBRANE AND HOLLOW FIBER TECHNOLOGY

Friedman and Gaden (1970) investigated the possibility of improving lactic acid productivity by incorporating a dialysis unit in the fermentation process for continuous removal of inhibitory or toxic metabolites. The pH was automatically controlled at 5.8. When *L. delbrueckii* was used, acid production was reported to be significantly higher in the dialysis culture system. The highest overall productivity of 8 g/Lh from 5% glucose (yeast extract 3%, $MgSO_4$ 0.6 g/L, $FeSO_4$ 0.03 g/L, $MnSO_4$ 0.03 g/L, Na-acetate 1 g/L, K_2HPO_4 0.5 g/L, KH_2PO_4 0.5 g/L) was obtained in contrast to 5 g/Lh in a conventional fermentation system without dialysis. Stieber et al. (1977) employed a dialysis unit for the continuous removal of ammonium lactate from the fermentation broth during lactic acid production from whey by *L. bulgaricus*. They reported a volumetric productivity of as high as 11.7 g/Lh at a 19-hour residence time and 97% conversion, contrasting this with the value 1.6 g/Lh reported by Keller and Gerhardt (1975) for conventional continuous fermentation at a 31-hour residence time. The process was later further improved by incorporating a cell recycle system (Stieber and Gerhardt, 1981). This resulted in a nearly complete conversion of lactose at a 2.9-hour residence time, in contrast to about 75% conversion at a 27.5-hour residence time without cell recycle. Setti (1974) reported an improved lactic acid yield and production rate by connecting the fermentor to a reverse osmosis unit, which allowed selective permeation of the product with an almost complete retention of substrate and a nearly pure product solution.

Royce et al. (1982) recently reported on the production of lactic acid by homofermentative *L. delbrueckii* NRRL B445 in a hollow fiber fermentor, with final cell densities in the fluid surrounding the fibers as high as 480 g (d.m.)/L and volumetric productivity as high as 100 g/Lh (fermentor volume 3.6 cm³, 300-fiber unit, Amicon Vitafiber® shell, Celanese Cellgard® microporous polypropylene hollow fibers type X-10 100 micron i.d., 150 micron o.d.). A densely packed cell mass occupied the entire volume around the fibers, which allowed the nutrients and products to diffuse freely. Cell yields were reported to be significantly lower than in batch fermentations.

III. IMMOBILIZED LIVING CELL FERMENTATIONS

A. Lactic Acid Production

Griffith and Compere (1975); Compere and Griffith, (1981) reported on a method to attach living microbial cells on Berl saddles coated with 10% (w/v) gelatin in the presence of a polyelectrolyte (1% w/v solution of Nalco 8172) and subsequently cross-linked by spraying with 5% (w/v) glutaraldehyde. Compere and Griffith (1975) applied the technique to produce lactic acid from acid whey in a column bioreactor, employing a mixed culture of lactic acid–producing bacteria. The lactic acid concentration was reported to increase from 1.4% to 2.1% in a single pass through the 5×183 cm column reactor at a 10- to 20-hour residence time. Lactic acid was finally separated from the fermentation broth by ion exchange at about 62% yield.

Divies (1977a) entrapped various microbial cells, including lactic acid bacteria, in polyacrylamide gel, employing essentially a method first applied to whole-cell immobilization by Mosbach and Mosbach (1966) and applied at the industrial scale in 1973 by Tanabe Seiyaku Company Ltd. in Japan for the production of L-aspartic acid from ammonium fumarate by polyacrylamide gel-immobilized *Escherichia coli* cells (Chibata, 1980). Divies suspended microbial cells in 100 ml of 0.05 M phosphate buffer of pH 7.0 at 15°C and added 8.2 g of the monomers acrylamide and N,N-methylene-*bis*-acrylamide (81% acrylamide) and 200 mg of ammonium persulfate as the polymerizing catalyst. The polymerization was complete in about 1–2 min. It may be estimated that a productivity of about 14 g/Lh (based on biocatalyst volume) at about a 1.5-hour residence time was obtained with *L. delbrueckii* ATCC 9649 on 10% (w/v) glucose (yeast extract 20 g/L, $K_2SO_4 \cdot 7H_2O$ 0.2 g/L, $MnSO_4$ 0.01 g/L, $FeCl_3$ 0.002 g/L, Na-acetate 1 g/L, KH_2PO_4 2 g/L) fed to a 150-ml (liquid + biocatalyst) stirred tank reactor.

Linko (1980) first reported the successful immobilization of *Lactobacillus lactis* cells into calcium alginate gel beads at the Second International Congress on Engineering and Food in 1979. The biocatalyst could be used for at least five batch cycles for lactic acid production from 2% (w/v) glucose medium containing small quantities of nutrients without any significant loss in activity. After testing of a number of lactic acid producing bacteria a strain of *L. del-*

brueckii, recently identified as *L. casei* ssp. *rhamnosis*, was found to be the best producer on a glucose medium. The organism was normally grown on MRS-medium (Man et al., 1960), on which it grew rapidly:

Glucose (or lactose)	20 g
Bacto-protease-peptone	10 g
Bacto-meat extract	10 g
Yeast extract	5 g
Na-acetate	5 g
Ammonium citrate	2 g
KH_2PO_4	2 g
$MgSO_4 \cdot 7H_2O$	0.1 g
$MnSO_4 \cdot 4H_2O$	0.05 g
Tween 80	1 ml
Deionized water to	1 L
pH 6.5, 37°C	

In about 20 hours the pH decreased to 4.05, and the cell growth stopped at about 12 g/L lactic acid content in the fermentation broth. Cells were harvested at this point for immobilization. The centrifuged cells were suspended in 6% sodium alginate (BDH, Poole, UK) and extruded by slight pressure into 0.5 M calcium chloride through a specially constructed device (Linko and Linko, 1984) of multiple hollow needles of 0.6 mm diameter to obtain biocatalyst beads of relatively uniform size (2 mm diameter). The beads were allowed to harden for about 30 min at room temperature. When the biocatalyst was used in a continuous packed-bed column reactor, a steady state was normally reached in about a week (Linko, 1981). The maximum lactic acid yield from 4.8% (w/v) glucose medium containing 1% (w/v) yeast extract and 4.8% (w/v) solid calcium carbonate as buffer (pH 5.5, 43°C) was about 93% of theoretical; and in general with this immobilized organism, about 90% of the lactic acid obtained was L-lactic acid (Linko, 1981; Stenroos et al., 1982b). Excellent operational stability of the biocatalyst was demonstrated by using the same reactor for 32 recycle batch runs during a total period of 157 days. At a 3-hour residence time per cycle the maximum final lactic acid yield reached in about 40 hours decreased from 97% (87% L-lactic acid) in the first run to 86% (78% L-lactic acid) in the last experiment. In comparison, 93–95% lactic acid yields have been reported for conventional batch fermentations of 4–6 days with free cells (Casida, 1964; Lockwood, 1979). Varying the ratio of bacterial cell mass (24% d.m.) to 6% sodium alginate from 1:2 to 1:5 had little effect on lactic acid production after the steady state conditions were reached. Typical biocatalyst cell concentration at steady state was about 10^9 cells/g.

In continuous column reactor operation, the L-lactic acid yield was little affected when the residence time was varied from 8 to 18 hours, with a maximum of about 43 g/L (about 83% of theoretical; 30 g of biocatalyst with 1.5 g d.m. of cells) at about a 12-hour residence time, whereas the production rate fell rapidly from about 175 mg/gh to about 10 mg/gh when the residence time was increased from 8 to 38 hours. The biocatalyst activity half-life in continuous operation was over 50 days, and after 30 days of continuous operation the production rate was still about 70% of the original. The same biocatalyst could be revived and reused several times, even after prolonged storage at +4–7°C. The activity decreased little during 55 days of cold storage, and was quickly revived to nearly original activity during subsequent operation employing substrate supplemented with nutrients (Stenroos et al., 1982a). Only about 10% of the activity was lost during 5 months of intermittent storage and operation.

The lactic acid yield of up to about 80% was somewhat lower by similarly immobilized *L. bulgaricus* on 5% (w/v) glucose medium in a recycle batch column reactor, and only up to about 60% on 5 to 9% lactose medium (Stenroos et al., 1982b). When dairy industry by-products of equivalent lactose level such as whey, whey-UF-permeate, and electrodialytically demineralized whey were used as substrate, lactic acid yields varied from 70% to 80%. However, in continuous column operation with immobilized *L. bulgaricus* cells the maximum lactic acid yield on 4.8% (w/v) lactose (containing 1% w/v yeast extract and 3% w/v $CaCO_3$, pH 6.0, 45°C) was only about 40% at a 12-hour residence time, significantly less than that obtained with *L. casei* ssp. *rhamnosus* on glucose. Furthermore, nearly all of the lactic acid produced by the immobilized *L. bulgaricus* in continuous operation was D-lactic acid.

B. Fermentation of Pentoses

Griffith and Compere (1977; Compere and Griffith, 1980) applied their gelatin-coated Berl saddle biocatalyst system to waste liquors of about 6% sugar content from pulp, paper, and fiberboard industries. Such wood molasses wastes were buffered with calcium hydroxide and pretreated with several enzyme preparations to lower the viscosity. No effect on lactic acid yield by the enzyme pretreatment was observed. The column reactor (about 5 × 183 cm), packed with the carrier particles, was inoculated with a mixed culture of commercial kefir lactobacilli and lactose-fermenting yeasts and yielded about 3% lactic acid with a residual sugar content of only 0.25%. The reactor was first operated for several weeks on whey as substrate, followed by wood molasses. The buffer concentration was reported to be critical for optimal lactic acid production. With wood molasses a by-product of hardwood cellulose manufacture (International Paper Company) of 6% (w/v) sugar content, approximately 3.1% (w/v) lactic acid was obtained by using 250 meq of calcium lime as buffer. Maximum lactic acid production level was obtained from Masonite Corporation pressed board molasses of about 8% (w/v) sugar, reaching as high as 5% (w/v) with about 0.75% (w/v) residual sugar level.

In our own recent experiments, pentose-rich hardwood molasses of 5% (w/v) sugar content was used as substrate for calcium alginate gel-entrapped *L. plantarum pentosus* ATCC 8041 cells in a slightly conical packed-bed reactor operated in a circulation batch mode (10 g biocatalyst/50 ml substrate containing 1% w/v yeast extract and 5% w/v CaCO₃, 13.6 ml/min, 30°C), and greater than 90% lactic acid yield was obtained in 24 hours. The yield decreased considerably with increasing substrate concentration.

Tipayang and Kozaki (1982) recently employed calcium alginate (3%) gel-entrapped *L. vaccinostercus* Kozaki and Okada sp. nov. for lactic acid production from 2% xylose containing 1% peptone and 0.5% yeast extract. The immobilized biocatalyst was reported to be very stable with no leakage of cells into the surrounding medium. Lactic acid production was carried out in a batch system under stationary conditions, and the immobilized cells were reported to produce more lactic acid than washed free cells. Lactic acid levels in the fermentation broth reached 0.9–1% (w/v) in 5 days, and the yield remained fairly constant for five repeated batch experiments.

C. Malolactic Fermentation

Divies (1977a) reported on the fermentation of D,L-malic acid to lactic acid by polyacrylamide gel-entrapped *L. casei* (INRA/Montpellier No. CAP_1C_1), cultured on the following medium:

Glucose	10 g
D,L-malic acid	8 g
Yeast extract	4 g
Difco tryptone	2 g
KH_2PO_4	2 g
Clarified tomato juice	80 ml
$K_2SO_4 \cdot 7H_2O$	0.2 g
$MnSO_4 \cdot 4H_2O$	0.01 g
$FeCl_3$	0.002 g
Water to	1 L
pH 4.7 (adjusted with 10 N sodium bicarbonate)	

The substrate solution was same as above, except that only 1 g/L glucose was used, 90 ml/L ethanol (95%) and 50 mg/L cycloheximide were added as preservatives, and the pH was adjusted to 3.5. In a 150-ml reactor (50 ml biocatalyst, 38 ml/h flow rate) the immobilized cells were able to produce lactic acid for 10 months at a rate of 1.9 g/Lh.

The decarboxylation of L-malic acid to L-lactic acid during wine fermentation is often desirable in order to decrease acidity and to prevent fermentation in the bottle. Furthermore, an improvement of the flavor profile by certain lactic acid bacteria and by *Leuconostoc oenos* has been reported. In conventional wine processing, this malolactic fermentation normally takes place spontaneously by bacteria either from grape skins or from vats, during storage of the new wine at slightly elevated temperatures. Recently, Gestrelius (1982) demonstrated that malolactic fermentation may also be carried out by employing calcium alginate gel-entrapped *L. oenos* cells in a continuous column reactor.

Gestrelius mixed one volume of wet cell sludge (about 15% d.m.) with 5–12 volumes of sterile 5% sodium alginate and dripped the suspension into sterile 2% calcium chloride. After hardening for 2 hours the biocatalyst beads were placed on a sterile filtered grape juice/glycerol medium (reconstituted grape juice with 30 mM malate, 10% ethanol, and 35% glycerol, pH 5.0), incubated overnight at 6°C, and stored at -20°C. The storage half-life of the biocatalyst at this temperature was about 3 months. However, in continuous column operation, residual malolactic activity did not significantly decrease during 6 months of operation. Excellent mechanical properties of the calcium alginate beads made reactor bed heights as high as 1 m possible. The operational half-life with grape juice containing 12% ethanol was about 40 days at 20°C, pH 3.5–3.8, and about 20% of the activity was retained during complete conversion of substrate containing 100 ppm of SO_2. A pilot plant reactor system consisting of three 16-L packed-bed reactors was constructed and used to treat new 1980 Monterey County Cabernet Sauvignon wine of 10.8% ethanol and 1.73 g/L malic acid (pH 3.4, reducing sugars 0.1%, tartaric acid 3.15 g/L, free/total SO_2 0/7 ppm), and nearly complete conversion of L-malic acid to lactic acid at a 2-hour residence time, 16–17°C could be demonstrated. The reactor was kept under 0.3 atm pressure to keep the carbon dioxide produced in solution.

D. Applications in Dairy Technology

Continuous production methods based on conventional fermentation technology have been developed for sour curd cheese and for yogurt. Linklater and Griffin (1971b) produced sour curd cheese from skim milk in a laboratory-scale two-stage process, in which the skim milk was first prefermented by *S. lactis* in a continuous stirred tank reactor to a pH of 5.4–6.0. Both maximum cell growth and maximum acid production rate took place at pH 5.4 at a 1.5-hour residence time. At pH 5.4 or less, casein precipitated in the reactor, taking after 100 hours about a quarter of the reactor volume. The preferment was passed through a plug-flow fermentor, with an overall residence time of about 5 hours, to obtain the desired end pH of 4.7. The cheese was finally obtained by centrifuging.

Driessen et al. (1977a, b) produced yogurt continuously from pasteurized skim milk using a mixed culture of *Streptococcus thermophilus* and *L. bacillus bulgaricus*, employing a process essentially similar to that presented by Linklater and Griffin (1971b). The maximum acid production rate was obtained at a first stage effluent of pH 5.5. However, at an effluent pH of 5.7 or less the final yogurt consistency was broken down by pumping and mixing actions. The best yogurt structure was obtained if the milk was cooled to 37°C before pumping to the plug-flow fermentor. More recently, MacBean et al. (1979) reported difficulties in stabilizing the flow in the second fermentation stage. A two-stage continuous yogurt fermentation process has also been commercialized (Van der Loo, 1980). In this process, milk is first homogenized and heat-treated at 85°C for 5 min for sufficient viscosity and pasteurized at 90°C for 10 s to prevent contamination. The effluent pH from the first fermentation stage is controlled at 5.7 ± 0.05, which is reached after about 30 min of residence time in a 2-m^3 vessel. Milk is then cooled down to 37°C for the second stage, during which the pH decreases to 4.30. The product is finally cooled down to 8°C.

Divies (1977b) suggested the use of either polyacrylamide or calcium alginate gel-entrapped mixed culture of *Streptococcus cremoris* and *L. bulgaricus* cells for continuous prefermentation of milk for subsequent yogurt manufacture. In one reported experiment the prefermentation took place at 40°C to reach pH 5.45 at about a 15-min residence time with calcium alginate gel-immobilized cells. Yogurt was reportedly also produced in one step with polyacrylamide gel-entrapped bacteria at 40°C, with the final pH of 4.8 obtained at a 1.5-hour residence time (product lactic acid concentration was about 8.3 g/L). According to Siess and Divies (1981), about 30% of the *L. casei* cells are destroyed during immobilization in polyacrylamide gel, but cell growth and division regenerates a stable population in the biocatalyst beads. Koistinen et al. (1983) employed calcium alginate (8%) gel-immobilized *S. cremoris* Hansen 1977 and *L. helveticus* cas v/a cells for prefermentation of pasteurized skim milk. Most experiments were performed in small 16-ml laboratory packed-bed column reactors with *L. helveticus*, owing to its better lactic acid productivity and better resistance against contamination and to the poor stability of calcium alginate gel with *S. cremoris*. Furthermore, according to Lawrence and Thomas (1979), a successful use of *Streptococci* in continuous fermentation of milk is unlikely because of poor stability of the organism. McKay et al. (1972) discovered that as much as 90% of *S. lactis* colonies were lactose negative after 165 hours of continuous culture in a chemostat operated at a 5.9-hour residence time, which Lawrence and Thomas (1979) explained by the loss of plasmids that control the lactic acid fermentation.

In our work with immobilized *L. helveticus,* the column reactor was thermostated at 40–42°C and operated continuously for 5 hours daily, after which the biocatalyst was removed, washed with tap water, and subsequently stored overnight in a column under 5% lactose solution circulation at room temperature. The reactor parts were washed with 1% sodium hydroxide, 1% nitric acid, and finally with water. The lactic acid productivity of a new reactor was always poor during the first 2 days, during which time there was a relatively stable productivity of 3–4 mg/gh (calculated on biocatalyst gel basis) at a 15-min residence time, with a maximum of 5.5 mg/gh. At greater than 20-min residence times the rapid decrease in pH inhibited further fermentation, and the precipitation of milk proteins on the gel surface is likely to result in a decrease in biocatalyst activity. At residence times of 15–20 min a decrease in skim milk pH by 0.6–0.8 units could be achieved without excessive casein precipitation. It was also discovered that immobilized *L. helveticus* appeared to produce relatively less D-acid than free cells, which produce L- and D-acids at a ratio of about 2:1; the immobilized cells exhibited a ratio of about 4:1.

IV. ANALYTICAL APPLICATIONS OF IMMOBILIZED CELLS

Matsunaga et al. (1978) employed *L. arabinosus* ATCC 8014 immobilized in agar gel for rapid nicotinic acid determination by using immobilized bacteria and a combined glass electrode to measure the lactic acid produced. Bacteria immobilized in agar gel gave a larger potential difference (ΔE) than cells immobilized either in polyacrylamide gel or in collagen, and the response was linear between nicotinic acid concentrations of about 5×10^{-8} to 5×10^{-6} g/ml. The optimal agar concentration for immobilization was reported to be 2% (w/v), inasmuch as a further increase in the agar concentration would be expected to decrease the pore size in the interstitial space, preventing the growth of bacteria. On the other hand, too low an agar concentration would allow the bacteria to leak out.

V. EPILOGUE

Although applications of acid fermentations by immobilized cells have been relatively little investigated until recently, it is likely that the interest in such applications will increase in the near future. The vast potential of immobilized lactic acid bacteria bioreactors both for lactic acid production and for food applications has been demonstrated. Nevertheless, more work is still needed to scale up for industrial production. Certain problems in large-scale reactor design need to be solved. However, the immobilized lactic acid bacteria system offers an attractive alternative to obtain relatively pure lactic acid high in L-lactic acid content, thus reducing purification costs. Particularly attractive appears to be the application in prefermentation for the manufacture of certain milk products.

REFERENCES

Casida, L. E., Jr. (1964) *Industrial Microbiology*, John Wiley and Sons, New York.
Chibata, I. (1980) in *Food Process Engineering*, Vol. II, *Enzyme Engineering in Food Processing*, pp. 1–26, Applied Science Publishers, London.
Childs, C. C., and Welshby, B. (1966) *Process Biochem. 1*(8), 441–444.
Compere, A. L., and Griffith, W. L. (1975) *Dev. Ind. Microbiol. 17*, 247–252.
Compere, A. L., and Griffith, W. L. (1980) *Tappi 63*(2), 101–104.
Compere, A. L., and Griffith, W. L. (1981) U.S. Patent 4,287,305.
Cox, C. G., and MacBean, R. D. (1977) *Aust. J. Dairy Technol. 32*(1), 13–22.
Czarnetzky, E. J. (1958) U.S. Patent 2,904,437.
Divies, C. (1977a) French Demande 2,320,349 and Ger. Offen. 2,633,076.
Divies, C. (1977b) French Demande 2,359,202.
Driessen, F. M., Ubbels, J., and Stadhouders, J. (1977a) *Biotechnol. Bioeng. 19*, 821–829.

Driessen, F. M., Ubbels, J., and Stadhouders, J. (1977b) *Biotechnol. Bioeng. 19*, 841–851.

Eveleigh, D. D. (1981) *Scientific American 245*(3), 120–130.

Friedman, M. R., and Gaden, E. L. (1970) *Biotechnol. Bioeng. 12*, 961–974.

Gestrelius, S. (1982) in *Enzyme Engineering*, Vol. VI (Chibata, I., Fukui, S., and Wingard, L. B., Jr., eds.), pp. 245–250, Plenum Press, New York.

Griffith, W. L., and Compere, A. L. (1975) *Dev. Ind. Microbiol. 17*, 241–246.

Griffith, W. L., and Compere, A. L. (1977) *Dev. Ind. Microbiol. 18*, 723–726.

Henderson, H. E., Crickenberger, R., Reddy, C. A., and Rossman, E. (1973) *Beef Cattle Research Report*, p. 14, Michigan State University Agricultural Experimental Station, Mich.

Jansen, H. C. (1945) Dutch Patent 57,848.

Keller, A. K., and Gerhardt, P. (1975) *Biotechnol. Bioeng. 17*, 997–1018.

Koistinen, T., Harju, M., Heikonen, M., and Linko, P. (1983) *Karjantuote 66*(9), 25–28.

Lawrence, R. C., and Thomas, T. D. (1979) in *Microbial Technology* (Bull, A. T., Ellwood, D. C., and Ratledge, C., eds.), pp. 187–219, Cambridge University Press, Cambridge, England.

Linklater, P. M., and Griffin, C. J. (1971a) *J. Dairy Res. 38*, 127–136.

Linklater, P. M., and Griffin, C. J. (1971b) *J. Dairy Res. 38*, 137–144.

Linko, P. (1980) in *Food Process Engineering*, Vol. II, *Enzyme Engineering in Food Processing*, pp. 27–39, Applied Science Publishers, London.

Linko, P. (1981) in *Advances in Biotechnology*, Vol. I (Moo-Young, M., Robinson, C. W., and Vezzina, C., eds.), pp. 711–716, Pergamon Press, New York.

Linko, P., and Linko, Y.-Y. (1984) CRC Critical Rev. Biotechnol. *1*(4), 289–338.

Lipinsky, E. S. (1981) *Science 212*, 1465–1471.

Lockwood, L. B. (1979) in *Microbial Technology*, Vol. I, 2nd ed. (Peppler, J., and Perlman, D., eds.), pp. 356–388, Academic Press, New York.

MacBean, R. D., Hall, R. J., and Linklater, P. M. (1979) *Biotechnol. Bioeng. 21*, 1517–1541.

Man, J. C., Rogosa, M., and Sharpe, M. E. (1960) *J. Appl. Bacteriol. 23*, 130–135.

Marshall, K. R. (1970) in *Proceedings of the Eighteenth International Dairy Congress*, International Dairy Science Association, Sydney, p. 447.

Matsunaga, T., Karube, I., and Suzuki, S. (1978) *Anal. Chim. Acta 99*, 233–239.

McKay, L. L., Baldwin, K. A., and Zottola, E. A. (1972) *Appl. Microbiol. 23*, 1090–1096.

Miall, L. M. (1978) in *Economic Microbiology*, Vol. II (Rose, A. H., ed.), pp. 47–119, Academic Press, New York.

Milko, E. S., Sperelup, O. V., and Rabotnova, I. L. (1966) *Ann. Allgem. Mikrobiol. 6*, 297–301.

Mosbach, K., and Mosbach, R. (1966) *Acta Chem. Scand. 20*, 2807–2810.

Perquin, L. H. C. (1945) Dutch Patent 58,545.

Royce, T. B. V., Blanch, H. W., and Wilke, C. R. (1982) *Biotechnol. Lett. 4*, 483–488.

Setti, D. (1974) in *Proceedings of the Fourth International Congress of Food Sciences and Technology*, International Union of Food Science Technology, Madrid, Vol. IV, pp. 289–295.

Siess, M. H., and Divies, C. (1981) *Eur. J. Appl. Microbiol. Biotechnol. 12*, 10–15.

Stenroos, S.-L., Linko, Y.-Y., and Linko, P. (1982a) *Biotechnol. Lett. 4*, 159–164.

Stenroos, S.-L., Linko, Y.-Y., Linko, P., Harju, M., and Heikonen, M. (1982b) in *Enzyme Engineering*, Vol. VI (Chibata, I., Fukui, S., and Wingard, L. B., Jr., eds.), pp. 299–301, Plenum Press, New York.

Stieber, R. W., and Gerhardt, P. (1979) *Biotechnol. Bioeng. Symp. 9*, 137–148.

Stieber, R. W., and Gerhardt, P. (1981) *Biotechnol. Bioeng. 23*, 523–534.

Stieber, R. W., Coulman, G. A., and Gerhardt, P. (1977) *Appl. Environ. Microbiol. 34*, 733–739.

Tipayang, P., and Kozaki, M. (1982) *J. Ferment. Technol. 60*, 595–598.

Van der Loo, L. G. W. (1980) *Deut. Milchwirtschaft 29*, 1199–1202.

Villet, R. H. (1979) *SERI/TR-332-336*, Vol. I, Solar Energy Research Institute, Golden, Colo.

Whittier, E. O., and Rogers, L. A. (1931) *Ind. Eng. Chem. 23*, 532–534.

Yamada, K. (1977) *Biotechnol. Bioeng. 19*, 1563–1621.

Zeikus, J. G. (1980) *Ann. Rev. Microbiol. 34*, 423–464.

3

Immobilized Biocatalysts to Produce Amino Acids and Other Organic Compounds

Ichiro Chibata
Tetsuya Tosa
Tadashi Sato

I. INTRODUCTION

Since the late 1970's, biotechnology has received increasing attention with respect to problems of natural resources and energy. Biotechnology is a comprehensive term that includes the techniques of fermentation technology, enzyme engineering, and genetic engineering. In this field, enzyme engineering, using immobilized enzymes and microbial cells, is one of the main focal topics.

The utilization of enzymes began early in the history of human beings, and it has gradually expanded to a variety of fields such as brewing, food production, textiles, tanning, and medicine. Furthermore, the recent developments in biochemistry, the clarification of mechanisms of enzyme reaction, the development of new enzyme sources, and especially progress in applied microbiology and genetic engineering have markedly accelerated the utilization of enzymes. However, enzymes are not always ideal catalysts for practical application because they are generally unstable and cannot be used in organic solvents or at elevated temperature. Conventionally, enzyme reactions have been carried out in batch processes, incubating a mixture of substrate and soluble enzyme, and it is technically very difficult to recover the active enzyme from the reaction mixture for reuse. Thus in order to use enzymes more economically and efficiently the techniques of immobilization of enzymes have been the subject of increased interest since the late 1960's. Many papers and reviews on this subject have been published.

Many useful organic compounds have been produced by fermentation utilizing the catalytic activities of multienzyme systems in microorganisms. In order to skip the procedure of extracting enzymes from microbial cells, direct immobilization of whole microbial cells has also been attempted since the early 1970's.

Furthermore, the reactions for production of useful compounds in microorganisms often require generation of ATP and coenzymes such as NAD, NADP, and Coenzyme A. When immobilized microbial cells are kept in the living state, they may be applied to such reactions. On the basis of these ideas, studies on immobilized living cells have been increasing.

This chapter reviews studies on the biosynthesis and bioconversion of some useful organic compounds by immobilized biocatalysts, namely, immobilized enzymes, immobilized microbial cells, immobilized plant cells, and immobilized animal cells.

II. PREPARATION OF IMMOBILIZED BIOCATALYSTS

A large number of immobilization methods for enzymes, microbial cells, plant cells, and animal cells have been published. These methods can be classified into three categories: carrier-binding, cross-linking, and entrapping.

A. Carrier-Binding

The carrier-binding method is based on linking enzymes or cells directly to water-insoluble carriers and can be further divided into three categories according to the binding mode: physical adsorption, ionic binding, and covalent binding. As carriers, water-insoluble polysaccharides (cellulose, dextran, and agarose derivatives), proteins (gelatin and albumin), synthetic polymers (ion-exchange resins and polyvinylchloride), inorganic materials (brick, sand, and porous glass), and others are used. This method is mainly applied for immobilization of enzymes. For immobilization of cells the method is considered not to be advantageous because leakage of the enzymes may easily occur due to autolysis during the enzyme reaction.

Immobilization of enzymes by covalent binding is carried out under relatively severe conditions in comparison with those of physical adsorption or ionic binding. Thus in the former case, conformational changes of the enzyme structure and partial destruction of the active center may occur. Accordingly, unless immobilization of an enzyme by covalent binding is carried out under well-controlled conditions, immobilized enzymes having high activity cannot be obtained. However, the binding forces between the enzyme and carrier are strong, and the enzyme cannot easily leak out from carriers even in the presence of substrates or salts at high concentration. However, when the activity of enzymes immobilized by covalent binding decrease during long-term operation, regeneration is impossible.

On the other hand, immobilization of enzymes by the ionic binding method can be achieved simply under mild conditions. Accordingly, preparations having relatively high activity are obtained. However, the binding forces between enzyme and carrier are weak in comparison with covalent binding. Therefore leakage of the enzyme from the carrier may occur with changes of ionic strength or pH of the substrate or product solution. This type of immobilized enzyme can be regenerated when the enzyme activity decreases after prolonged operation. Thus the ionic binding method can be advantageous in comparison with the covalent binding method, particularly when expensive carriers or enzymes are used.

B. Cross-Linking

This immobilization method is based on the formation of chemical bonds, as in the covalent binding method, but water-insoluble carriers are not used. The immobilization of enzymes or cells is performed by the formation of intermolecular cross-linkages between the enzyme molecules or the cells, by means of bifunctional or multifunctional reagents.

As cross-linking reagents, such compounds as glutaraldehyde, bisisoocyanate derivative, and bisdiazobenzidine have been employed.

The cross-linking reactions are carried out under relatively severe conditions, as in the case of the covalent-binding method.

Few papers have appeared on this method so far. However, it is possible that suitable cross-linking reagents for immobilization of enzymes or cells will be found in the future.

C. Entrapping

The entrapping method is based on confining enzymes or cells in the lattice of a polymer matrix or enclosing them in semipermeable membranes.

This method differs from the covalent-binding and cross-linking methods in that the enzyme or cell itself does not bind to the gel matrix or membrane. Thus this method may have wide applicability. Many papers on the immobilization of microbial cells, plant cells, and animal cells by entrapment have been published. For this method the following matrices are among those that have been employed: collagen, gelatin, agar, alginate, carrageenan, cellulose triacetate, polyacrylamide, epoxy resin, photo-cross-linkable resin, polyester, polystyrene, and polyurethanes. Among these matrices, polyacrylamide gel, calcium alginate, and carrageenan have been most extensively used for immobilization of many kinds of enzymes and cells.

Recently, immobilized living cells have been studied for the production of useful compounds. In this case, carrageenan gel is useful. x-Carrageenan becomes a gel by cooling, and its gelation occurs also by contacting with a solution containing a gel-inducing reagent such as K^+, NH_4^+, Ca^{2+}, and amines. If a suitable gel-inducing reagent is selected for immobilization of individual microbial cells, the immobilization can be performed under very mild conditions and without the use of chemicals that may kill the cells. In addition, if the immobilized cells are suspended in physiological saline, the gel dissolves rapidly, and a cell suspension can be obtained. This is advantageous for investigating characteristics of microbial cells after immobilization. For example, immobilized cells are washed thoroughly with sterilized potassium chloride solution and are suspended in sterilized physiological saline to dissolve the x-carrageenan gel. The number of living cells can then be counted by the serial diluted drop-plate method.

Enzymes can also be encapsulated within semipermeable microcapsules or within liposomes, and microencapsulated enzymes have received attention for their potential use in medical applications.

III. CHEMICAL PROCESSES USING IMMOBILIZED BIOCATALYSTS

Applications of immobilized biocatalysts are extending to a variety of fields. In this section, examples of immobilized biocatalysts applied to biosynthesis and bioconversion of organic compounds are reviewed.

A. Production of Amino Acids

Amino acids are widely used in the food, feed, medicine, and cosmetic industries and also as starting materials for synthetic chemicals. For food and nutritional applications, only the L-isomers of the amino acids are active. There are many chemical synthetic methods for amino acid production, but the products are mixtures of the L- and D-isomers. To obtain the L-amino acid from the chemically synthesized D,L-form, optical resolution is necessary. Although these mixtures can be resolved optically by chemical or enzymatic methods, the resolution steps are generally complicated. Thus immobilized enzyme technology has been applied to the resolution of racemic mixtures. On the other hand, biosynthesis of amino acids by microorganisms or isolated enzyme systems leads exclusively to production of the L-isomer. For this reason, biological synthesis is generally the preferred production method. Here also, immobilized technology has been applied to develop a more efficient process for producing optically active amino acids.

1) Optical Resolution of D,L-Amino Acids.   To obtain natural L-amino acids
from the chemically synthesized D,L-form, optical resolution is necessary.
Among the many optical resolution methods the enzymatic method using mold
aminoacylase developed by us is one of the most advantageous procedures,
yielding optically pure L-amino acids. A chemically synthesized acyl-D,L-amino
acid is asymmetrically hydrolyzed by aminoacylase to give L-amino acid and
unhydrolyzed acyl-D-amino acid as follows:

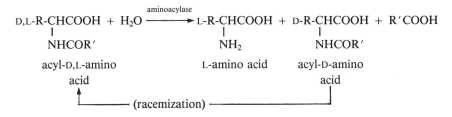

The two products are easily separated by the difference in their solubilities.
Acyl-D-amino acid is racemized and reused for the resolution procedure.

From 1954 to 1969 this enzymatic resolution method was employed by
Tanabe Seiyaku Co. Ltd. for the industrial production of several L-amino
acids. The enzyme reaction was carried out batchwise by incubating a mixture
containing substrate and soluble enzyme. However, this procedure had some
disadvantages inherent to a batch process using soluble enzyme.

For instance, in order to isolate the L-amino acid from the reaction mixture
we had to remove enzyme protein by pH and/or heat treatments. This resulted
in uneconomical use of enzyme. In addition, since complicated purification
procedures were necessary to remove contaminating proteins and coloring
substances, the yield of L-amino acid was lowered. Also, much labor was necessary for batch operation.

To overcome these drawbacks and to improve the procedure, we studied
extensively the continuous optical resolution of D,L-amino acids using immobilized aminoacylase. Mold aminoacylase was immobilized by a variety of
methods. As a result, relatively active and stable immobilized aminoacylases
were obtained by ionic binding to DEAE-Sephadex (Tosa et al., 1966, 1969),
covalent binding to iodoacetyl cellulose (Sato et al., 1971), and entrapment in
polyacrylamide gel lattice (Mori et al., 1972). To select the most suitable preparation for industrial purposes, the properties of these three immobilized
aminoacylases were compared (Chibata et al., 1972). Finally, we chose aminoacylase immobilized by ionic binding to DEAE-Sephadex for industrial purposes because the preparation is easy, the activity is stable and high, and the
regeneration of deteriorated preparations is possible.

Further, we carried out chemical engineering studies and designed an enzyme reactor system for the production of L-amino acids with immobilized
aminoacylase (Tosa et al., 1971; Chibata et al., 1972). In the immobilized enzyme process the overall production cost was more than 40% lower than that

of the conventional batch process using the soluble enzyme. Savings of enzyme and labor costs are the main contributors, as well as the increase of product yield due to easy isolation of L-amino acids from the reaction mixture. Since 1969 we have been operating several enzyme reactions in our plants for the production of L-amino acids, including L-methionine, L-valine, and L-phenyl-alanine, representing the first industrial application of immobilized enzymes.

Mold aminoacylase from *Aspergillus* sp. was also immobilized, by cova-lent binding to alkylaminosilanized porous glass with glutaraldehyde or to the diazonium derivative of arylaminosilanized porous glass, and these immobi-lized aminoacylases were used for continuous preparation of L-amino acids from acetyl-D,L-amino acids (Weetall and Detar, 1974). Bacterial and animal aminoacylases have been also immobilized. Bacterial aminoacylase was cova-lently bound to diazotized polyaminostyrene (Mitz and Summaria, 1961; Lilly et al., 1965). The aminoacylase from pig kidney was immobilized by ionic binding to DEAE-cellulose (Barth and Masková, 1971) and by entrapping in cellulose nitrate (Leuschner, 1966).

Besides these immobilized enzyme methods, *Aspergillus ochraceus* cells having aminoacylase activity were immobilized by cross-linking with egg al-bumin and glutaraldehyde, and the continuous optical resolution of acetyl-D,L-methionine was investigated by using these immobilized cells (Hirano et al., 1977). These preparations were employed batchwise and at a la-boratory scale for the preparation of L-amino acids from acyl-D,L-amino acids.

As well as aminoacylase, leucine aminopeptidase immobilized by covalent binding to CNBr-activated Sepharose was used for preparation of L-amino acids from D,L-amino acids (Koelsch, 1972).

2. Biosynthesis of Optically Active Amino Acids.   Many papers on the bio-synthesis of optically active amino acids using immobilized biocatalysts have been published (Table 3.1). In this section our industrial applications of im-mobilized biocatalysts for production of L-alanine and L-aspartic acid are described.

a. L-Alanine.   L-Alanine is a useful amino acid not only as a medicine but also as a food additive because of its good taste. It has been produced industrially from L-aspartic acid in our plant since 1965 by a batchwise enzyme reaction using L-aspartate β-decarboxylase from *Pseudomonas dacunhae* (Chibata et al., 1965).

To develop a more efficient process for producing L-alanine, we studied continuous L-alanine production from L-aspartic acid using *P. dacunhae* cells immobilized with $\varkappa$-carrageenan (Yamamoto et al., 1980; Takamatsu et al., 1981).

In this continuous production system, one of the problems is the evolution of CO_2 gas during the L-aspartate β-decarboxylase reaction. It is not efficient

TABLE 3.1 Synthesis of Optically Active Amino Acids by Immobilized Biocatalysts

Amino Acid	Raw Materials	Enzymes or Microbial Cells	Matrices for Immobilization	Reference
L-Alanine	L-Aspartic acid	*Pseudomonas dacunhae* (L-aspartate β-decarboxylase)	Carrageenan	Yamamoto et al. (1980)
	Ammonium fumarate	*Escherichia coli* (aspartase) and *Pseudomonas dacunhae* (L-aspartate β-decarboxylase)	Carrageenan	Sato et al. (1982) Takamatsu et al. (1982)
		Aspartase and L-aspartate β-decarboxylase	Ultrafiltration membrane	Jandel et al. (1982) Wandrey et al. (1982)
	Glucose	*Corynebacterium dismutans* (multienzymes)	Polyacrylamide	Sarkar and Mayaudon (1983)
			Carrageenan	Sarkar and Mayaudon (1983)
L-Arginine	Glucose	*Serratia marcescens* (multienzymes)	Carrageenan	Fujimura et al. (1983)
L-Aspartic acid	Ammonium fumarate	*Escherichia coli* (aspartase)	Polyacrylamide	Chibata et al. (1974c)
		Aspartase	Carrageenan	Sato et al. (1979)
			Polyurethane	Fusee et al. (1981)
			Polyacrylamide	Tosa et al. (1973)
			Duolite A-7	Yokote et al. (1978) Kimura et al. (1981)
			Cellulose triacetate	Pittalis et al. (1979)
ε-Aminocaproic acid	ε-Aminocaproic acid cyclic dimer	*Achromobacter guttatus* (cyclic dimer hydrolase)	Polyacrylamide	Kinoshita et al. (1975)
L-Citrulline	L-Arginine	*Pseudomonas putida* (L-arginine deiminase)	Polyacrylamide	Yamamoto et al. (1974a)

(continued)

43

TABLE 3.1 *(continued)*

Amino Acid	Raw Materials	Enzymes or Microbial Cells	Matrices for Immobilization	Reference
L-Glutamic acid	Glucose	*Brevibacterium flavum* (multienzymes)	Collagen	Constantinides et al. (1981)
		Corynebacterium glutamicum (multienzymes)	Polyacrylamide	Slowinski and Charm (1973)
L-Isoleucine	Glucose	*Serratia marcescens* (multienzymes)	Carrageenan	Wada et al. (1980)
L-Lysine	Diaminopimelic acid	*Microbacterium ammoniaphilum* (diaminopimelic acid decarboxylase)	Polyacrylamide	Kanemitsu (1975)
	D,L-α-Amino-ε-caprolactam	L-α-Amino-ε-caprolactam hydrolase and α-amino-ε-caprolactam racemase	DEAE-Sephadex	Fukumura (1975)
D-α-Phenylglycine	Phenylhydantoin	*Bacillus* sp. (hydantoinase)	Polyacrylamide	Yamada et al. (1980b)
D-α-Hydroxyphenylglycine	*p*-Hydroxyphenylhydantoin	*Bacillus* sp. (hydantoinase)	Polyacrylamide	Yamada et al. (1980b)

Product	Substrate	Enzyme / Microorganism	Support	Reference
L-Tryptophan	Indol and L-serine	Tryptophan synthase *Escherichia coli* (tryptophan synthase)	Cellulose triacetate Polyacrylamide Chitosan	Dinelli (1972) Zaffaroni et al. (1974) Chibata et al. (1974a) Bang et al. (1983) Verlop and Klein (1981)
L-Tryptophan	Indol, pyruvate, and NH$_3$	Tryptophanase *Escherichia coli* (tryptophanase)	CNBr-activated Sepharose Polyacrylamide	Fukui et al. (1975a) Fukui et al. (1975b) Maréchal et al. (1979)
5-Hydroxytryptophan	5-Hydroxyindol and L-serine	*Escherichia coli* (tryptophan synthase)	Polyacrylamide	Chibata et al. (1974b).
L-Tyrosine	Phenol, pyruvate, and NH$_3$	β-Tyrosinase *Erwinia herbicola* (β-tyrosinase)	CNBr-activated Sepharose Collagen and glutaraldehyde	Axén and Ernback (1971) Fukui et al. (1975a) Fukui et al. (1975b) Yamada et al. (1978)
L-DOPA	Pyrocatechol, pyruvate, and NH$_3$	*Erwinia herbicola* (β-tyrosinase)	Collagen and dialdehyde-starch	Yamada et al. (1978)

to perform this reaction using a conventional column system at normal pressure because the evolution of CO_2 gas makes it difficult to obtain complete plug-flow of substrate solution during the reaction. Therefore we investigated the most advantageous reaction system using immobilized *P. dacunhae* for continuous production of L-alanine and designed a closed column reactor that performs the enzyme reaction at high pressure, for example, 10 kg/cm². By using this reactor, in which liberated CO_2 gas is dissolved into the reaction mixture, complete plug-flow of substrate solution is obtained, and the pH of reaction mixture is not much changed.

The efficiency of immobilized cells for production of L-alanine in the closed column system at high pressure is much higher than in the conventional column system at normal pressure. The stability of immobilized cells in the two systems is similar.

We reasoned that if immobilized *Escherichia coli* having aspartase activity and immobilized *P. dacunhae* having L-aspartate β-decarboxylase were employed in a single reactor, L-alanine might be produced more efficiently from ammonium fumarate (Sato et al., 1982) according to the following equation:

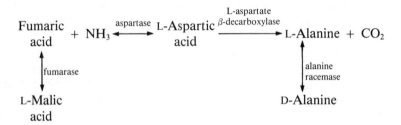

However, since these two microorganisms have alanine racemase and fumarase activities, D-alanine and L-malic acid were formed as by-products. Thus for effective production of L-alanine, the actions of these undesirable enzymes had to be removed. Therefore we tried various procedures for removal of these enzymes (Takamatsu et al., 1982). It was found that when *E. coli* cells were treated at pH 5.0, 45°C for 1 h, and *P. dacunhae* cells at pH 4.75, 30°C, for 1 h before immobilization with $\varkappa$-carrageenan, the aspartase of *E. coli* and the L-aspartate β-decarboxylase of *P. dacunhae* were not decreased, but the alanine racemase and fumarase of the two microorganisms were almost completely inactivated.

We then designed an enzyme reactor system for production of L-alanine from ammonium fumarate using the two immobilized cell types (Takamatsu et al., 1983). We industrialized this system in 1982, and we are satisfied with both the economical efficiency and the quality of the product.

Continuous production of L-alanine from ammonium fumarate also was carried out in a two-stage membrane reactor containing partially purified aspartase and L-aspartate β-decarboxylase (Jandel et al., 1982; Wandrey et al., 1982).

Recently, *Corynebacterium dismutans* was immobilized in a polyacrylamide gel lattice, and the immobilized cells were used for synthesis of ^{14}C-L-alanine from ^{14}C-glucose (Sarkar and Mayaudon, 1983).

b. L-Aspartic Acid.   L-Aspartic acid is widely used not only as a medicine but also as a food additive and has been produced commercially by fermentative or enzymic batch process from ammonium fumarate using the action of aspartase. In order to improve the productivity of this system we studied the continuous production of L-aspartic acid using immobilized aspartase (Tosa et al., 1973). Since the aspartase is an intracellular enzyme, it was necessary to extract the enzyme from microbial cells before immobilization. Extracted intracellular enzymes are generally unstable, and most of the immobilization methods we tried resulted in low activity and poor yield. Although entrapment in polyacrylamide gel lattice gave relatively active immobilized aspartase, its operational stability was not adequate for the industrial production of L-aspartic acid. We reasoned that if whole microbial cells having the enzyme activity could be immobilized directly, these disadvantages might be overcome. Therefore we studied the immobilization of whole microbial cells. Whole cells of *E. coli* having high aspartase activity were immobilized by various methods; the most active immobilized *E. coli* cells were obtained by entrapping the cells in polyacrylamide gel lattice (Chibata et al., 1974a).

A column packed with the immobilized *E. coli* cells had good operational stability; its half-life was 120 days at 37°C (Tosa et al., 1974; Sato et al., 1975).

For the industrial application of this technique we carried out a chemical engineering analysis of the continuous enzyme reaction using a column packed with the immobilized *E. coli* cells, and the aspartase reactor system was designed (Sato et al., 1975). Since this aspartase reaction is exothermic, the column reactor used for industrial production of L-aspartic acid was designed as a multistage system with a radiator.

This immobilized cell system has been operating industrially since 1973 at Tanabe Seiyaku Co. Ltd., Japan. The overall production cost by this system was reduced to about 60% of that of the conventional batchwise reaction using intact cells, owing to the marked increase of productivity of L-aspartic acid per unit of cells, reduction of labor costs due to automation, and an increase in the yield of L-aspartic acid. Furthermore, the procedure employing immobilized cells is advantageous from the standpoint of waste treatment.

For further improvement of this process we developed the use of x-carrageenan as an immobilization matrix for *E. coli* cells. The efficiency of *E. coli* cells immobilized with polyacrylamide and x-carrageenan in L-aspartic acid production was compared. *E. coli* cells immobilized with x-carrageenan and treated with glutaraldehyde and hexamethylenediamine showed the highest productivity, and the half-life was 680 days at 37°C (Sato et al., 1979). Therefore in 1978 we changed the conventional polyacrylamide gel method to the carrageenan method for industrial production of L-aspartic acid. If a 1000-liter column is used, the theoretical yield of L-aspartic acid is 3.4 tons/day and 100 tons/month.

Several papers by other investigators have been published on the continuous production of L-aspartic acid using immobilized enzymes and immobilized microbial cells. The continuous production of L-aspartic acid from ammonium fumarate was investigated at a laboratory scale using *E. coli* cells immobilized with polyurethane (Fusee et al., 1981). Aspartase extracted from *E. coli* cells was immobilized in the presence of the substrate to a weakly basic anion exchange resin, Duolite A-7, by ionic binding, and the immobilized enzyme was used for continuous production of L-aspartic acid from ammonium fumarate (Yokote et al., 1978). Aspartase extracted from *Escherichia intermedia* was entrapped in cellulose triacetate porous fiber (Pittalis et al., 1979), and the enzyme from a thermophilic bacterium, *Bacillus aminogenes*, was immobilized ionically on Duolite A-7 and then treated with glutaraldehyde (Kimura et al., 1981).

c. L-Glutamic Acid.   Generally, L-glutamic acid is produced industrially by chemical synthesis or fermentation. The microorganism commonly employed is a biotin-requiring strain of *Corynebacterium glutamicum*. Whole cells of *C. glutamicum* were immobilized by entrapping in polyacrylamide gel lattice, and the production of L-glutamic acid from glucose was investigated (Slowinski and Charm, 1973). Immobilized cells corresponding to 2.8 mg (dry weight) were suspended in 40 ml of a fermentation medium containing glucose as the carbon source, and the suspension was incubated at 30°C with stirring. After 144 hours of incubation, 15 mg/ml of L-glutamic acid accumulated in the reaction medium. The productivity of L-glutamic acid by immobilized cells was said to be superior in comparison with that of intact cells.

Also, living cells of *Brevibacterium flavum* immobilized in collagen were used for the continuous production of L-glutamic acid in a recycle reactor system (Constantinides et al., 1981).

d. L-Tryptophan.   L-Tryptophan can be produced from indole, pyruvate, and ammonia by a biosynthetic method involving tryptophanase. The enzyme was immobilized by covalent binding to CNBr-activated Sepharose (Fukui et al., 1975a, b). The immobilized enzyme was used for production of L-tryptophan from indole, pyruvate, and ammonia. L-Tryptophan was synthesized in higher yields by a repeated batch process. However, a gradual decrease of activity was observed during repeated use in the absence of coenzyme, pyridoxal phosphate. A partially purified tryptophan synthase from *E. coli* was immobilized by entrapment in cellulose triacetate fibers for production of L-tryptophan from indole and L-serine in batch and continuous processes (Dinelli, 1972; Zaffaroni et al., 1974). At high indole concentration the enzyme activity was reduced owing to adsorption of indole to the fibers, leading to an increased diffusional barrier.

Also, we immobilized whole cells of *E. coli* having tryptophan synthase activity into a polyacrylamide gel lattice and investigated the synthesis of L-tryptophan from indole and L-serine (Chibata et al., 1974a) or 5-hydroxytryptophan from 5-hydroxyindole and L-serine (Chibata et al., 1974b). The yields of tryptophan and 5-hydroxytryptophan were 86% and 72%, respectively.

e. L-Tyrosine.   In analogy with tryptophanase, tyrosine can be synthesized from phenol, pyruvate, and ammonia by the action of β-tyrosinase. Also, the enzyme can be used to produce 3,4-dihydroxy-L-phenylalanine (L-DOPA), useful for treating Parkinson's disease, from pyrocatechol, pyruvate, and ammonia.

β-Tyrosinase from *Escherichia intermedia* was immobilized by covalent binding to CNBr-activated Sepharose (Axén and Ernback, 1971; Fukui et al., 1975a, b), and the immobilized enzyme preparation was used for the continuous production of L-tyrosine and L-DOPA.

Also, *Erwinia herbicola* having β-tyrosinase activity was immobilized by cross-linking with collagen and glutaraldehyde or dialdehyde starch, and the immobilized cells were used for production of L-tyrosine from phenol, pyruvate, and ammonia and production of L-DOPA from pyrocatechol, pyruvate, and ammonia, respectively (Yamada et al., 1978).

f. L-Lysine.   L-Lysine is an essential amino acid and is used as a medicine, as a food additive, and in animal feed.

In 1974 an interesting report on the preparation of L-lysine using two kinds of immobilized enzymes was published by Tore Co. Ltd., Japan (Fukumura, 1975). Cyclohexane, a by-product of nylon production, was used as a starting material and was converted to D,L-α-amino-ε-caprolactam by a synthetic chemical method. L-α-Amino-ε-caprolactam hydrolase and α-amino-ε-caprolactam racemase were immobilized by ionic binding to DEAE-Sephadex, and the two immobilized enzymes were allowed to react with D,L-α-amino-ε-caprolactam simultaneously. L-α-Amino-ε-caprolactam was hydrolyzed to give L-lysine, and the remaining D-α-amino-ε-caprolactam was racemized to the D,L-form by the action of racemase and again hydrolyzed to L-lysine. The D,L-form was finally converted to L-lysine. Further, *Microbacterium ammoniaphilum* having diaminopimelic acid decarboxylase activity was immobilized in a polyacrylamide gel lattice, and the immobilized cells were used for production of L-lysine from diaminopimelic acid (Kanemitsu, 1975).

g. Other Amino Acids.   Besides those described above, other amino acids have been produced by immobilized biocatalysts (Table 3.1). For example, we studied continuous production of L-citrulline from L-arginine using immobilized *Pseudomonas putida* having L-arginine deiminase activity (Yamamoto et al., 1974a). Also, whole cells of *Achromobacter guttatus* having cyclic dimer hydrolase were immobilized by entrapment in polyacrylamide gel, and the immobilized cells were used for production of ε-aminocaproic acid from ε-aminocaproic acid cyclic dimer (Kinoshita et al., 1975).

Further, D-α-phenylglycine and D-α-hydroxyphenylglycine, which are important components of the semisynthetic penicillins and cephalosporins, were produced from phenylhydantoin and *p*-hydroxyphenylhydantoin by immobilized *Bacillus* sp. having hydantoinase (Yamada et al., 1980b). Recently, we studied continuous production of L-isoleucine and L-arginine from glucose using immobilized living cell systems (Wada et al., 1980; Fujimura et al., 1983).

B. Production of Antibiotics

Immobilized enzymes and microbial cells are very useful in the production of antibiotics; the immobilized biocatalysts have been utilized in some cases for the commercial production of penicillins and cephalosporins.

1. Penicillins and 6-Aminopenicillanic Acid.   There are relatively few reports on the enzymic acylation of 6-aminopenicillanic acid (6-APA) for production of penicillins (Table 3.2). The reasons include the relatively low yield of acylated product and the availability of simple chemical methods for acylation. Low yields are attributable to significant product or substrate inhibition and the reversibility of the reaction.

Penicillin amidase from *E. coli*, entrapped in cellulose triacetate fibers, was used in a continuous flow reaction to produce ampicillin or amoxycillin from 6-APA and D-phenylglycine methyl ester or D-*p*-hydroxyphenylglycine methyl ester, respectively (Marconi et al., 1975). In all experiments the conversion was less than 50%. Also, succinoylated penicillin amidase adsorbed on DEAE-Sephadex was used for production of ampicillin in a stirred batch reactor; the yield of ampicillin was 67% (Kamogashira et al., 1972a, b).

In another process, immobilized whole cells were used to synthesize ampicillin from 6-APA and D-phenylglycine methyl ester. Whole cells of *Achromobacter* sp. or *Bacillus megaterium* were immobilized by ionic binding on DEAE-cellulose (Fujii et al., 1973a, b). These immobilized cells were used in a column; the yield of ampicillin was 54%. Mycelium and protoplasts of *Penicillium chrysogenum* were immobilized with polyacrylamide (Morikawa et al., 1979a) and calcium alginate (Kurzatkowski et al., 1982) for studies on the synthesis of penicillin G from glucose.

6-APA is an important intermediate compound for the production of semisynthetic penicillins and is produced commercially on a large scale from penicillin G (benzyl penicillin) or penicillin V (phenoxymethyl penicillin), both of which are produced by fermentation, by deacylation using the action of penicillin amidase. This enzymatic process is widely applied, and the conventional process based on soluble enzyme has been largely replaced by immobilized biocatalyst processes.

Partially purified penicillin amidase from *E. coli*, entrapped in cellulose triacetate fibers, was used for production of 6-APA from penicillin G in a column as part of a recirculation batch reactor with continuous titration of liberated acid (Dinelli, 1972; Marconi et al., 1973; Giacobbe et al., 1977). In this system, 6% substrate solution was hydrolyzed to at least 90% during a 1.5- to 2-h retention time; the half-life of the column was about 2 months at 37°C.

Further, purified penicillin amidase from *E. coli*, immobilized by covalent binding to CNBr-activated Sephadex, was used in a batch process for the industrial production of 6-APA from penicillin G (Delin et al., 1973; Lagerlöf et al., 1976). The immobilized enzyme could be used for more than 100 batches without addition of fresh enzyme, and 6-APA of 98% purity was obtained in

90% yield. Also, penicillin amidase from *E. coli*, immobilized covalently to bromoacetyl cellulose (Savidge et al., 1969) or polymethacrylate resin with glutaraldehyde (Savidge et al., 1974), was used in an ultrafiltration reactor to hydrolyze penicillin G or penicillin V. 6-APA could be obtained in about 90% yield.

Penicillin amidase was also immobilized by covalent binding to a chloro-*s*-triazine derivative of DEAE-cellulose (Self et al., 1969), by physical adsorption to bentonite (Ryu et al., 1972), by ionic binding to DEAE-Sephadex (Kamogashira et al., 1972a, b) and by entrapping in a polyacrylamide gel lattice (Huper, 1973).

Immobilized whole cells also have been used to hydrolyze penicillins. We studied a continuous method for the production of 6-APA from penicillin G using *E. coli* cells immobilized with polyacrylamide (Sato et al., 1976). The half-life of the column under continuous operation was estimated to be 42 days at 30°C and 17 days at 40°C. Pure 6-APA could be produced in 78% yield. Further, by cloning the penicillin amidase gene of *E. coli* ATCC 11105 on multicopy plasmids a new hybrid strain, *E. coli* 5K (PHM 12), having high enzyme activity was obtained (Mayer et al., 1980). The whole cells of the new *E. coli* strain were immobilized in calcium alginate or an epoxymatrix obtained by polycondensation of a water-soluble epoxy-precursor and polyfunctional amine (Klein and Wagner, 1980). This is considered to be the first promising example combining the techniques of genetic engineering and enzyme engineering.

2. Cephalosporins, 7-Aminocephalosporanic Acid, and 7-Aminodesacetoxy-cephalosporanic Acid.   Cephalosporins can be synthesized from penicillins or produced by direct fermentation by using *Cephalosporium acremonium*. The fermentation product is cephalosporin C, which contains the nucleus, 7-aminocephalosporanic acid (7-ACA), and the side chain, α-aminoadipic acid. Cephalosporins can also be produced chemically from penicillins by expanding the five-membered thiazolidine ring of penicillin to the six-membered dihydrothiazine ring of cephalosporin containing the nucleus 7-aminodesacetoxy cephalosporanic acid (7-ADCA).

Cephalosporin amidases from various microorganisms were immobilized by various methods, and the immobilized enzyme systems have been used for the production of various cephalosporin derivatives. The enzyme from *E. coli*, entrapped in cellulose triacetate fibers, was used for the production of cephaloxin from 7-ADCA and D-phenylglycine methyl ester (Marconi et al., 1975). In this case, 75% of the 7-ADCA was converted to cephalexin in one hour at 25°C. When the reaction was carried out for a longer time, the yield of product decreased. This was probably due to the enzyme catalyzing the reverse reaction.

The enzyme from *Bacillus megaterium*, adsorbed on alumina (Fujii et al., 1974) or celite (Fujii et al., 1973b), was also employed to carry out the reverse reaction, that is, the acylation of 7-ACA or 7-ADCA. The commercially im-

TABLE 3.2 Synthesis of Antibiotics by Immobilized Biocatalysts

Antibiotic	Raw Materials	Enzymes or Microbial Cells	Matrices for Immobilization	Reference
Amoxycillin	6-APA and p-hydroxyphenylglycine methyl ester	Penicillin amidase	Cellulose triacetate	Marconi et al. (1975)
Ampicillin	6-APA and D-phenylglycine methyl ester	*Bacillus megaterium* (penicillin amidase)	DEAE-cellulose	Fujii et al. (1973a, b)
		Achromobacter aceris (penicillin amidase)	DEAE-cellulose	Fujii et al. (1973a, b)
		Achromobacter liquidum (penicillin amidase)	DEAE-cellulose	Fujii et al. (1973a, b)
		Kluyvera citrophila (penicillin amidase)	Polyacrylamide	Morikawa et al. (1980a)
		Penicillin amidase	Cellulose triacetate	Marconi et al. (1975)
		Succinolylated penicillin amidase	DEAE-Sephadex	Kamogashira et al. (1972a, b)
Bacitracin	Nutrient medium (starch, peptone, meat extract)	*Bacillus* sp. (multienzymes)	Polyacrylamide	Morikawa et al. (1979b) Morikawa et al. (1980b)
Cephalexin	7-ADCA and D-phenylglycine methyl ester	*Achromobacter* sp. (cepharosporin amidase)	DEAE-cellulose	Fujii et al. (1973a)
		Achromobacter sp. (cepharosporin amidase)	Hydroxylapatite	Fujii et al. (1973a)
		Bacillus megaterium (cepharosporin amidase)	Celite	Fujii et al. (1973a)
		Cepharosporin amidase	Cellulose triacetate	Marconi et al. (1975)

52

Product	Substrate	Organism (enzyme)	Support	Reference
Cephalexin	7-ADCA and D-phenylglycine methyl ester	*Xanthomonas citri* (cepharosporin amidase)	Polyacrylamide	Kim et al. (1983)
Cephaloglycine	7-ACA and D-phenylglycine methyl ester	Cepharosporin amidase	Alumina	Fujii et al. (1974)
Cephalosporin C	3-(*N*-Morpholino) propane sulfonic acid	*Streptomyces clavuligeus* (multienzymes)	Polyacrylamide	Freeman and Aharonowitz (1981)
Cephalothin	7-ACA and 2-thiophene acetic acid methyl ester	Cepharosporin amidase	Celite	Fujii et al. (1973b)
Nikkomycin	Nutrient medium (malt extract, peptone, starch, mannitol)	*Streptomyces tendae* (multienzymes)	Calcium alginate	Veelken and Pape (1982)
Patulin	Glucose	*Penicillium urticae* (multienzymes)	Carrageenan	Deo and Gaucher (1983)
Penicillin G	Glucose	*Penicillium chrysogenum* (multienzymes)	Polyacrylamide	Morikawa et al. (1979a)
			Calcium alginate	Kurzatkowski et al. (1982)
Tylosin	Nutrient medium (malt extract, peptone, starch, mannitol)	*Streptomyces* sp. (multienzymes)	Calcium alginate	Veelken and Pape (1982)

portant cephalothin and cephaloglycine were synthesized from 7-ACA and 2-thiophene acetic acid or D-phenylglycine methyl ester by continuously passing the substrate solution through the immobilized enzyme column.

Further, several immobilized whole cell systems were investigated for production of cephalexin. Acetone-dried cells of *Achromobacter* sp., adsorbed on DEAE-cellulose or hydroxylapatite, were used for the production of cephalexin (Fujii et al., 1973a). Cephalosporin C was synthesized from 3-(*N*-morpholino) propane sulfonic acid by whole cells of *Streptomyces clavuligerus* immobilized with polyacrylamide (Freeman and Aharonowitz, 1981).

For enzymatic deacylation of cephalosporins, most studies have been carried out on the 7-ADCA nucleus. Substrates for the reaction are usually either 7-phenylacetamidodesacetoxy-cephalosporanic acid (phenylacetyl-7-ADCA) or 7-phenoxyacetamidodesacetoxy cephalosporanic acid (phenoxyacetyl-7-ADCA). These substrates are obtained easily by the ring expansion reaction from penicillin G and penicillin V, respectively.

The enzyme from *B. megaterium* was adsorbed on celite, and the immobilized enzyme was used for the production of 7-ADCA from phenylacetyl-7-ADCA (Fujii et al., 1976). Crystalline 7-ADCA (1.13 kg corresponding to 85% yield) was produced during four days of continuous operation of a column packed with the immobilized enzyme (10 ℓ).

For the deacylation of the same substrate to 7-ADCA the enzyme from *Arthrobacter viscous* was immobilized by adsorption on hydroxylapatite (Takeda et al., 1975).

3. Other Antibiotics.   Few reports have been published on the utilization of immobilized cells for the production of secondary metabolites such as antibiotics (Table 3.2). Whole cells of *Bacillus* sp., bacitracin-producing bacteria, were immobilized in a polyacrylamide gel lattice, and the immobilized living cells were used for the production of bacitracin in batch and in continuous culture systems (Morikawa et al., 1979b, 1980b). Also, the macrolide antibiotic, tylosin, and nucleoside peptide antibiotic, nikkomycin, were produced by calcium alginate-immobilized living cells of *Streptomyces* sp. and *Streptomyces tendae*, respectively (Veelken and Pape, 1982). Continuous production of these antibiotics was carried out in an air-bubbled reactor. Further, conidia of *Penicillium urticae*, immobilized in *x*-carrageenan beads, were germinated in situ to form a patulin-producing cell mass by incubating these beads in a growth-supporting medium, and the immobilized cells were used for the production of patulin from glucose (Deo and Gaucher, 1983).

C. Organic Acids
Organic acids are widely used in food and medicine, and some of them are produced by conventional fermentation. For improvement of productivities, immobilized microbial cells have been investigated.

1. Malic Acid.   L-Malic acid is an essential compound in cellular metabolism and is mainly used in the pharmaceutical field. We studied a process for the continuous production of L-malic acid from fumaric acid (Yamamoto et al., 1976, 1977). *Brevibacterium ammoniagenes* was chosen as a bacterium having high fumarase activity and was immobilized with polyacrylamide gel. However, when immobilized *B. ammoniagenes* was used for the production of L-malic acid from fumaric acid, succinic acid was formed as a by-product. Success in the industrial production of pure L-malic acid depends upon the prevention of succinic acid formation during the enzyme reaction because it is difficult to separate this by-product from malic acid. Thus attempts were made to suppress the side reaction. By treating the immobilized cells with bile extract the formation of succinic acid was suppressed.

We studied the conditions for the production of L-malic acid using a column packed with cells immobilized by polyacrylamide gel and treated with bile extract. This production system was industrialized in 1974. As in the case of L-aspartic acid production we investigated the carrageenan method to improve the productivity of L-malic acid (Takata et al., 1979, 1980). After screening of various microorganisms having higher fumarase activity, *Brevibacterium flavum* was found to show higher enzyme activity after immobilization with x-carrageenan than that of the formerly used *B. ammoniagenes*. Therefore the polyacrylamide method was changed to the carrageenan method in 1977.

Further, we found that the heat stability of fumarase activity of *B. flavum* increased when immobilized with x-carrageenan in the presence of polyethyleneimine (Takata et al., 1982). The productivity of *B. flavum* immobilized with x-carrageenan and polyethyleneimine increased to two times that of the cells immobilized with x-carrageenan only. Thus in 1980 the industrial production system of L-malic acid was changed to this improved method.

Besides our studies, the thermophilic bacterium *Thermus rubens* nov sp. was immobilized ionically on Duolite for the production of L-malic acid from fumaric acid (Ado et al., 1982).

2. Urocanic Acid.   Urocanic acid is used as a sun-screening agent in the pharmaceutical and cosmetic fields and has been produced from L-histidine by the action of L-histidine ammonia lyase. We immobilized *Achromobacter liquidum* cells having high L-histidine ammonia lyase activity by the polyacrylamide gel method and investigated the continuous reaction using the immobilized cells packed into a column (Yamamoto et al., 1974b). However, *A. liquidum* was found to have urocanase activity, which converts urocanic acid to imidazolone propionic acid. Thus for effective production of urocanic acid it was necessary to prevent the action of this enzyme. Various procedures were investigated, and it was found that when *A. liquidum* was heated at 70°C for 30 min, the L-histidine ammonia lyase activity was not decreased, but urocanase was completely inactivated. Thus the simple heat treatment before immobilization of the cells completely suppressed the conversion of urocanic acid to imidazole propionic acid by urocanase. When L-histidine solution containing Mg^{2+} was passed through a column packed with immobilized *A. liquidum* cells, the amino acid was completely converted to urocanic acid. Urocanic acid of high purity was obtained from the effluent of the column in good yield without re-crystallization by merely adjusting the pH of the effluent to around 4.7. The immobilized cell column was very stable, and its half-life was 180 days at 37°C.

Whole cells of *Micrococcus luteus* containing L-histidine ammonia lyase activity were immobilized by covalent coupling to CM-cellulose with carbodi-imide reagent, and the immobilized cells were used for continuous production of urocanic acid from L-histidine (Jack and Zajic, 1977). Also, *Pseudomonas fluorescens* having L-histidine ammonia lyase was used in hollow fibers for the production of urocanic acid (Kan and Shuler, 1978).

3. α-Keto Acids.   Recent papers indicate the potential use of α-keto acids for therapy for chronic uremia. Some α-keto acids are prepared from the corres-ponding amino acids by the action of amino acid oxidase. As shown in Table 3.3, two papers have been published on the use of immobilized microbial cells for the amino acid oxidase reaction. *Trigonopsis variabilis* having high D-amino acid oxidase activity was immobilized with calcium alginate, and the immobilized cells were used for the production of α-keto acid from the D-amino acid (Brodelius et al., 1980). The enzyme showed relatively high activ-ity toward the following D-amino acids: valine, leucine, isoleucine, methionine, phenylalanine, tyrosine, tryptophan, and histidine.

Further, whole cells of *Chlorella vulgaris* and *Anacystis nidulans* contain-ing L-amino acid oxidase activity, which was increased by illumination with red light, were coimmobilized with agarose (Wikström et al., 1982). The coim-mobilized cells produced O_2 photosynthetically within the beads. Illuminated cells produced up to ten times more keto acid than did nonilluminated cells.

4. 2-Ketogulonic Acid.   2-Ketogulonic acid is an important intermediate for the synthesis of vitamin C and can be produced by microbial oxidation from L-sorbose via L-sorbosone as an intermediate. L-Sorbose is transformed to L-sorbosone by L-sorbose dehydrogenase, and the L-sorbosone is converted to 2-ketogulonic acid by L-sorbosone oxidase.

Whole cells of *Gluconobacter melanogenus* immobilized with polyacryl-amide gel were effective in converting L-sorbose to L-sorbosone (Martin and Perlman, 1976a). The rate of conversion of sorbosone to 2-ketogulonic acid, however, was limited by L-sorbosone oxidase. Consequently, the process was improved by coimmobilizing *G. melanogenus* and *Pseudomonas syringus* cells having high sorbosone oxidase activity into a polyacrylamide gel lattice for direct oxidation of L-sorbose to 2-ketogulonic acid (Martin and Perlman, 1976b).

5. Other Organic Acids.   As shown in Table 3.3, the immobilized living cell systems can be applied to aerobic multistep reactions such as the production of acetic acid and citric acid. Whole cells of *Acetobacter aceti* were immobilized with porous ceramic, and the immobilized living cells were used for the production of acetic acid in a simple medium containing glucose (Ghommidh et al., 1981; Ghommidh and Navarro, 1982). In addition, citric acid was produced from glucose by using living cells of *Aspergillus niger* immobilized with calcium alginate (P. Linko, 1981). 12-Ketochenodeoxycholic acid, used chemotherapeutically for the solubilization of cholesterol gallstones, was produced from dehydrocholic acid using living cells of *Brevibacterium fuscum* immobilized with x-carrageenan (Sawada et al., 1981). For the production of lactic acid, whole cells of *Lactobacillus lactis* (P. Linko, 1981) or *Lactobacillus delbrueckii* (Stenroos et al., 1982) were immobilized with calcium alginate, and *L. delbrueckii* cells were used in hollow fibers (Roy et al., 1982).

D. Production of Alcohols

Recently, the production of alcohols, mainly ethanol, using immobilized living microbial cells has been the subject of increased interest, and many papers have been published (Table 3.4). This immobilized living cell system appears to be a promising technique for the industrial production of ethanol.

1. Ethanol.   Ethanol is produced from glucose via a series of multistep enzyme reactions involving ATP and NADH regeneration systems. We have studied ethanol production as a model system using immobilized living cells and found that ethanol can be continuously and efficiently produced by using living yeast cells immobilized with carrageenan (Wada et al., 1979). The immobilization of the yeast, *Saccharomyces carlsbergensis*, was carried out as follows. Without harvesting of the yeast cells, precultured broth was mixed directly with carrageenan solution and immobilized in bead form. The gels, containing a small amount of cells (3.5×10^6 cells/ml of gel), were incubated in complete nutritional medium on a rotary shaker at 30°C. After 60 hours of incubation the number of living cells in the gel increased by 1000-fold (5.4×10^9 cells/ml of gel). The number of living cells per unit volume of gel was about ten times higher than that of the full growth of free cells in ordinary liquid culture, and they formed a thin condensed layer near the surface of the beads. The yeast-cell layer may be formed depending on the diffusion of nutrients into gel. That is, the cell layer is formed as a result of selection of suitable environmental conditions by microbial cells.

Ethanol production was studied in a column packed with immobilized living yeast cells. In continuous production of ethanol the steady state was maintained for longer than 90 days. For constant production of high concentration of ethanol, it was necessary to increase the concentration of glucose stepwise.

TABLE 3.3 Synthesis of Organic Acids by Immobilized Biocatalysts

Organic Acid	Raw Materials	Enzymes or Microbial Cells	Matrices for Immobilization	Reference
Acetic acid	Glucose	Acetobacter aceti (multienzymes)	Porous ceramic	Ghommidh et al. (1981)
				Ghommidh and Navarro (1982)
		Acetobacter sp. (multienzymes)	Hydrous titanium	Kennedy et al. (1976)
Citric acid	Glucose	Aspergillus niger (multienzymes)	Calcium alginate	P. Linko (1981)
α,ω-Dodecanedioic acid	α,ω-Dodecanediol	Candida tropicalis (oxidase)	Carrageenan	Yi and Rehm (1982)
Gluconic acid	Glucose	Aspergillus niger (glucose oxidase)	Calcium alginate	P. Linko (1981)
α-Keto acid	D-Amino acid	Trigonopsis variabilis (D-amino acid oxidase)	Calcium alginate	Brodelius et al. (1980)
	L-Amino acid	Chlorella vulgaris and Anacystis nidulans (L-amino acid oxidase)	Agarose	Wikström et al. (1982)
12-Ketochenodeoxycholic acid	Dehydrocholic acid	Brevibacterium fuscum (reductase)	Carrageenan	Sawada et al. (1981)
2-Ketogluconic acid	Glucose	Serratia marcescens (multienzymes)	Collagen	Venkatasubramanian et al. (1978)
2-Ketoglonic acid	L-Sorbose	Gluconobacter melanogenus (L-sorbose dehydrogenase) and Pseudomonas syringus (L-sorbosone oxidase)	Polyacrylamide	Martin and Perlman (1976b)

Product	Substrate	Microorganism (enzyme)	Immobilization method	Reference
Lactic acid	Glucose	*Lactobacillus lactis* (multienzymes)	Calcium alginate	P. Linko (1981)
		Lactobacillus delbrueckii (multienzymes)	Calcium alginate	Stenroos et al. (1982)
			Hollow fiber	Roy et al. (1982)
		Lactobacillus sp. (multienzymes)	Calcium alginate	Tipayang and Kozaki (1982)
	Sugar	Mixed culture of lactobacilli and yeasts (multienzymes)	Gelatin	Compere and Griffith (1975)
L-Malic acid	Fumaric acid	*Brevibacterium ammoniuagenes* (fumarase)	Polyacrylamide	Yamamoto et al. (1976)
				Yamamoto et al. (1977)
		Brevibacterium flavum (fumarase)	Carrageenan	Takata et al. (1979)
		Thermus rubens nov. sp. (fumarase)	Duolite A-7	Takata et al. (1980)
				Ado et al. (1982)
d-Tartaric acid	cis-Epoxysuccinic acid	*Achromobacter sp.* (D-tartaric epoxydase)	Polyacrylamide	Kawabata and Ichikura (1977)
α,ω-Tridecanedioic	α,ω-Tridecaediol	*Candida tropicalis (oxidase)*	Carrageenan	Yi and Rehm (1982)
Urocanic acid	L-Histidine	*Achromobacter liquidum* (L–histidine ammonia lyase)	Polyacrylamide	Yamamoto et al. (1974b)
		Micrococcus luteus (L–histidine ammonia lyase)	Carbodiimide activated CM-cellulose	Jack and Zajic (1977)
		Pseudomonas fluorescens (L–histidine ammonia lyase)	Hollow fiber	Kan and Shuler (1978)

TABLE 3.4 Production of Alcohols by Immobilized Biocatalysts

Alcohol	Microbial Cells	Matrices for Immobilization	Reference
Ethanol	Saccharomyces bayanus	Carrageenan	Amin and Verachtert (1982)
	Saccharomyces carlsbergensis	Carrageenan	Wada et al. (1979)
	Saccharomyces cerevisiae	Calcium alginate	Kierstan and Bucke (1977)
			Larsson and Mosbach (1979)
			Y.-Y. Linko and Linko (1981)
			Y.-Y. Linko et al. (1981)
			Veliky and Williams (1981)
			Cho et al. (1983)
			Lee et al. (1983)
		Gelatin	Sivaraman et al. (1982)
		Pectin	Navarro et al. (1983)
		ECTEOLA-cellulose film	Lueng et al. (1983)
		Hollow fiber	Inloes et al. (1983)
	Saccharomyces formosensis	Calcium alginate	Fukushima and Hanai (1982)
	Pachysolen tannophilus	Calcium alginate	Slininger et al. (1982)
	Zymomonas mobilis	Borosilicate glass	Arcuri et al. (1980)
		Calcium alginate	Grote et al. (1980)
			Margaritis and Wallace (1982)
		Carrageenan	Grote et al. (1980)
			Amin and Verachtert (1982)
		Ion exchange resin	Krug and Daugulis (1983)
Isopropanol	Clostridium butyricum	Calcium alginate	Krouwel et al. (1980)
			Krouwel et al. (1983)
n-Butanol	Clostridium butyricum	Calcium alginate	Krouwel et al. (1980)
			Krouwel et al. (1983)
2,3-Butanediol	Enterobacter aerogenes	Carrageenan	Chua et al. (1980)
Glycerol	Saccharomyces cerevisiae	Carrageenan	Bisping and Rehm (1982)

Besides our studies, several papers have been published on ethanol production using immobilized living yeast cells (Table 3.4). A large amount of cells of *Saccharomyces cerevisiae* entrapped in calcium alginate gel was used for production of ethanol from glucose (Kierstan and Bucke, 1977). However, favorable results were not obtained owing to a low level of producing activity and stability. Also, *S. cerevisiae* cells immobilized in calcium alginate gel were used for continuous production of ethanol from cane molasses (Y. Y. Linko and Linko, 1981). Further, *Zymomonas mobilis* having several times higher ethanol productivity than yeast were immobilized in calcium alginate gel and x-carrageenan gel. Both types of immobilized cells showed high ethanol productivity, but there was a decline of 30% in activity after 800 hours of operation.

Ethanol production by the immobilized biocatalyst system is being further investigated to determine the conditions of optimum operation and to scale up the reaction. If a novel strain can be constructed that has resistance to higher concentrations of ethanol and/or glucose and that grows well in the immobilization matrix, the immobilized biocatalyst system will become a more promising method for ethanol production.

2. Other Alcohols. Immobilized biocatalyst systems have been investigated for the production of *n*-butanol, isopropanol and 2,3-butanediol. Continuous production of *n*-butanol and isopropanol using *Clostridium butylicum* entrapped in calcium alginate gel was studied (Krouwel et al., 1980, 1983). The immobilized cells were placed in a conical column to avoid problems associated with gas evolution. The productivity was about four times higher than that of a batch fermentation using free cells.

2,3-Butanediol was produced by *Enterobacter aerogenes* immobilized with x-carrageenan (Chua et al., 1980). When the immobilized cells were grown in a nutrient medium for one day, maximum activity was obtained, and the steady state was maintained for 10 days.

Living cells of *S. cerevisiae* immobilized with x-carrageenan could be shifted from ethanol production to glycerol production when sodium sulfite was added to the fermentation broth (Bisping and Rehm, 1982). The addition of sulfite caused an excess formation of NADH, which is used to reduce dihydroxyacetone phosphate to glycerol phosphate.

E. Production of Steroids

Microorganisms catalyze numerous biotransformations of steroids, and the use of immobilized biocatalysts has been reported for these reactions.

The synthesis of hydrocortisone and prednisolone from Reichstein's compound S was studied by using biocatalysts immobilized by entrapment with polyacrylamide (Mosbach and Larsson, 1970; Larsson et al., 1976). Mycelia of *Curvularia lunata* were used for the first reaction, 11 β-hydroxylation; for the second reaction, Δ^1-dehydrogenation, whole cells of *Corynebacterium simplex* were employed. By combining both types of immobilized cells, prednisolone was produced directly from Reichstein's compound S.

Whole cells of *C. simplex* were entrapped in collagen membranes for the production of prednisolone from hydrocortisone (Constantinides, 1980). Small chips of the collagen–cell complex were packed in a column and used for continuous operation. Further, spores of *C. lunata* were entrapped with photo-cross-linkable resin prepolymer and were grown into mycelial forms in a nutrient medium containing Reichstein's compound S as an inducer (Sonomoto et al., 1981). The immobilized mycelia showed higher 11 β-hydroxylation activity. In addition, *Arthrobacter globiformis* cells immobilized with polyacrylamide were used for the transformation of hydrocortisone in a batchwise system (Koshcheenko et al., 1981).

Steroid transformation reactions have been carried out in organic solvents using immobilized microbial cells. *Arthrobacter simplex* cells immobilized with urethane polymer and *Nocardia rhodochrous* cells immobilized with photo-cross-linkable resin were used in nonaqueous solvents for conversion of hydrocortisone and testosterone (Sonomoto et al., 1980; Omata et al., 1980; Fukui et al., 1980). In the future, immobilized systems are expected to be used extensively for reactions in organic solvents.

F. Other Useful Organic Compounds

Besides the studies cited above, many papers have been published on the production of useful organic compounds such as coenzymes and vitamins.

We immobilized *Achromobacter aceris* cells having high NAD kinase activity in a polyacrylamide gel and used them for continuous production of NADP from NAD and ATP (Uchida et al., 1978). In this case the utilization of ATP was not economical. For further improvement of NADP production we have studied the utilization of polyphosphate NAD kinase (Murata et al., 1979a). *Brevibacterium ammoniagenes* cells having high polyphosphate kinase activity were immobilized in polyacrylamide gel. By using a column packed with the immobilized cells, highly pure NADP was continuously produced from NAD and metaphosphate, the phosphoryl donor, in high yield. In addition, *S. cerevisiae* cells for high ATP regenerating activity and *B. ammoniagenes* cells for high NAD kinase activity were coimmobilized by entrapment in polyacrylamide gel (Murata et al., 1981) or by microcapsulation with cellulose acetate-butylate (Ado et al., 1980), and the coimmobilized cells were also used for production of NADP.

For production of flavine adenine dinucleotide (FAD) from flavin mononucleotide and ATP, whole cells of *Arthrobacter oxydans* having high FAD pyrophosphorylase activity were immobilized in films of polyvinylalcohol cross-linked with tetraethylsilicate (Yamada et al., 1980a). By the same immobilization procedure, whole cells of *P. fluorescens* having high pyridoxine 5′-phosphate oxidase activity were immobilized and used for production of pyridoxal 5′-phosphate from pyridoxine 5′-phosphate (Yamada et al., 1980).

Continuous production of Coenzyme A from panthothenic acid, L-cysteine, and ATP was investigated by using *B. ammoniagenes* cells immobilized with polyacrylamide gel (Shimizu et al., 1975). The reaction consisted of the following five steps: pantothenic acid ⟶ phosphopantothenic acid ⟶ phosphopantothenoyl cysteine ⟶ phosphopantetheine ⟶ dephosphocoenzyme A ⟶ Coenzyme A. Further, a system for production of Coenzyme A was investigated by using *S. cerevisiae* cells and *B. ammoniagenes* cells coimmobilized by microencapsulation with ethylcellulose (Samejima et al., 1978).

For production of glucose-6-phosphate and glucose-1-phosphate from glucose and *p*-nitrophenylphosphate, whole cells of *Escherichia freundii* having acid phosphatase activity were immobilized in a polyacrylamide gel lattice (Saif et al., 1975). On the other hand, continuous production of glucose-6-phosphate from glucose and metaphosphate was investigated by using polyphosphate glucokinase activity of *Achromobacter butyri* cells immobilized with polyacrylamide gel (Murata et al., 1979b).

Vitamin B_{12} production using immobilized living cells was studied (Yongsmith et al., 1982). *Propionibacterium* sp. cells were entrapped with urethane prepolymer, photo-cross-linkable prepolymer, ᴋ-carrageenan, calcium alginate, and agar. Among these matrices, urethane prepolymer was the most suitable for vitamin B_{12} production by immobilized cells.

Further, whole cells of *Enterobacter aerogenes*, which has transglycosylation activity, were immobilized with a hydrophilic photo-cross-linkable resin, and the immobilized cells were used to produce adenine arabinoside from uracil arabinoside and adenine in an appropriate water-organic solvent system (Yokozeki et al., 1982).

The production of dihydroxyacetone, used as a pharmaceutical intermediate and as a sun tanning agent, was studied by employing the oxidation of glycerol using *Acetobacter xylinum* cells immobilized with polyacrylamide gel (Nabe et al., 1979).

The production of glutathione from glutamate, cysteine, glycine, and ATP with immobilized biocatalysts has been reported. Glutathione is biosynthesized in reactions requiring ATP catalyzed by γ-glutamylcysteine synthetase and glutathione synthetase. Therefore the bioreaction system used for glutathione production consists of a sequential synthetic enzyme process and an ATP regeneration process. To establish an efficient glutathione production system, we investigated the possibility of utilizing the acetate kinase reaction as an ATP regeneration system.

When *S. cerevisiae* cells, for high ATP regeneration activity, and *E. coli* cells, for high glutathione synthetase activity, were coimmobilized in a polyacrylamide gel lattice, the immobilized cells showed high productivity (Murata et al., 1981).

Furthermore, attempts to produce extracellular enzymes by immobilized microbial cells were reported. Namely, production of α-amylase was attempted in a batch system using whole cells of *Bacillus subtilis* immobilized in a polyacrylamide gel lattice (Kokubu et al., 1978). α-Amylase production by the immobilized cells was about three times greater than that by washed cells at optimum conditions. Similarly, immobilized *Streptomyces fradiae* cells were used for protease production (Kokubu et al., 1981). Cellulase production by *Trichoderma reesei* cells immobilized with x-carrageenan was studied in continuous culture (Frein et al., 1982).

More recently, *B. subtilis* cells carrying plasmids encoding rat proinsulin were immobilized in agarose beads, and the immobilized cells were used for continuous production of proinsulin in a small model of a continuous stirred tank reactor (Mosbach et al., 1983). Continuous formation of proinsulin took place over a period of 80 h. This is interesting as an example in which the advantages of immobilized cell technology were applied to microorganisms modified by recombinant DNA techniques to produce a variety of eukaryotic proteins such as hormones.

In addition to immobilization of microbial cells, studies on immobilization of plant and animal cells have started. For example, plant cells of *Morinda*, *Catharanthus*, and *Digitalis* have been immobilized in gel beads of calcium alginate, and the immobilized plant cells were used for production and transformation of alkaloids. Immobilized *Catharanthus roseus* and *Morinda citrifolia* cells were used for synthesis of adjmalicine isomers from tryptamine and secologanin (Veliky and Jones, 1981) and for de novo synthesis of anthraquinones (Brodelius et al., 1979), respectively. Also, in hydroxylation of alkaloids, *Daacrs carota* and *Digitalis lanata* immobilized in calcium alginate were used for the production of 5-β-hydroxygitoxigenin from gitoxigenin (Brodelius et al., 1979) and for the production of digoxin from digitoxin (Veliky and Jones, 1981).

As an example of the use of animal cells, the posterior silk glands of the silkworm were immobilized in polyacrylamide gel, and the immobilized organ was used for production of silk protein in the presence of amino acids and energy sources (Ikariyama et al., 1979). Further, prostaglandin synthesis from arachidonic acid was carried out by using ram seminal microsomes immobilized with photo-cross-linkable resin (Ahern et al., 1983). In addition, animal cells immobilized in agarose beads were used for production of monoclonal antibodies and other biomolecules (Nilsson et al., 1983). Continuous production of interleukin-2 using lymphoblastoid MLA 144 cells immobilized in agarose beads was investigated. When the immobilized cells were kept in spinner flasks, continuous production of interleukin-2 occurred over the 13-day period tested. The same entrapped hybridoma cells were also tested for their capacity to form monoclonal IgG 2A antibodies against herpes simplex type 2 glycoprotein. Daily removal and assay of the supernatant for a period of 1 week showed a more or less unchanged production capacity.

Although the papers related to immobilization of plant and animal cells are still very few, they are expected to increase rapidly in the near future.

IV. CONCLUSIONS

Besides the use of immobilized biocatalysts for production of useful organic compounds discussed above, immobilized biocatalyst systems have been applied in a variety of fields and play a very important role in enzyme engineering in the field of biotechnology. The use of immobilized living cell systems especially are now expanding to include not only microbial cells but also plant and animal cells, and it is expected to be an important technique in biotechnology.

Recently, many discussions and reports have been published on the applications of genetic engineering. Genetic engineering, more specifically, recombinant DNA techniques, and enzyme engineering are not competitive technologies; they should cooperate with each other to assure further progress. Fundamentally, genetic engineering cannot be an independent production technology. To be an efficient production technology it should be combined with fermentation technology, enzyme engineering, and also isolation and purification process technologies.

If a novel microorganism with desired characteristics is produced by genetic engineering and is used in the form of immobilized cells, this will be a very promising technology.

Therefore we are convinced that if cooperation among scientists and engineers in a variety of fields related to biotechnologies such as genetic engineering, fermentation technology, enzyme engineering, and separation process technology is accelerated, biotechnology will contribute greatly to the future welfare of human beings.

REFERENCES

Ado, Y., Kawamoto, T., Masunaga, I., Takayama, K., Takasawa, S., and Kimura, K. (1982) *Enzyme Eng. 6*, 303–304.

Ado, Y., Kimura, K., and Samejima, H. (1980) *Enzyme Eng. 5*, 295–304.

Ahern, T. J., Katoh, S., and Sada, E. (1983) *Biotechnol. Bioeng. 25*, 881–885.

Amin, G., and Verachtert, H. (1982) *Eur. J. Appl. Microbiol. Biotechnol. 14*, 59–63.

Arcuri, E. J., Worden, R. M., and Shumate, I. I. (1980) *Biotechnol. Lett. 2*, 499–504.

Axén, R., and Ernback, S. (1971) *Eur. J. Biochem. 18*, 351–360.

Bang, W.-G., Behrendt, U., Lang, S., and Wagner, F. (1983) *Biotechnol. Bioeng. 25*, 1013–1025.

Barth, T., and Masková, H. (1971) *Collection Czech. Chem. Comm. 36*, 2398–2400.

Bisping, B., and Rehm, H. J. (1982) *Eur. J. Appl. Microbiol. Biotechnol. 14*, 136–139.

Brodelius, P., Deus, B., Mosbach, K., and Zenk, M. H. (1979) *FEBS Lett. 103*, 93–97.

Brodelius, P., Hägerdal, B., and Mosbach, K. (1980) *Enzyme Eng. 5*, 383–387.

Chibata, I., Kakimoto, T., and Kato, J. (1965) *J. Appl. Microbiol. 13*, 638–645.

Chibata, I., Tosa, T., Sato, T., Mori, T., and Matsuo, Y. (1972) *Proc. IV. IFS: Ferment. Technol. Today*, 383–389.

Chibata, I., Kakimoto, T., and Nabe, K. (1974a) Japanese Patent Kokai, 74-81591.

Chibata, I., Kakimoto, T., and Nabe, K. (1974b) Japanese Patent Kokai, 74-81590.

Chibata, I., Tosa, T., and Sato, T. (1974c) *Appl. Microbiol. 27*, 878–885.

Cho, G. H., Choi, C. Y., Choi, Y. D., and Han, M. H. (1982) *J. Chem. Tech. Biotechnol. 32*, 959–967.

Chua, J. W., Erarslan, A., Kinoshita, S., and Taguchi, H. (1980) *J. Ferment. Technol. 58*, 123–127.

Compere, A. L., and Griffith, W. L. (1975) *Dev. Ind. Microbiol. 17*, 241–246.

Constantinides, A. (1980) *Biotechnol. Bioeng. 22*, 119–136.

Constantinides, A., Bhatia, D., and Vieth, W. R. (1981) *Biotechnol. Bioeng. 23*, 899–916.

Delin, P. S., Ekström, B., Sjöberg, B., and Thelin, K. H. (1973) U.S. Patent 3736230.

Deo, Y. M., and Gaucher, G. M. (1983) *Biotechnol. Lett. 5*, 125–130.

Dinelli, D. (1972) *Process Biochem. 7*, 9–12.

Freeman, A., and Aharonowitz, Y. (1981) *Biotechnol. Bioeng. 23*, 2747–2759.

Frein, E. M., Montenecourt, B. S., and Eveleigh, D. E. (1982) *Biotechnol. Lett. 4*, 287–292.

Fujii, T., Matsumoto, K., Shibuya, Y., Hanamitsu, K., Yamaguchi, T., Watanabe, T., and Abe, S. (1973a) Japanese Patent Kokai, 73-75792.

Fujii, T., Matsumoto, K., Shibuya, Y., Hanamitsu, K., Yamaguchi, T., Watanabe, T., and Abe, S. (1973b) Belgian Patent 803832.

Fujii, T., Matsumoto, K., Shibuya, Y., Hanamitsu, K., Yamaguchi, T., Watanabe, T., and Abe, S. (1974) Japanese Patent 74-47594.

Fujii, T., Matsumoto, K., and Watanabe, T. (1976) *Process Biochem. 11*, 21–24.

Fujimura, M., Kato, J., Tosa, T., and Chibata, I. (1984) *Appl. Microbiol. Biotechnol. 19*, 79–84.

Fukui, S., Ahmed, S. A., Omata, T., and Tanaka, A. (1980) *Eur. J. Appl. Microbiol. 10*, 289–301.

Fukui, S., Ikeda, S., Fujimura, M., Yamada, H., and Kumagai, H. (1975a) *Eur. J. Biochem. 51*, 155–164.

Fukui, S., Ikeda, S., Fujimura, M., Yamada, H., and Kumagai, H. (1975b) *Eur. J. Appl. Microbiol. 1*, 25–39.

Fukumura, T. (1975) Japanese Patent 75-15795.

Fukushima, S., and Hanai, S. (1982) *Enzyme Eng. 6*, 347–348.

Fusee, M. C., Swann, W. E., and Calton, G. J. (1981) *Appl. Environ. Microbiol. 42*, 672–676.

Ghommidh, C., and Navarro, J. M. (1982) *Biotechnol. Bioeng. 24*, 1991–1999.

Ghommidh, C., Navarro, J. M., and Durand, G. (1981) *Biotechnol. Lett. 3*, 93–98.

Giacobbe, F., Iasonna, A., and Cecere, F. (1977) *J. Solid-Phase Biochem. 2*, 195–202.

Grote, W., Lee, K. J., and Rogers, P. L. (1980) *Biotechnol. Lett. 2*, 481–486.

Hirano, K., Karube, I., and Suzuki, S. (1977) *Biotechnol. Bioeng. 19*, 311–321.

Huper, F. (1973) German Patent 2157972.

Ikariyama, Y., Aizawa, M., and Suzuki, S. (1979) *J. Solid-Phase Biochem. 4*, 279–288.

Inloes, D. S., Taylor, D. P., Cohen, S. N., Michaels, A. S., and Robertson, C. R. (1983) *Appl. Environ. Microbiol. 46*, 264–278.

Jack, T. R., and Zajic, J. E. (1977) *Biotechnol. Bioeng. 19*, 631–648.

Jandel, A.-S., Husted, H., and Wandrey, C. (1982) *Eur. J. Appl. Microbiol. Biotechnol. 15*, 59–63.

Kamogashira, T., Kawaguchi, T., Miyazaki, W., and Doi, T. (1972a) Japanese Patent 72-28190.

Kamogashira, T., Mihara, S., Tamaoka, H., and Doi, T. (1972b) Japanese Patent 72-28187.

Kan, J. K., and Shuler, M. L. (1978) *Biotechnol. Bioeng. 20*, 217–230.

Kanemitsu, O. (1975) Japanese Patent Kokai, 75-132181.

Kawabata, Y., and Ichikura, S. (1977) Japanese Patent Kokai, 77-102496.

Kennedy, J. F., Barker, S. A., and Humphreys, J. D. (1976) *Nature 261*, 242–244.

Kierstan, M., and Bucke, C. (1977) *Biotechnol. Bioeng. 19*, 387–397.

Kim, I. H., Nam, D. H., and Ryu, D. D. Y. (1983) *Appl. Biochem. Biotechnol. 8*, 195–202.

Kimura, K., Takayama, K., Ado, T., Kawamoto, T., and Masunaga, I. (1981) Japanese Patent 81-75097.

Kinoshita, S., Muranaka, M., and Okada, H. (1975) *J. Ferment. Technol. 53*, 223–229.

Klein, J., and Wagner, F. (1980) *Enzyme Eng. 5*, 335–345.

Koelsch, R. (1972) *Enzymologia 42*, 257–267.

Kokubu, T., Karube, I., and Suzuki, S. (1978) *Eur. J. Appl. Microbiol. Biotechnol. 5*, 233–240.

Kokubu, T., Karube, I., and Suzuki, S. (1981) *Biotechnol. Bioeng. 23*, 29–39.

Koshcheenko, K. A., Sukhodolskaya, G. V., Tyurin, V. S., and Skryabin, G. K. (1981) *Eur. J. Appl. Microbiol. Biotechnol. 12*, 161–169.

Krouwel, P. G., Groot, W. J., Kossen, N. W. F., and van der Lean, C. G. (1983) *Enzyme Microb. Technol. 5*, 46–54.

Krouwel, P. G., van der Laan, W. F. M., and Kossen, N. W. F. (1980) *Biotechnol. Lett. 2*, 253–258.

Krug, T. A., and Daugulis, A. J. (1983) *Biotechnol. Lett. 5*, 159–164.

Kurzatkowski, W., Kurytowicz, W., and Paszkiewicz, A. (1982) *Eur. J. Appl. Microbiol. Biotechnol. 15*, 211–213.

Lagerlöf, E., Nathoust-Westfelt, L., Ekström, B., and Sjöberg, B. (1976) in *Methods in Enzymology*, Vol. XLIV (Mosbach, K., ed.), pp. 759–768, Academic Press, New York.

Larsson, P.-O., and Mosbach, K. (1979) *Biotechnol. Lett. 1*, 501–506.

Larsson, P.-O., Ohlson, S., and Mosbach, K. (1976) *Nature 263*, 796.

Lee, T. H., Ahn, J. C., and Ryu, D. D. Y. (1983) *Enzyme Microb. Technol. 5*, 41–45.

Leuschner, F. (1966) German Patent 1227855.

Lilly, M. D., Money, C., Hornby, W. E., and Crook, E. M. (1965) *Biochem. J. 95*, 45.

Linko, P. (1981) in *Advances in Biotechnology*, Vol. I (Moo-Young, M., Robinson, C. W., and Vezina, C., eds.), pp. 711–716, Pergamon Press, New York.

Linko, Y.-Y., and Linko, P. (1981) *Biotechnol. Lett. 3*, 21–26.

Linko, Y.-Y., Jalanka, H., and Linko, P. (1981) *Biotechnol. Lett. 3*, 263–268.

Lueng, K.-L., Joshi, S., and Yamazaki, H. (1983) *Enzyme Microb. Technol. J. Antibiot. 5*, 181–184.

Marconi, W., Bartoli, F., Cecere, F., Galli, G., and Morisi, F. (1975) *Agr. Biol. Chem. 39*, 277–279.

Marconi, W., Cecere, F., Morisi, F., Della, P. G., and Rappuoli, B. (1973) *26*, 228–232.

Maréchal, P. D.-L., Calderon-Seguin, R., Vandecasteele, J. P., and Azerad, R. (1979) *Eur. J. Appl. Microbiol. Biotechnol. 7*, 33–44.

Margaritis, A., and Wallace, J. B. (1982) *Biotechnol. Bioeng. Symp. 12*, 147–159.

Martin, C. K. A., and Perlman, D. (1976a) *Biotechnol. Bioeng. 18*, 217–237.

Martin, C. K. A., and Perlman, D. (1976b) *Eur. J. Appl. Microbiol. Biotechnol. 3*, 91–95.

Mayer, H., Collins, J., and Wagner, F. (1980) *Enzyme Eng. 5*, 61–69.

Mitz, M. A., and Summaria, L. J. (1961) *Nature 189*, 576–577.

Mori, T., Sato, T., Tosa, T., and Chibata, I. (1972) *Enzymologia 43*, 213–226.

Morikawa, Y., Karube, I., and Suzuki, S. (1979a) *Biotechnol. Bioeng. 21*, 261–270.

Morikawa, Y., Karube, I., and Suzuki, S. (1980a) *Eur. J. Appl. Microbiol. Biotechnol. 10*, 23–30.

Morikawa, Y., Karube, I., and Suzuki, S. (1980b) *Biotechnol. Bioeng. 22*, 1015–1023.

Morikawa, Y., Ochiai, K., Karube, I., and Suzuki, S. (1979b) *Antimicrob. Agents Chemother. 15*, 126–130.

Mosbach, K., Birnbaum, S., Hardy, K., Davies, J., and Bülow, L. (1983) *Nature 302*, 543–545.

Mosbach, K., and Larsson, P.-O. (1970) *Biotechnol. Bioeng. 12*, 19–27.

Murata, K., Kato, J., and Chibata, I. (1979a) *Biotechnol. Bioeng. 22*, 887–895.

Murata, K., Tani, K., Kato, J., and Chibata, I. (1981) *Enzyme Microb. Technol. 3*, 233–242.

Murata, K., Uchida, T., Tani, K., Kato, J., and Chibata, I. (1979b) *Eur. J. Appl. Microbiol. Biotechnol. 7*, 45–52.

Nabe, K., Izuo, N., Yamada, S., and Chibata, I. (1979) *Appl. Environ. Microbiol. 38*, 1056–1060.

Navarro, A. R., Rubio, M. C., and Callieri, D. A. S. (1983) *Eur. J. Appl. Microbiol. Biotechnol. 17*, 148–151.

Nilsson, K., Scheirer, W., Merten, O. W., Östberg, L., Liehl, E., and Katinger, H. W. D. (1983) *Nature 302*, 629–630.

Omata, T., Tanaka, A., and Fukui, S. (1980) *J. Ferment. Technol. 58*, 339–343.

Pittalis, F., Bartoli, F., and Morisi, F. (1979) *Enzyme Microb. Technol. 1*, 189–192.

Roy, T. B. V., Blanch, H. W., and Wilke, C. R. (1982) *Biotechnol. Lett. 4*, 483–488.

Ryu, D. D. Y., Bruno, C. F., Lee, B. K., and Venkatasubramanian, K. (1972) *Proc. IV. IFS: Ferment. Technol. Today*, 307–314.

Saif, S. R., Tani, Y., and Ogata, Y. (1975) *J. Ferment. Technol. 53*, 380–385.

Samejima, H., Kimura, K., Ado, Y., Suzuki, Y., and Tadokoro, T. (1978) *Enzyme Eng. 4*, 237–244.

Sarkar, J. M., and Mayaudon, J. (1983) *Biotechnol. Lett. 5*, 201–206.

Sato, T., Mori, T., Tosa, T., and Chibata, I. (1971) *Arch. Biochem. Biophys. 147*, 788–796.

Sato, T., Mori, T., Tosa, T., Chibata, I., Furui, M., Yamashita, K., and Sumi, A. (1975) *Biotechnol. Bioeng. 18*, 1797–1804.

Sato, T., Nishida, Y., Tosa, T., and Chibata, I. (1979) *Biochim. Biophys. Acta 570*, 179–186.

Sato, T., Takamatsu, S., Yamamoto, K., Umemura, I., Tosa, T., and Chibata, I. (1982) *Enzyme Eng. 6*, 271–272.

Sato, T., Tosa, T., and Chibata, I. (1976) *Eur. J. Appl. Microbiol. 2*, 153–160.

Savidge, T., Powell, L. W., and Carrington, T. R. (1969) German Patent 1917057.

Savidge, T., Powell, L. W., and Warren, K. B. (1974) German Patent 2336829.

Sawada, H., Kinoshita, S., Yoshida, T., and Taguchi, H. (1981) *J. Ferment. Technol. 59*, 111–114.

Self, D. A., Kay, G., and Lilly, M. D. (1969) *Biotechnol. Bioeng. 11*, 337–348.

Shimizu, S., Morioka, H., Tani, Y., and Ogata, K. (1975) *J. Ferment. Technol. 53*, 77–83.

Sivaraman, H., Rao, B. S., Pundle, A. V., and Sivaraman, C. (1982) *Biotechnol. Lett. 4*, 359–364.

Slininger, P. J., Bothast, R. J., Black, L. T., and McChee, J. E. (1982) *Biotechnol. Bioeng. 24*, 2241–2251.

Slowinski, W., and Charm, S. E. (1973) *Biotechnol. Bioeng. 15*, 973–979.

Sonomoto, K., Hog, M. M., Tanaka, A., and Fukui, S. (1981) *J. Ferment. Technol. 59*, 465–469.

Sonomoto, K., Jin, I. N., Tanaka, A., and Fukui, S. (1980) *Agr. Biol. Chem. 44*, 1119–1126.

Stenroos, S.-L., Linko, Y.-Y., and Linko, P. (1982) *Biotechnol. Lett. 4*, 159–164.

Takamatsu, S., Tosa, T., and Chibata, I. (1983) *Nippon Kagaku Kaishi 9*, 1369–1376.

Takamatsu, S., Umemura, I., Yamamoto, K., Sato, T., Tosa, T., and Chibata, I. (1982) *Eur. J. Appl. Microbiol. Biotechnol. 15*, 147–152.

Takamatsu, S., Yamamoto, K., Tosa, T., and Chibata, I. (1981) *J. Ferment. Technol. 59*, 489–493.

Takata, I., Kayashima, K., Tosa, T., and Chibata, I. (1982) *J. Ferment. Technol. 60*, 431–437.

Takata, I., Yamamoto, K., Tosa, T., and Chibata, I. (1979) *Eur. J. Appl. Microbiol. Biotechnol. 7*, 161–172.

Takata, I., Yamamoto, K., Tosa, T., and Chibata, I. (1980) *Enzyme Microb. Technol. 2*, 30–36.

Takeda, H., Matsumoto, I., and Matsuda, K. (1975) Japanese Patent 75-3588.

Tipayang, P., and Kozaki, M. (1982) *J. Ferment. Technol. 60*, 595–598.

Tosa, T., Mori, T., and Chibata, I. (1969) *Agr. Biol. Chem. 33*, 1053–1059.

Tosa, T., Mori, T., and Chibata, I. (1971) *J. Ferment. Technol. 49*, 522–528.

Tosa, T., Mori, T., Fuse, N., and Chibata, I. (1966) *Enzymologia 31*, 214–224.

Tosa, T., Sato, T., Mori, T., and Chibata, I. (1974) *Appl. Microbiol. 27*, 886–889.

Tosa, T., Sato, T., Mori, T., Matsuo, Y., and Chibata, I. (1973) *Biotechnol. Bioeng. 15*, 69–84.

Uchida, T., Watanabe, T., Kato, J., and Chibata, I. (1978) *Biotechnol. Bioeng. 20*, 255–266.

Veelken, M., and Pape, H. (1982) *Eur. J. Appl. Microbiol. Biotechnol. 15*, 206–210.

Veliky, I. A., and Jones, A. (1981) *Biotechnol. Lett. 3*, 551–554.

Veliky, I. A., and Williams, R. E. (1981) *Biotechnol. Lett. 3*, 275–280.

Venkatasubramanian, K., Constantinides, A., and Vieth, W. R. (1978) *Enzyme Eng. 3*, 29–41.

Verlop, K.-D., and Klein, J. (1981) *Biotechnol. Lett. 3*, 9–14.

Wada, M., Kato, J., and Chibata, I. (1979) *Eur. J. Appl. Microbiol. Biotechnol. 8*, 241–247.

Wada, M., Uchida, T., Kato, J., and Chibata, I. (1980) *Biotechnol. Bioeng. 22*, 1175–1188.

Wandrey, C., Wichmann, R., and Jandel, A.-S. (1982) *Enzyme Eng. 6*, 61–67.

Weetall, H. H., and Detar, C. C. (1974) *Biotechnol. Bioeng. 16*, 1537–1544.

Wikström, P., Szwajcer, E., Brodelius, P., Nilsson, K., and Mosbach, K. (1982) *Biotechnol. Lett. 4*, 153–158.

Yamada, H., Yamada, K., Kumagai, H., Hino, T., and Okamura, S. (1978) *Enzyme Eng. 3*, 57–62.

Yamada, H., Shimizu, H., Tani, Y., and Hino, T. (1980a) *Enzyme Eng. 5*, 405–411.

Yamada, H., Shimizu, S., Shimada, H., Tani, Y., Takahashi, S., and Ohashi, T. (1980b) *Biochimie 62*, 395.

Yamamoto, K., Sato, T., Tosa, T., and Chibata, I. (1974a) *Biotechnol. Bioeng. 16*, 1589–1599.

Yamamoto, K., Sato, T., Tosa, T., and Chibata, I. (1974b) *Biotechnol. Bioeng. 16*, 1601–1610.

Yamamoto, K., Tosa, T., and Chibata, I. (1980) *Biotechnol. Bioeng. 22*, 2045–2054.

Yamamoto, K., Tosa, T., Yamashita, K., and Chibata, I. (1976) *Eur. J. Appl. Microbiol. 3*, 169–183.

Yamamoto, K., Tosa, T., Yamashita, K., and Chibata, I. (1977) *Biotechnol. Bioeng. 19*, 1101–1114.

Yi, Z.-H., and Rehm, H. J. (1982) *Eur. J. Appl. Microbiol. Biotechnol. 16*, 1–4.

Yokote, Y., Maeda, S., Yabushita, H., Noguchi, S., Kimura, K., and Samejima, H. (1978) *J. Solid-Phase Biochem. 3*, 247–261.

Yokozeki, K., Yamanaka, S., Utagawa, T., Takinami, K., Hirose, Y., Tanaka, A., Sonomoto, K., and Fukui, S. (1982) *Eur. J. Appl. Microbiol. Biotechnol. 14*, 225–231.

Yongsmith, B., Sonomoto, K., Tanaka, A., and Fukui, S. (1982) *Eur. J. Appl. Microbiol. Biotechnol. 16*, 70–74.

Zaffaroni, P., Vitobello, V., Cecere, F., Giacomozzi, E., and Morisi, F. (1974) *Agr. Biol. Chem. 38*, 1335–1342.

Production of 6-APA with Immobilized Cells: A Catalyst Development

Joachim Klein
Fritz Wagner
Burkhard Kressdorf
Reinhard Müller
Handoko Tjokrosoeharto
Klaus-Dieter Vorlop

I. INTRODUCTION

Enzymatic cleavage of penicillins, especially penicillin G, is now the most widely used route for the industrial production of 6-aminopenicillanic acid. Another important aspect of this process is the very early introduction of the concept of immobilized enzymes (Vandamme and Voets, 1974); the many contributions in the literature are a demonstration of the significance of this development in enzyme technology (Balasingham et al., 1972; Warburton et al., 1973). In addition to immobilized enzymes, immobilized cells also have been tested (Sato et al., 1976), but it was generally felt that the much lower activity of the latter system was due to systematic deficiencies like a lower level of immobilized active protein and an additional diffusion barrier caused by the cell wall. Good yields of the purified enzyme, its satisfactory operational stability, and the development of improved carrier materials were additional factors in favor of the immobilized enzyme method for this monoenzyme catalyzed reaction (Sauber and Krämer, 1982).

However, on the basis of systematic research and development with regard to microbial strain selection, genetic engineering, optimization of fermentation technology, composition and structure of polymeric carriers, and effective immobilization methods we were finally able to prepare immobilized cell catalysts, which are comparable in catalytic activity to immobilized enzyme systems. While separate aspects of this work have been published elsewhere (Klein and Wagner, 1980; Klein and Vorlop, 1983a, b), it is the intention of this chapter to give a comprehensive overview and combined case history to illustrate the strategy to be used for the development of a highly active and competitive immobilized cell biocatalyst.

The structural formulae of the components involved in this reaction are presented in Fig. 4.1. The main reaction path is the cleavage of penicillin G to the main product 6-aminopenicillanic acid (6-APA) and the stoichiometrically coupled coproduct phenylacetic acid, catalyzed by the enzyme penicillin acylase. The most important side reaction is the cleavage of the β-lactam ring—catalyzed enzymatically by β-lactamase or chemically by OH^{θ}- or Me^{2+}-ions—to form penicillenic G-acid or, after decarboxylation, penillenic G-acid. The main reaction is an equilibrium reaction that, depending on the substrate

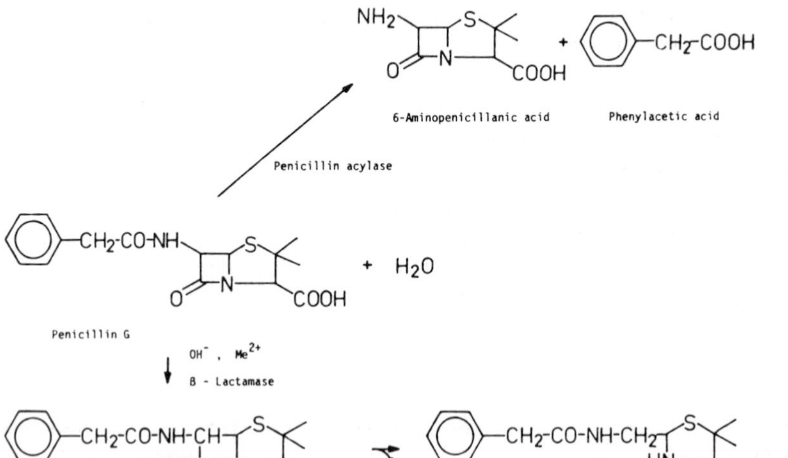

FIGURE 4.1   Reaction scheme for the cleavage of penicillin G to 6-APA, including the main side reactions.

concentration or temperature and pH, can be shifted more or less quantitatively to the 6-APA-side (Svedas et al., 1980). In any practical application the side reaction has to be kept below 1% of the penicillin G consumption.

The kinetics of the cleavage reaction to 6-APA have been studied extensively by using the isolated enzyme or penicillin acylase-containing whole cells. Following Warburton et al. (1973) the rate equation can be written as

$$v = \frac{v_{max}}{\left(1 + \dfrac{K_M^S}{c_S}\right)\left[1 + \dfrac{c_A}{K_I^A}\right] + \dfrac{c_P}{K_I^P}\left[1 + \dfrac{K_M^S}{c_S}\right] + \dfrac{c_A \cdot c_P}{c_S} \cdot \dfrac{K_M^S}{K_I^A K_I^P} + \dfrac{c_S}{K_I^S}} \qquad (4.1)$$

This model includes a competitive product inhibition by phenylacetic acid (K_I^A), a noncompetitive product inhibition by 6-APA (K_I^P), and a substrate inhibition by penicillin G (K_I^S). K_M^S, c_S, c_A, and c_P stand for the Michaelis constant, the substrate concentration, the phenylacetic acid concentration, and the 6-APA concentration, respectively.

This equation holds in general for the purified enzyme as well as for the whole cell enclosed enzyme, with a significant difference, however, when the absolute values of the kinetic parameters are considered. Table 4.1 gives a comparison of typical Michaelis and inhibition constants. The higher K_M-values for the whole cells compared to the enzyme are usually attributed to the additional diffusion barrier originating from the cell wall.

The kinetic analysis is usually based on acid/base titration under pH-static conditions. Sodium hydroxide or ammonium hydroxide is used to neutralize the acidic coproduct phenylacetic acid; the equimolar consumption of the base can directly be converted to 6-APA formation per unit time and thus to the activity of the enzyme preparation.

The purity of the product or the possible formation of by-products can best be controlled by HPLC analysis (Klein and Vorlop, 1984).

Two cell lines have been used in the course of this study: the highly active wild-type *Escherichia coli* ATCC 11105 and the superproducing strain *E. coli* 5K (pHM12), which has been constructed by genetic engineering methods. It should be mentioned that this work includes the first report on the combination of genetic engineering techniques and cell immobilization techniques given in the literature (Klein and Wagner, 1980).

TABLE 4.1 Comparison of Kinetic Constants of Penicillin Acylase from *E. coli* ATCC 11105

	$T(°C)$	pH	K_M^S	K_I^S	K_S^A	K_I^P	Reference
Free enzyme	25	8.1	0.02	—	0.2	15	Kutzbach and Rauenbusch (1974)
Whole cell	37	7.8	7.8	280	10	25	Washausen (1979)

II. STRAIN IMPROVEMENT

A. Increase of Acylase Activity of *E. coli* ATCC 11105

A commercially interesting immobilized cell system for the production of 6-APA requires a higher specific acylase activity of the whole cell than has so far been available with *E. coli* ATCC 11105. Since the induced acylase synthesis is strongly repressed by carbohydrates like glucose, a medium with 1% yeast extract, 1% tryptone, and 0.5% beef extract as C- and N-sources and 0.25% NaCl is the most favorable one for the enzyme formation. Under controlled conditions in a 10-l bioreactor a strong increase of penicillin G acylase activity was observed only in the case of temperature shift from 27°C to 24°C at an optical density (OD_{546nm}) of 6.0 (Table 4.2) (Klein and Wagner, 1980). The high activity of 69 U/g wet cells is attainable under carefully controlled conditions. However, the productivity of immobilized ATCC 11105, starting from 69 U/g bacterial wet weight, is not sufficient in comparison to immobilized enzyme preparations. Therefore a genetic engineering program was initiated.

B. Acylase Gene Transfer on a Multi Copy Plasmid

First, the acylase gene has to be isolated from the *E. coli* ATCC 11105 genome without positive selection pressure for *E. coli* recipients containing a genome fragment. With the cosmid cloning system used in this study, hybrid plasmids, which include large foreign DNA fragments, are introduced with higher efficiency into the recipient bacteria than the original cosmids. From the proline and leucine auxotroph recipients there are pro$^+$ and leu$^+$ clones at a frequency of about 0.2%. However, the acylase gene, which should also be found in a screening of some 500 clones, was obtained only once in some 10,000 clones by an overlay test with the 6-APA sensitive *Serratia marcescens*. The

TABLE 4.2   Influence of Temperature Shift on Penicillin G-Acylase Activity at Different Growth Phases of *E. coli* ATCC 11105

Temperature Shift (°C)	OD_{546nm}	Specific Activity (µ/g)
27 to 24	3.5	34
27 to 24	5.0	41
27 to 24	5.5	44
27 to 24	6.0	69
27 to 24	6.5	41
27 to 24	7.0	31
27 to 24	7.5	30
24 to 27	5.0	32
24 to 27	6.0	36
24 to 27	7.0	34

cosmid hybrid expresses acylase at a level equivalent to that of *E. coli* ATCC 11105, and it was also inducible.

By subcloning onto high-copy plasmids like pBR 322 the level of penicillin G acylase could be increased. Furthermore, the expression is constitutive because the structural gene was likely separated from regulatory gene regions by this fragment reduction. Half of the activity was lost by additional deletions of the subcloned fragment (Mayer et al., 1979). Afterwards, a further increase of enzyme production and a lowered sensitivity to catabolite repression were reached by UV mutagenesis. *E. coli* 5K (pHM12) was obtained having fivefold higher activity compared to the induced wild-type strain (Mayer et al., 1980).

C. Improvement of Cultivation Conditions for the Hybrid Strain *E. coli* 5K (pHM12)

1. Hybrid Strain Stability.   First, for plasmid-containing strains, growth conditions must be found that guarantee plasmid preservation. Even multicopy plasmids, whose distribution is not guaranteed by special mechanisms, are lost statistically, and furthermore through mutations from some cells of a growing culture. Such plasmid-free cells generally overgrow the hybrid strain, since the expression of the plasmid-coded genes at the cost of the host metabolism has no selective advantage. Also, the hybrid strain 5K (pHM12) is overgrown by plasmid-free 5K cells. The plasmid-coded tetracycline resistance can be used to force plasmid preservation by the addition of 4 mg/l of this antibiotic (Fig. 4.2). By this means acylase formation is also guaranteed on the basis of the observed stringent coupling of both properties. Tetracycline is necessary only

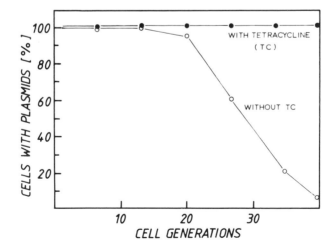

FIGURE 4.2   Effect of tetracycline (4 mg/l) on the appearance of plasmid-free cells.

for precultures, since it lasts at least ten generations, until the occurrence of plasmid-free cells leads to a decrease of acylase activity.

2. Improvement of Acylase Formation.   From the physical parameters it is easy to investigate the suitable temperature for acylase formation (Table 4.3). The partial reduction of the basal medium developed for *E. coli* ATCC 11105 shows that yeast extract is particularly important for acylase production. The same specific activity is attainable with technical-grade nutritive substances if cheaper technical yeast extract is used as the only C- and N-source. Since acylase formation is strongly influenced by pH, the effect of nutritional additions could be reliably investigated only if the course of pH was similar, based on high buffer capacity in the medium. Low concentrations of additional quickly metabolized carbon sources cause a slight increase of specific activity and enhanced utilization of the yeast extract medium (Table 4.4). At higher concentrations, acylase formation is lowered by catabolite repression (Tjokrosoeharto, 1983).

On the basis of these results, quantitative nutritional improvement was carried out first by the method of Greek-Latin squares and afterwards with the gradient process according to Box and Wilson (Table 4.5).

With cultivation in a bioreactor the course of acylase formation can be examined exactly. The specific activity increases parallel to cell growth, with a certain delay (Fig. 4.3), but the results of quantitative nutritional improvement in shake flasks were not transferable. Also the improvement in a chemostat

TABLE 4.3   Influence of Incubation Temperature, Basal Medium, O, 2% Fructose

Temperature (°C)	Activity (U/g bww)	Yield (g bww/l)	pH
20	136	13	7.2
24	183	17	8.1
27	233	17	8.4
30	170	16	8.4
33	57	14	8.5
37	9	16	8.7

TABLE 4.4   Effect of Additional Carbon Sources, 3% Technical Yeast Extract

Addition	pH	Yield (gbww/l)	Specific Activity (U/g bww)	Volume Activity (U/l)
Control	7.8	12.9	145	1871
Maltose	7.7	12.6	170	2142
Sucrose	7.8	13.7	138	1904
Lactose	7.6	13.8	156	2153
Fructose	7.7	14.7	153	2249
Sorbit	7.4	14.2	145	2059
D,L-Malate	7.7	13.8	163	2249

TABLE 4.5   Quantitative Improvement of the Culture Medium: Influence of the Contribution of the Variable Part (%)

Medium	1	2	3	4	5	6	7	8	9	THE-1
Technical Yeast Extract	2.1	1.9	1.7	1.5	1.3	1.1	0.9	0.7	0.5	3.0
NH_4CL	0.28	0.27	0.26	0.25	0.24	0.23	0.22	0.21	0.20	0.0
D,L-Malate	0.34	0.31	0.28	0.25	0.22	0.19	0.16	0.13	0.10	0.0
Specific Activ. (U/g bww)	242	239	237	248	258	234	255	252	246	208
Bacteria wet weight (g/l)	9.4	9.2	7.9	8.8	9.1	9.9	7.7	6.0	5.5	7.6

with the help of the pulse method is not applicable despite the apparent coupling of growth and acylase formation. Additions like fructose, which show a positive effect in a final steady state, lead first to a decrease of specific activity. The possibility of obtaining a steady state shows also that the hybrid strain has a stability sufficient for industrial utilization. With the chemostat the influence of pH can be investigated exactly. Figure 4.4 confirms results from batch fermentations, that a pH of 6.8 causes the highest acylase activity.

Compared to the studies in continuous culture the influence of oxygen concentration during incubation is greater for batch fermentations. Oxygen should be the growth-limiting substrate during the logarithmic growth phase. Limited aeration thus means that μ_{max} remains lower than would be possible in this medium (Fig. 4.5).

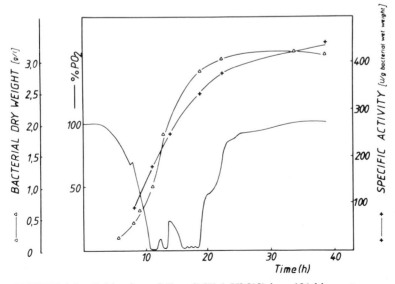

FIGURE 4.3   Cultivation of *E. coli* 5K (pHM12) in a 121-bioreactor.

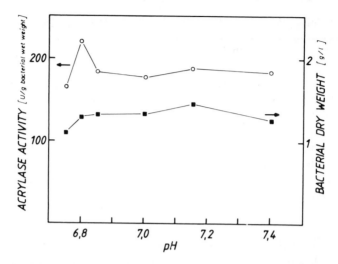

FIGURE 4.4   Penicillin G acylase activity in dependence of the pH value.

Since the results from the quantitative nutritional improvement are not transferable, the influence of additions must also be tested in batch fermentations. The addition of 0.25% D,L-malate, which was the most suitable, caused a slight increase of specific activity and cell yield (related to the organic dry weight of technical yeast extract), from $\gamma_{x/s} = 0.13$ to $\gamma_{x/s} = 0.15$. In the

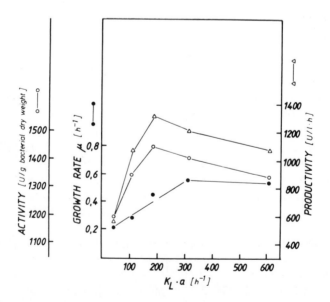

FIGURE 4.5   Influence of the $K_L a$ value on specific growth rate and penicillin G acylase specific activity and productivity.

shake flask experiments, $\gamma_{x/s}$ was increased much more, from 0.12 up to 0.22, by the quantitative improvement. Some experiments to reduce the observed catabolite repression through higher amounts of quickly metabolized carbon sources by feeding were not successful, although this method should be suitable on the basis of positive results with additional fructose in continuous culture. These feeding experiments were not continued because it seemed undesirable to develop too complicated an industrial fermentation process.

In summarizing these results it can be stated that application of recombinant DNA techniques and optimization of growth conditions with respect to penicillin acylase formation led to a remarkable enhancement of specific penicillin acylase activity from 12 U/g wet cells to 420 U/g wet cells, including stable and economic fermentation conditions.

III. PREPARATION OF IMMOBILIZED CELL BIOCATALYSTS

In the course of our work on the development of procedures for entrapment of whole cells in polymeric carriers (Klein and Wagner, 1979), several chemically and structurally different matrices were obtained. In relation to this study, two development steps were involved: screening for the most effective type of carrier and optimization of the catalyst preparation on the basis of a highly active strain.

A. Screening of Immobilization Procedures

Polymeric carriers can be prepared starting from: (1) monomeric, (2) oligomeric, and (3) polymeric precursors.

When our program was started, only one report on the immobilization of penicillin acylase-containing *E. coli* cells was available (Sato et al., 1976). This described the polyacrylamide block polymerization technique, a type 1 procedure. Owing to constraints in cell loading and control of particle shape and size, only a very low activity of 0.33 U/g wet catalyst was obtained.

Our first approach was the application of an alternative monomer, namely, methacrylamide, to improve the mechanical stability of the matrix. A suspension polymerization technique was applied (Klein and Schara, 1980) to obtain catalyst particles of controlled size and shape. A specific activity of 1.1 U/g wet catalyst was a certain improvement, but the problems of limited cell capacity remained.

The second approach involved carrier preparation from oligomeric precursors based on epoxy resins. Water-soluble or water-dispersable two-component systems are available that show a curing reaction at lower temperatures ($T < 30°C$) and moderate pH (pH < 9.0) in the presence of water.

In a first set of experiments the two-component system was directly mixed with the cell mass, and a dense block was obtained, which had to be ground to obtain catalyst particles of irregular shape and size. Particles of very small size showed high specific activity; but owing to the very limited porosity of the material, this activity was drastically decreased with increasing particle size because of diffusional transport limitation.

In the second set of experiments a scheme for the preparation of spherical epoxy beads was developed in which the control of geometry and porosity was introduced by an ionotrope gelation process. Figure 4.6 is a graphic representation of this procedure. At comparable particle size and cell loading the positive influence of porosity on the specific activity of the catalyst could be demonstrated (see Table 4.6) (Klein and Eng, 1979).

Another type of polycondensation network from oligomeric precursors is polyurethane. Foam-type or gel-type structures can be prepared; but owing principally to low levels of cell loading, only low specific activities could be obtained (Klein and Kluge, 1981). The third approach involves network formation from polymeric precursors; in this context the method of ionotropic gelation is of primary importance. Calcium alginate, the most widely used system, has been tested, and catalyst particles with rather high specific activity were obtained. Limits exist, however, because of the poor mechanical stability due to the ionic nature of substrate and products.

Finally, a simple precipitation technique was tested with "Eudragit" preparations, but only a lower level of cell loading, and thus specific activity, could be realized.

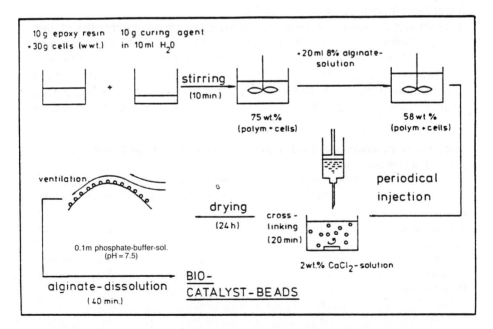

FIGURE 4.6   Schematic diagram for the preparation of epoxy beads.

TABLE 4.6 Catalytic Activity of Immobilized _E. coli_ ATCC 11105 Whole Cells for Epoxy Carrier Systems

Geometry	Loading (g wet cells/g wet catalyst)	(%)	(%)	Activity (U/ml catalyst)
Resin particles				
$\theta \sim 0,1$ mm	1.24	70	40	4.8
$\theta \sim 3\text{-}4$ mm	1.24	70	11	0.98
Porous beads				
$\theta \sim 3$ mm	1.13	67	28	2.3

Reproduced by permission from _Biotechnology Letters_.

These results (Klein and Wagner, 1980) are summarized in Table 4.7; two procedures — namely, the epoxy and the ionotropic gelation matrix — can be selected for further optimization studies, while the others can be eliminated, mainly owing to constraints in cell loading capacity. From a comparison of the epoxy system and the ionotropic gel system the following can be stated: Both matrices can be prepared under conditions of high cell loading, and thus high values of specific activity can be achieved. While the stability of the epoxy matrix is superior to the alginate preparation — owing to swelling and abrasion — the opposite is true for yield of activity as related to toxic inactivation in the course of catalyst preparation.

B. Optimization of Catalyst Preparation for the Immobilization of _E. coli_ 5K (pHM12)

There was a 20-fold increase of activity as compared to earlier reports, but with the limiting specific activity of the wild strain _E. coli_ ATCC 11105 an upper limit of catalyst activity on the order of 10 U/g catalyst has to be accepted.

Progress can be expected only if two problems can be solved: The first one is the development of cell strains with much higher specific penicillin acylase activity, and successful work in this direction has been covered in Section II of

TABLE 4.7 Comparison of the Catalytic Performance of Immobilized _E. coli_ ATCC 11105 Whole Cells

Polymer	Particle Diameter (mm)	Cell Loading (g wet cells/ml cat.)	Activity (U/ml cat.)	(%)
Eudragit®	5	0.14	2.1	65
Calcium alginate	2	0.5	4.4	60
Epoxide				
Crash	0.1	0.7	4.8	40
Beads	3	0.67	2.3	28
Polyurethane				
Gel	1	0.05	0.5	43
Foam	5	0.09	0.56	27
PMAAm	0.5	0.15	1.07	47

this chapter. The second problem is immobilization without additional loss of efficiency.

When cells with different activity are immobilized in epoxy beads under otherwise identical conditions, the absolute activity increases, but the overall yield of activity decreases.

While the cell activity was improved by a factor of 10, the absolute activity could be brought only to a level of 10 U/g catalyst, and the activity yield dropped to 10%. This was confirmed by studying different particle sizes down to 0.5 mm, and the very high activity level of the *E. coli* 5K (pHM12) cells could be expressed only if particles in the size range of <0.1 mm were prepared. On the other hand, the yield of activity of 45% clearly showed the limits caused by toxic deactivation typical for this type of carrier material.

To obtain catalyst particles that combine attractive activity yields with particle sizes of practical value, that is, on the order of >0.2 mm, the ionotropic gelation methods seem to offer the greater potential. In this respect, one has to realize that in addition to the calcium alginate method a variety of polymer-counterion systems exist (Klein and Manecke, 1982), allowing adaption to the specific immobilization problem. In our case the systems Chitosan/polyphosphate (Vorlop and Klein, 1981, 1983) and $\varkappa$-Carrageenan/K$^\oplus$ showed the most promising properties.

The chitosan procedure can be described in detail as follows: 9 g *E. coli* cells (wet weight) are mixed with 9 ml H$_2$O and suspended in 82 g of a sterilized 1.3% chitosan acetate solution at pH = 5.6 (chitosan from Chugai Boyeki Europe Office, Düsseldorf, F.R.G.). This suspension is injected dropwise into a 2% Na-tripolyphosphate solution (pH = 8.3) and stirred gently for 4 hours. Bead formation is accompanied by shrinking: starting from an initial 9% cell loading, the catalyst bead finally has a cell loading of 30%.

Some typical activity data of *E. coli* 5K (pHM12)/Chitosan biocatalysts are given in Table 4.8, which demonstrates the success in obtaining immobilized cell preparations with very high activity, which up to now could be reached only with immobilized enzyme catalysts (Klein and Vorlop, 1984).

TABLE 4.8   Activity of Chitosan Biocatalysts for Various Particle Diameter and Cell Loading

Particle Diameter (mm)	Cell Loading Start (%)	End (%)	Activity (U/g wet catalyst)
Free cells	—	—	175
~0.6	20	>50	60
<0.25	20	>50	101
Free cells	—	—	245
~0.8	9	~30	49

IV. EFFICIENCY OF IMMOBILIZED CELL CATALYSTS

As was mentioned in Section III in a qualitative fashion, the activity of the immobilized cell preparations is not simply and directly related to the free cell activity but depends also on certain parameters, like cell loading, porosity, or particle size. This is a well-known fact in heterogeneous catalysis in general, where the coupling of a reaction with transport steps has to be taken into account. If only pore diffusion resistance, and not surface boundary layer resistance, is considered, a dimensionless number, usually called Thiele-modulus Φ, can be used to treat quantitatively for the transport-reaction coupling phenomenon. Assuming the validity of Michaelis-Menten rate equation—which can be used for simple enzymatic reactions in whole cells too—the following expression for the Thiele modulus has been derived:

$$\Phi = \frac{r_0}{3} \sqrt{\frac{V'}{2\,K_M^S \cdot D_S}} \left(\frac{C_S}{K_M^S + C_S}\right) \left[\frac{C_S}{K_M^S} - \ln\left(1 + \frac{C_S}{K_M^S}\right)\right]^{-1/2} \quad (4.2)$$

where r_0 is the particle radius, V' the rate of reaction, K_M^S the Michaelis constant, C_S the substrate concentration, and D_S the effective substrate diffusivity in the porous catalyst particle. On the other hand, the effectiveness factor η is defined as the ratio of the effective reaction rate v_e to the maximum reaction rate v_{max} that could be observed without transport limitation:

$$\eta = \frac{v_e}{v_{max}} \quad (4.3)$$

Based on numerical calculations, typical functions

$$\eta = f(\Phi) \quad (4.4)$$

have been developed that interrelate the two dimensionless parameters and can be checked experimentally. This model has been used for the cleavage reaction of penicillin G to 6-APA with immobilized *E. coli* cells for epoxy (Klein and Vorlop, 1983a) and polymethacrylamide beads (Washausen, 1979).

A more careful inspection of the situation reveals the fact that the rate expression, incorporated in Eq. (4.2), is a simplified form of Eq. (4.1). Based on Eq. (4.1) and a complete material balance, an exact mathematical model has been developed and used in numerical calculation procedures. Two objectives have been considered, namely, (1) the calculation of the pH profile in the catalyst bead and (2) the calculation of the catalytic effectiveness itself (Klein and Vorlop 1983b, 1984).

The material balance in the interior of the catalyst bead is given by the following set of differential equations, keeping furthermore Eq. (4.1) in mind:

(Penicillin G)
$$D_S \cdot \left[\frac{d^2c_S}{d\tau^2} + \frac{2}{\tau} \cdot \frac{dc_S}{d\tau} \right] = V \tag{4.5}$$

(6-APA)
$$D_P \cdot \left[\frac{d^2c_P}{d\tau^2} + \frac{2}{\tau} \frac{dc_P}{d\tau} \right] = -V \tag{4.6}$$

(Phenylacetic acid)
$$D_{A^-} \cdot \left[\frac{d^2c_{A^-}}{d\tau^2} + \frac{2}{\tau} \frac{dc_{A^-}}{d\tau} \right] = -V \tag{4.7}$$

D_S, D_P, and D_{A^-} are the effective intraparticle diffusivities of penicillin G, 6-APA, and phenylacetic acid. To calculate the pH value at any point in the catalyst particle, the buffering capacity of the solution was also taken into account.

The solution of the differential equations (4.5), (4.6), and (4.7) is based on the following general equation:

$$\eta = \frac{\int_\sigma^{\tau_a} V(C) \cdot \tau^2 \cdot d\tau}{V(C_0) \cdot \int^{\tau_a} \tau^2 \cdot d\tau} = \frac{\int_\sigma^{\tau_a} V(C) \cdot \tau^2 \cdot d\tau}{V(C_a) \cdot \frac{1}{3}\tau_\sigma^3} \tag{4.8}$$

The method of Runge and Kutta, with an optimization of pH and substrate concentration at the catalyst surface, has been chosen as solution procedure. So the calculation of concentration and pH profiles in the catalyst bead as well as the catalytic effectiveness factor η was possible. A comparison of experimentally determined and calculated values of the effectiveness factor is given in Table 4.9. A satisfactory agreement of theory and experiment can be observed.

TABLE 4.9 Comparison of Calculated and Measured Activity of Carrageenan Biocatalysts (25% Cell Loading)

Particle Diameter (mm)	Measured (U/g wet catalyst)	(%)	Calculated (U/g wet catalyst)	(%)
~4	11	20	16	30
~2.8	17	29	24	45
~1.7	25	47	33	61
~1	34	64	44	81
Free cells (glutaraldehyde treated)	215	100	215	100
Free cells	247	—	—	—

As a consequence of the formation of phenylacetic acid, a pH gradient within the catalyst beads has to be expected. It becomes obvious that around the bead center a pH below 5.2 exists, whereas the pH in the solution was 8.2. The pH profile was more or less pronounced depending on the buffer concentration. The activity of the biocatalysts in a buffered medium was higher than in the unbuffered one (Klein and Vorlop, 1984). For an effective process a buffered reaction solution should be used.

V. PROCESS CHARACTERISTICS

In the course of the development of a practical process for 6-APA production from penicillin G with immobilized cells the following facts have to be noted.

1. The conversion has to be as close as possible to 100% (e.g., 98%) combined with a product yield at the same level.
2. Penicillin G solutions of a certain concentration (7.5–10%) should be used.
3. During the reaction, intensive mixing has to be provided to control the pH exactly by alkali addition and to prevent external diffusion limitations.

The simplest and most effective solution to this problem would be the discontinuous, well-stirred batch reactor, as was shown by a computer simulation (Klein and Vorlop, 1984). In this case a penicillin G solution of a certain concentration (7.5–10%) is brought to complete conversion, and the catalyst can be reused after separation from the solution. Short reaction times (on the order of 2–3 hours) at moderate temperatures ($T = 28$–$37°C$) have to be realized to minimize the formation of unwanted by-products. By using *E. coli* 5K (pHM12) cells entrapped in chitosan beads it was possible to fulfill these requirements.

VI. OPERATIONAL STABILITY

The most important point of an industrial biocatalyst, besides the activity, is operational stability. The operational stability includes the catalytic and the mechanical stability (abrasion) of the catalyst. Therefore we tested the repeated use of our immobilized cell biocatalysts, especially the chitosan biocatalysts, in a discontinuous stirred tank reactor.

Penicillin G solutions of 7.5% ($37°C$) were converted to 99%. The product solution was separated, and a new penicillin G solution was employed. This was done by an automatic system with pH control, a filling device for the penicillin G solution, and a drainage appliance for the product solution. As an example, two types of chitosan biocatalysts were tested:

1. Fresh *E. coli* 5K (pHM12) cells were immobilized.
2. Glutaraldehyde treated cells were immobilized.

The results shown in Fig. 4.7 clearly indicate the better operational stability of pretreated immobilized cells. After an initial activity loss this preparation showed very good operational stability. The activity loss after ten batches was < 3%. Figure 4.8 shows the titration graphs (consumed alkali) belonging to it. The untreated cells lost half the activity after only two to three batches, so further experiments were carried out with pretreated cells only.

After optimization of the immobilization procedure the chitosan biocatalysts ($\theta \sim$ 0.6 mm, 40–50% cell loading) converted 10% penicillin G solutions to ~99% (37°C, pH 8.1) after a reaction time of ~2 hours. The product yield was ~98%; only 0.8–1.4% by-products were formed, which could be monitored by HPLC. After 70 reaction cycles (batches) an activity loss of only 10% was observed (Fig. 4.9). So the long time stability is quite comparable with that of commercial immobilized enzymes (Röhm, 1983; Boehringer, 1983).

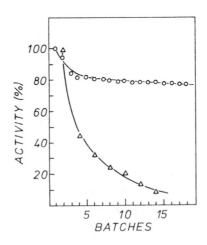

FIGURE 4.7 Operational stability of chitosan biocatalysts with a high penicillin acylase activity. $\bigcirc$ = *E. coli* 5K (pHM12) cells treated with glutaraldehyde; $\triangle$ = nontreated cells.

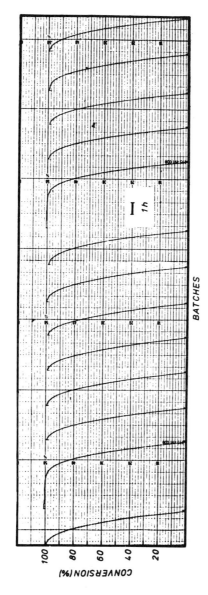

FIGURE 4.8 Titration graphs (alkali consumption) for the discontinuous penicillin G cleavage.

87

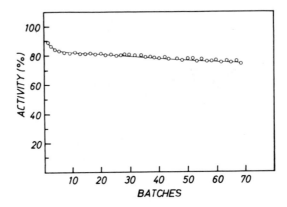

FIGURE 4.9   Operational stability of chitosan biocatalysts.

VII. CONCLUSIONS

A comparison of activities of immobilized cell biocatalysts with penicillin G acylase as reported by different authors and by our group is summarized in Table 4.10. The highest activity was shown by the chitosan catalyst. This highly active catalyst resulted from systematic research and development with regard to genetic engineering, optimization of fermentation technology, and effective cell immobilization.

It is of interest to compare both the immobilized penicillin acylase and immobilized cells containing the enzyme. A few years ago there was a large difference between the activities of the immobilized enzyme preparations and the immobilized cells. The activity of the inexpensive chitosan cell biocatalyst is now quite comparable to the activities of the expensive immobilized enzymes (Table 4.11). The long time stability is also quite comparable with that of immobilized enzymes, as shown in Section VI.

The *E. coli* 5K (pHM12) cells can also be used for the cleavage of penicillin V, deacetoxycephalosporin G, etc. (Mayer et al., 1981). The inexpensive chitosan cell catalyst thus can be employed for the production of 6-aminopenicillanic acid (6-APA) or 7-aminodeacetoxycephalosporin acid (7-ADCA).

This work clearly indicates that the activities of immobilized highly active cells, which can be obtained by genetic engineering, are comparable with those of immobilized enzyme preparations. Thus the combination of genetic engineering and whole cell immobilization will lead to an increase in industrial use of immobilized whole cells.

TABLE 4.10 Activities of Immobilized Cell Biocatalysts with Penicillin G Acylase Activity

Support	Particle Diameter (mm)	Temperature (°C)	Activity (U/g wet cat)	Activity (U/g dry cat)	Activity Yield (%)	Reference
Polyacrylamide	—	37	0.33	—	—	Sato et al. (1976)
Carrageenan	—	40	?	6.57	—	Casas and Quintero (1981)
Agar-glutardialdehyde	—	40	5	—	25	Zhang (1982)
Polymethacrylamide beads	~0.25	37	0.88	—	40	Washausen (1979)
Epoxy beads	~3	37	2.3	—	23	Klein and Eng (1979)
	~0.1	37	4.8	—	40	
Epoxy crash	0.05–0.1	37	41.3	—	32	Klein and Wagner (1980)
Epoxy beads	~2	37	10	—	13	Eng (1980)
Silicone rubber	1.5–2	37	19	—	16	Kressdorf (personal communication, 1983)
Eudragit E 30 D beads	0.4–0.7	28	15	—	24	Vorlop (1984)
Kaurit crash	0.05–0.1	28	48	—	30	Vorlop (1984)
Mg Pectinate beads	1–1.2	37	17	—	45	Vorlop (1984)
dried	0.7–0.9	37	42	—	36	
Carrageenan beads	~4	37	11	—	20	Klein and Vorlop (1984)
Carrageenan beads	~1	37	34	—	64	
dried	<1	37	47	—	26	
Chitosan beads	~0.6	37	60	—	64	Klein and Vorlop (1984)
Chitosan beads	<0.25	37	101	—	>90	

TABLE 4.11 Activities of Immobilized Enzyme Preparations

Support	Particle Diameter (mm)	Temperature (°C)	Activity (U/g wet cat)	Activity (U/g dry cat)	Activity Yield (%)	Reference
Amberlite XAD7	0.4–0.8	37	70	–	77	Carleysmith et al. (1980)
CM-Cellulose	–	37	272	–	82	Carleysmith et al. (1980)
Eupergit C	0.1–0.2	37	100–350	–	70	Röhm (1983)
Sephadex G 200	–	37	225	–	–	Lagerlöf et al. (1976)
DEAE-Cellulose	–	37	35	–	–	Warburton et al. (1972)
Celluloseacetate	–	37	–	146	35	Dinelli et al. (1976)
Polyacrylamide	0.1–0.63	28	12–16	140–160	–	Boehringer (1983)

REFERENCES

Balasingham, K., Warburton, D., Dunnill, P., and Lilly, M. D. (1972) *Biochim. Biophys. Acta 276*, 250–256.

Boehringer Mannheim, 6800 Mannheim 31, F.R.G. product information (1983).

Carleysmith, S. W., Dunnill, P., and Lilly, M. D. (1980) *Biotechnol. Bioeng. 22*, 735–756.

Casas, L. T., and Quintero, R. R. (1981) poster presentation at Second European Congress of Biotechnology Eastbourne, U.K. European Federation of Biotechnology, Sponsor.

Dinelli, D., Marconi, W., Morisi, F. (1976) *Methods Enzymol. 44*, 227–243.

Eng, H. (1980) Ph.D. dissertation, Technical University of Braunschweig, Braunschweig, F.R.G.

Klein, J., and Eng, H. (1979) *Biotechnol. Lett. 1*, 171–176.

Klein, J., and Kluge, J. (1981) *Biotechnol. Lett. 3*, 65–70.

Klein, J., and Manecke, G. (1982) *Enzyme Eng. 6*, 181–189.

Klein, J., and Schara, P. (1980) *J. Solid-Phase Biochem. 5*(2), 61–78.

Klein, J., and Vorlop, K. D. (1983a) *ACS Sympos. Ser. 207*, 377–392.

Klein, J., and Vorlop, K. D. (1983b) *Chem. Ing. Tech. 55*(12), 976–977.

Klein, J., and Vorlop, K. D. (1984) *Ger. Chem. Eng. 7*, 233–240.

Klein, J., and Wagner, F. (1979) *Dechema Monogr. 84*, 265–325.

Klein, J., and Wagner, F. (1980) *Enzyme Eng. 5*, 335–345.

Kutzbach, C., and Rauenbusch, E. (1974) *Hoppe-Seyler's Z. Physiol. Chem. 354*, 45–53.

Lagerlöf, E., Nathorst-Westfeld, L., Ekström, B., and Sjöberg, B. (1976) *Methods Enzymol. 44*, 759–768.

Mayer, H., Collins, J., and Wagner, F. (1979) in *Plasmids of Medical, Environmental and Commercial Importance* (Timmis, K. N., and Pühler, A., eds.), pp. 459–470, Elsevier North-Holland Biochemical Press, Amsterdam.

Mayer, H., Collins, J., and Wagner, F. (1980) *Enzyme Eng. 5*, 61–69.

Mayer, H., Jaakkola, P., Schömer, U., Segner, A., Kemmer, T., and Wagner, F. (1981) in *Advances in Biotechnology*, Vol. I (Moo-Young, M., Robinson, C. W., and Vezina, C., eds.), pp. 83–86, Pergamon Press, Canada.

Röhm Pharma GmbH, 6100 Darmstadt 1. F.R.G., production information (1983).

Sato, T., Tosa, T., and Chibata, I. (1976) *Eur. J. Appl. Microbiol. 2*, 153–160.

Sauber, K., and Krämer, D. M. (1982) *Enzyme Eng. 6*, 235–236.

Svedas, V. K., Margolin, A. L., Berezin, I. V. (1980) *Enzyme Microbiol. Technol. 2*, 138–144.

Tjokrosoeharto, H. (1983) Ph.D. dissertation, Technical University of Braunschweig, Braunschweig, F.R.G.

Vandamme, E. J., and Voets, J. P. (1974) *Adv. Appl. Microbiol. 17*, 311–369.

Vorlop, K. D. (1984) Ph.D. dissertation, Technical University of Braunschweig, Braunschweig, F.R.G.

Vorlop, K. D., and Klein, J. (1981) *Biotechnol. Lett. 3*, 9–14.

Vorlop, K. D., and Klein, J. (1983) in *Enzyme Technology* (Lafferty, R. M., ed.), pp. 219–235, Springer, New York.

Warburton, D., Balasingham, K., Dunnill, R., and Lilly, M. D. (1972) *Biochim. Biophys. Acta 284*, 278–284.

Warburton, D., Dunnill, P., and Lilly, M. D. (1973) *Biotechnol. Bioeng. 15*, 13–25.

Washausen, P. (1979) Ph.D. dissertation, Technical University of Braunschweig, Braunschweig, F.R.G.

Zhang, S. Z. (1982) *Enzyme Eng. 6*, 265–270.

Enzymatic Processes for High-Fructose Corn Syrup

Sidney A. Barker
Graham S. Petch

I. INTRODUCTION

The conception and establishment of the enzymatic conversion of glucose to fructose by immobilized glucose isomerase on a huge industrial scale have been the first real success story of the newer biotechnology. While first conceived as a means of making glucose production more economically viable and at the same time providing a flexible alternative to sugar, they have already enabled certain nations (for example, the United States) to use much more of their home-grown sources of starch (e.g., maize), thereby encouraging home agriculture and cutting imports. Such developments have been crippled politically in the European Economic Community, but Japan and certain Third World countries are rapidly establishing and moving to the nominal goal of producing at least a third of their nation's sweetener requirements as high-fructose corn syrup (HFCS) or its equivalent. As such a goal is achieved, the natural trend will be to use the industry as a feedstock, from renewable sources, of chemical products easily derivable in high yield from fructose. Hydroxymethylfurfuraldehyde and chloromethylfurfuraldehyde are prime examples of such products.

However, there are two basic requirements that should be implemented in the current processes. The first is further improvement in the quality of the glucose feedstock that could be derived by an improved continuous process, largely based on immobilized amyloglucosidase (1,4-α-D-glucan glucohydrolase). Such an achievement would permit the whole process of HFCS production to be continuous. Further separation of HFCS (42% fructose) has created new markets filled by enriched fructose corn syrup (EFCS, 55% fructose) and even very enriched fructose corn syrups (VEFCS, 90% fructose). However, the second requirement in the industry is the true mating of the separation process with the reaction step not only for glucose to fructose conversion to overcome its natural equilibrium barrier, but also in the conversion of starch to glucose to overcome the reversion reaction that leads to the synthesis of isomaltose and other products from the true end-product glucose, thereby lowering the quality of the feed.

The reluctance of HFCS manufacturers to switch to continuous enzyme processes throughout is to be appreciated in the context of their heavy capital investment in their current plants and their wet milling end to produce the starch from the cereal. It is realistic to expect that there are currently under development processes that collapse into one the selective extraction of the starch and the means for its subsequent breakdown to glucose. Further, such developments are proceeding in parallel with the selective extraction of cellulose from wood with the means for its subsequent breakdown to glucose. The less demanding parameters for chemical industry feedstocks in comparison with those of the food industry also means that competition from chemical catalysis will be that much fiercer. It is thus probable that enhanced enzyme stability would become a primary goal despite the outstanding achievements that have been made.

Most immobilized glucose isomerase used currently also has the capacity to convert xylose to xylulose, a route not commercially exploited. Xylulose tends to be much more unstable than fructose, but this problem can be overcome, opening up a further avenue of valuable chemical intermediates.

At least four different types of glucose isomerase enzyme exist in microorganisms:

1. This type requires xylose in the growth medium and arsenate for effective conversion of glucose to fructose. It also has associated xylose isomerase activity.
2. This type is similar to 1 but without xylose isomerase activity, yet the same enzyme can isomerize glucose 6-phosphate to fructose 6-phosphate in the absence of phosphate.
3. These do not require arsenate, but their activity is enhanced by manganese and cobalt ions. Both glucose and xylose isomerase activity are exhibited at pH 6–7, and there is even a low level of ribose isomerase activity.

4. These resemble type 3 but are much more thermostable; and although some types have high pH optima (pH 9.3–9.5), undesirable from the commercial standpoint, there are a wide variety that can operate effectively at pH 7.5. They were the source of the first industrially viable glucose isomerases. Variations within the group include subtypes that do not require either xylose or cobalt in the culture medium for their production. Others require no cobalt for activity and stability. Another, *Bacillus coagulans*, required cobalt ions to isomerize glucose and ribose but manganese ions to isomerize xylose. The two major enzyme manufacturers in the field use *B. coagulans* and *Actinoplanes missouriensis*.

II. GLUCOSE FEEDSTOCK PREPARATION

Recent work (Benson et al., 1982) has highlighted the troublesome enzyme, transglucosidase, that often accompanies amyloglucosidase (exo-1,4-α-D-glucosidase; EC 3.2.1.3) production from *Aspergillus niger*, which is the preferred industrial enzyme source for glucose feedstock preparation from starch dextrins. This enzyme is a 1,4-α-D-glycosyltransferase (EC 2.4.1.24) and a distinct molecular species (Pazur and Ando, 1959). It is unfortunate that *Rhizopus* and *Endomycopsis* have inferior quality amyloglucosidases, since neither of them produce appreciable amounts of transglucosidase (Aunstrup, 1972).

Transglucosidase causes the synthesis of isomaltose and panose, both containing a 1,6-α-linkage; if not removed, it adds to the amyloglucosidase reversion reaction with glucose that produces as major products maltose and isomaltose (Underkofler et al., 1965). To assay transglucosidase, advantage is taken of its slow hydrolysis of methyl-α-D-glucoside, itself not attacked by amyloglucosidase. Maltose itself is unsuitable since both enzymes have overlapping specificities with it. Kathrein (1963) has patented a method for the removal of transglucosidase from amyloglucosidase involving the addition of 1% (w/v) magnesium oxide to the enzyme preparation at pH 3 and 30°C. After stirring for 1 hour the pH is readjusted to 4.0, and the suspension is filtered. Optimally, transglucosidase is assayed with 2% methyl-α-D-glucopyranoside at 40°C and pH 5.0 in 0.1 M acetate buffer with determination of glucose after 60 minutes. It is concluded that removal of transglucosidase activity from amyloglucosidase is beneficial in increasing the yield of glucose from starch, although complete conversion is prevented by the reversion reaction of the amyloglucosidase. For controlling this reaction the simultaneous presence of pullulanase (EC 3.2.1.41) is beneficial (Hurst, 1970). Pullulanase is an enzyme that attacks the linear polysaccharide pullulan, which is comprised of maltotriose as a repeating unit through α-1:6-glucosidic linkages. From pullulan, maltotriose is the major product.

The patent literature cites several examples of the combined use of pullulanase or amylo-1,6-glucosidase and other enzymes from fungi and barley malt to produce maltose syrups from corn starch. Using Maltrin 15 dextrin (D.E. 15.8) as feedstock, Bohnenkamp and Reilly (1980), with a combination of immobilized amyloglucosidase (*Aspergillus niger*; Novo) and immobilized fungal amylase (*Aspergillus oryzae*; maltose major product but some glucose), found the conditions for maximum maltose production at pH 4.8 and then for later conversion to glucose.

The effect of pullulanase on starch itself is demonstrated in a study of breadfruit starch by Loos et al. (1981). This starch contains 18.2% amylose; and when debranched by pullulanase, the proportions of various chain length fractions were >60 (30%), 38–60 (27%), and 15 (43%). If the pullulanase product was then treated with β-amylase, 94% was convertible to maltose compared to action on the original starch of 58%. It was noted, however, that even after the sequential action of pullulanase and β-amylase, some 5% still had a degree of polymerization > 60. In another sequence of enzymes, β-amylase afforded a β-limit dextrin that again resisted to a small extent (2%, degree of polymerization > 60 $\equiv$ 5% of original starch) the sequential action of pullulanase and β-amylase.

Modern amylomaize starches offer real problems in preparing from them glucose feeds for glucose isomerase reactors. The granules of high amylose corn starches are resistant to α-amylase attack and gelatinize at high temperatures and over a wide range. Discrete fractions of high-amylose corn meal have unusual resistance to heat and enzymes. Knutson et al. (1982) reported that the rate of hydrolysis of amylomaize starch by α-amylase decreases with increased amylose content and that the greatest amylose content is found in the smallest granules, which have the greatest surface area. Fully gelatinized granules (15 min at 100°C), 50% gelatinized granules (15 min at 80°C) and raw granules of amylomaize were studied.

Bacillus subtilis α-amylase was long the industrial enzyme of choice for the conversion of starch to the starch dextrins, which is the feedstock for subsequent amyloglucosidase action. Its relative cheapness and swiftness of action do not justify its immobilization. With the amylomaize starch problem of much higher gelation points, thermally stable α-amylases have been introduced. One interesting novel modification of corn starch with immobilized *B. subtilis* α-amylase that has been claimed is that the modified starches produced had a reduced tendency to retrograde (Hofreiter et al., 1978). The extent of retrogradation of starch pastes is related inversely to temperature and directly to concentration. Amylose, the linear component of starch, is chiefly responsible. In their studies the α-amylase was bound to a phenol-formaldehyde resin (Duolite S-761); adsorption of 97.4% of the enzyme was recorded, which was fixed with glutaraldehyde.

Immobilized *A. niger* amyloglucosidase (Lee et al., 1976b) has been used with a starch dextrin feed and compared with the soluble form of the enzyme under the same conditions. With a larger average molecular weight substrate there were lower overall rates of hydrolysis. The maltose concentration during the bulk of the reaction and the maximum glucose concentration attainable were lower with the immobilized enzyme. Attention was drawn to the fact that amyloglucosidase produces β-D-glucose. One molecule of this with another of either α- or β-glucose quickly forms maltose in small concentrations but more slowly produces higher concentrations of nigerose (α 1:3 link) and isomaltose (α 1:6 link). The greater yields of glucose with the soluble enzyme were attributed to the glucose concentration's being higher in the pores of the enzyme carrier than in the bulk solution. Higher glucose concentrations favor the reversion reaction. When Maltrin 15 feed was used at 30% w/v at pH 4.5 and 55°C with Novo amyloglucosidase immobilized on porous ZrO_2-coated glass, the maximum glucose concentrations were 94.3% (200 + mesh), 93.8% (40–80 mesh), and 91.2% (18–20 mesh), while soluble enzyme gave 94.1% glucose.

Allen et al. (1979) claimed that an immobilized amyloglucosidase on alumina fluid-bed reactor gave higher yields of glucose from 30% w/v dextrin feeds than a comparable fixed-bed reactor because it permitted the use of very small catalyst particles (down to 50 μm). Other advantages included freedom from plugging and high pressure drops in the fluidized-bed reactor. The amyloglucosidase was from *A. niger* (Wallerstein Co.) and in the immobilized form had excellent thermal stability over 4200 hours. At that time (1979), Allen et al. calculated that the fluidized-bed reactor could reduce by as much as 33% the processing cost of saccharifying low DE corn starch. The normalized residence times (total catalyst activity/mass flow rate) required to achieve a 92–93% ($\pm$ 3%) yield of glucose at 40°C and 60°C were 6.5 $\times$ 10^3 units min/g and 4 $\times$ 10^3 units min/g, respectively.

III. GLUCOSE ISOMERASE IMMOBILIZATION

Actinoplanes missouriensis glucose isomerase is an intracellular enzyme with optimum pH 7.0 and a K_m for glucose of 1.33M. Provided that the proper amount of magnesium is present, cobalt ion is not required (Gong et al., 1980). Purification was achieved by using DEAE cellulose, which also could be used for the glucose isomerase immobilization. The enzyme in solution was stabilized by 40% glycerol. No inducer is required for its culture.

Bacillus sp. 103 (*Novo*) glucose isomerase can be entrapped in cellulose triacetate (Adler et al., 1979). The instability of this enzyme was largely attributed to a protease that could be dealt with by hollow fiber treatment, thermal denaturation, or inactivation with an inhibitor. The stability so achieved corresponded to a 200-hour half-life for free enzyme in batch reactors at 60°C and a 100-hour half-life entrapped in cellulose acetate fibers in continuous reactors. Activity was measured at pH 6.8 with magnesium (0.1 M) and cobalt (0.001 M) ions.

Arthrobacter cells containing glucose isomerase are immobilized by the successive addition of a strong anionic polyelectrolyte (Primafloc C-7) and strong cationic polyelectrolyte (Primafloc A-10, both Rohm and Haas). After extruding and drying at 56°C for 24 h the catalyst particles are 10^{-3} m diameter and 4×10^{-3} min length. Boersma et al. (1979) reported that the K_m for action on glucose was 310 (20) and that on fructose was 340 (30) mol/m^3 using pH 8 and 60°C.

Bacillus coagulans Novo glucose isomerase has been immobilized on an anion exchange material (Huitron and Limon-Lason, 1978), such as DEAE-cellulose. This caused a shift of optimum pH from 7.2 to 6.8 but with retention of the same order of enzyme stability. For the immobilized enzyme the K_m was 0.25 M with 0.01 M Mg^{2+} and 0.19 M with 0.005 M Mg^{2+}. The enzyme was unstable above 70°C.

Streptomyces phaeochromogenus glucose isomerase does not require arsenate ion and has optimal activity at pH 7 and 70°C (Ryu and Chung, 1977) with 0.01 M $MgSO_4.7H_2O$ in the feed. The enzyme in cell form was produced by Godo Shusei Co. Japan and when heat-treated at 80° for 90 min intracellular fixation of the glucose isomerase was effected. In the reactor it had a half-life of 94 hours with a 2.3 M glucose feed. Its K_m for the forward reaction was 0.238 M and the reverse reaction 0.49 M with a maximum theoretical conversion of glucose to fructose of 57%. Under the conditions of the reaction, a reactor operation time of less than 3 hours gave no significant coloration or side products. One significant feature that emerged was that the reaction rate is only 10% of the capacity rate when the reaction mixture passes 40% of the reactor length or the time equivalent. Enzyme loading was important in that if increased by 100% the productivity increase was only about 60%. The productivity in a plug flow type continuous reactor was initially 26% better than in a continuous stirred tank type. This advantage decreased to 15% after 5 days of operation.

Lee et al. (1976a) immobilized two batches of partially purified glucose isomerase from *Streptomyces* sp. The support was porous glass (96% silica, ZrO_2 coated, 40–80 mesh) using silanization of the glass followed by glutaraldehyde coupling of the enzyme. Enzyme activity yield was 56%. The reactor was operated at 55°C, and 0.1 M Mg^{2+} ions gave maximal activation (threefold increase of glucose isomerase activity). Cobalt ions gave a maximum increase of 40% at 10^{-4} M. The optimum pH was 7. At 50°C the half-life of the enzyme was 240 days for the first 30 days; at 60°C it was only 15 days. There was a marked dependence of immobilized enzyme life on substrate concentration, the half-life at 1.5 M fructose being eight times that extrapolated to zero substrate level.

Stanley et al. (1976) immobilized *S. phaeochromogenes* glucose isomerase on chitin using glutaraldehyde as a cross-linking agent. However, glutaraldehyde was found to inhibit glucose isomerase activity, so the minimum effective level (0.07%) for enzyme fixation was determined. The column was operated at 60°C for 19 days at pH 7 before the reactor was recharged with enzyme. It was suspected that the enzyme was slowly desorbing from the column.

IV. GLUCOSE ISOMERASE REACTORS

While the immobilization of glucose isomerase has been a major success story, it should be appreciated that a wide variety of factors such as thermal denaturation, pH effects, shear effects, pore blockage, inactivators present in the feedstock, microbial contamination, or support erosion can cause enzyme deactivation. Park et al. (1981) outlined three control requirements of an enzyme reactor. These involved (1) the replacement of enzyme when the enzyme activity had fallen to a certain level, (2) the decrease in feed flow rate to compensate for loss of enzyme activity and (3) the potential to change the temperature and/or pH to compensate enzyme activity loss. Using Novo-immobilized glucose isomerase, Park et al. (1981) reported that maximum productivity was achieved with an increasing temperature profile. As was expected, fractional conversions of glucose to fructose at high temperature (71.5°) decrease more rapidly than those at lower temperatures (56.5°), although the initial conversions are higher. At 71.5° the enzyme was inactive in 150 hours.

Kikkert et al. (1981) supplied K_m and V_m values of glucose isomerase reported in the literature comprising both immobilized and nonimmobilized systems. In the temperature range of 50–65°C the maximum K_r for an ICI whole cell immobilized product was reached at pH 8.3 with a thermostated stirred tank reactor.

V. GLUCOSE-FRUCTOSE EQUILIBRIUM DISPLACEMENT

The achievement of a high-fructose corn syrup (HFCS, i.e., 42% fructose) with the same sweetness as sugar is relatively simple given that the equilibrium of glucose and fructose produced by immobilized glucose isomerase has a maximum theoretical fructose level of ~57% (Ryu and Chung, 1977). The real problems emerge in producing the enriched fructose corn syrup (EFCS, 55% fructose) and very enriched fructose corn syrup (VEFCS, 90% fructose) at commercial concentrations. Of course, these syrups can be produced by the addition of further amounts of fructose (or the removal of glucose), but the most effective method is the displacement of the equilibrium in favor of fructose.

The use of oxyanions of borate for the displacement of the alkali-catalyzed equilibration of glucose/fructose has been known for a long time, several authors having described this phenomenon (Takasaki, 1966; Mendicino, 1960; S. A. Barker et al., 1973). The alkali-catalyzed equilibration gives rise to more side products, in particular mannose and psicose, than does the glucose isomerase reaction. Increasing the fructose yields by the addition of borate oxyanions to glucose isomerase reactions was first reported by Takasaki (1971, 1972). The borate complex of fructose is more strongly bound than the borate complex of glucose under equivalent conditions with the result that fructose is effectively removed from the equilibrium, thus increasing the yield. Takasaki used soluble glucose isomerase isolated from cell-free extracts of *Streptomyces* sp. and demonstrated that the isomerization ratio could be raised from ~54% to 87% fructose by the addition of varying boron-derived oxyanions to the reaction system. A dependence of the isomerization ratio (usually expressed as a percentage based on fructose) on the mole ratio of boron compared to glucose was noted.

In more recent work carried out using immobilized *Arthrobacter* glucose isomerase (Saleh, 1978) the relation of the isomerization ratio to borate complexing and the utilization of commercially viable concentrations were studied in detail. It was demonstrated that the isomerization ratio could be predicted from the measured stoichiometry and pH dependence of the borate–glucose and borate–fructose complex formation.

The addition of either tetraborate or metaborate as the complexing agent was found to give similar conversions to fructose under similar conditions. The stoichiometry of the complexes, under enzymic reaction conditions, were found to be 2:1 fructose:boron and 1:1 glucose:boron. Generally, the optimum conversion of glucose to fructose was achieved when the borate concentration was equivalent to that required to complex with all the carbohydrate present if it had been isomerized to fructose (i.e., 2:1 carbohydrate:boron). Thus if the oxyanion (borate) concentration is kept constant and the glucose concentration in the feed solution for the isomerization is increased, the isomerization ratio is decreased (Table 5.1). If the borate concentration was increased in order to complex with the carbohydrate present at higher concentrations, then the immobilized glucose isomerase became deactivated. A similarly rapid loss of enzyme activity occurred when excess borate was used.

The production of fructose is also pH-dependent, since the tendency to produce the glucose–borate complex increases with an increase in pH. How-

TABLE 5.1   Yield of D-Fructose in an Immobilized Glucose Isomerase Reactor Operating at 60°C on Various Glucose Feeds Containing Borate

Tetraborate Concentration (M, to boron)	Initial Glucose Concentration (M)	Final pH	Isomerization Ratio
0.6	1.11	8.0	84.5
0.6	1.66	6.7	76.5
0.6	2.22	7.1	71.0
0.9	1.66	7.9	82.0
0.9	2.22	7.8	75.5
0.9	1.11	7.2	65.5
1.2	2.22	8.2	75.5
1.2	1.66	7.2	61.0
1.2	1.11	7.7	41.5

From Saleh (1978).

ever, simply lowering the pH to a level at which the glucose–borate complex is not formed results in less tendency to form the fructose–borate complex. A further problem with the use of borate and the subsequent formation of the fructose–borate complex is that the pH of this complex is more acidic than either the parent acid or the glucose–borate complex. Thus with more fructose–borate complex being formed throughout the reaction, the pH of the reaction mixture falls continually. Since the pH of the solution affects the equilibrium composition, there is a requirement for continual pH adjustment during the enzymic isomerization. This continual adjustment of pH cannot easily be performed in many immobilized enzyme reactors. However, for high conversions to fructose it is the final pH of the reaction mixture that affects the yield. The final isomerization ratio obtained is therefore dependent upon a number of related factors: the stoichiometry of complex formation, the degree of complexing at a particular pH, and the change in degree of complexing as a result of composition-dependent pH changes. The exact borate salt used is not critical, since the borate species present in a solution is a consequence of the pH and concentration (provided that sufficient equilibration time is allowed). There would appear to be one exception to this; the use of boron oxide (B_2O_3) does not appear to be feasible, since the stoichiometry of the complex formation appears to be 1:1 (carbohydrate:boron) for both glucose and fructose. This results in lower optimal conversions to fructose.

The toxicity of borate oxyanions in general must preclude their application in glucose isomerization processes for the production of high-fructose syrups that are to be used in food production. It would also appear that the borate oxyanion has serious effects on the activity of the enzyme whether by denaturation or inhibition.

High yields of fructose from high initial carbohydrate concentrations can be obtained by changing oxyanions. The germanate oxyanion has been used in conjunction with immobilized glucose isomerase to complex the fructose (S. A. Barker et al., 1978; Pelmore, 1978). The fructose and glucose complexes of germanate show similar stoichiometry to those of borate (i.e., 2:1 fructose:germanate and 1:1 glucose:germanate). For a given pH and equivalent conditions the difference between the proportion of glucose and fructose complexed is greater for germanate than for borate (e.g., at pH 7.0, germanate exhibits 100% complex formation with fructose and 2% complex formation with glucose compared with borate, which exhibits 98% complex formation with fructose and 18% complex formation with glucose). This is the property that is largely responsible for the higher yields of fructose obtainable from the germanate-containing reaction system. A further benefit from the utilization of germanate was that it did not affect the immobilized glucose isomerase even at the higher concentrations used, which allowed maximum exploitation of complex formation. An excess of germanate over the stoichiometric equivalent was not used, since it was unnecessary in order to obtain maximum utilization of the differential complexing ability and limited initial solubility of germanium dioxide to produce germanates precluded this.

The maximum yield of fructose obtained in the presence of germanate was higher than that achieved with borate (93% and 87%, respectively, at low initial glucose concentrations). In contrast to the case when borate was used at high initial glucose concentrations (e.g., 40% w/v) the use of germanate (1.2 M) produces high conversions to fructose with an isomerization ratio of 94% (compared with 75.5% for borate). The data in Table 5.2 show that, as with borate, increasing the glucose concentration without increasing the germanate concentration results in lower yields of fructose. The risk of alkali-catalyzed isomerization products, such as mannose, is negligible owing to the final pH of such a solution being 7.7.

The use of oxyanions, particularly germanate, readily allows the production of EFCS or even VEFCS without the requirement of additional expensive separations and subsequent enrichments of HFCS. It should also be noted that if the production of fructose from corn starch is carried out within a single reaction vessel, the displacement of the glucose–fructose equilibrium, using germanate, will also displace the starch–glucose equilibrium. This displacement of the starch–glucose equilibrium should further increase yields of glucose and hence fructose from the corn starch.

VI. SEPARATION OF GLUCOSE AND FRUCTOSE

The separation of fructose from glucose–fructose mixtures has been studied extensively, since this allows the enrichment of glucose–fructose mixtures with fructose produced. Ideally, both the fructose and the glucose should be recovered from a solution allowing the recycling of glucose into the glucose isomerase process, producing yet more fructose. This requirement makes redundant processes that selectively destroy glucose.

The separation of glucose and fructose can be achieved readily by the precipitation of the sodium chloride–glucose double-salt from solutions containing fructose, glucose, and a stoichiometric amount of sodium chloride

TABLE 5.2 Yields of Fructose in an Immobilized Glucose Isomerase Reactor Operating at 60°C on Various Glucose Feeds Containing Germanate

Germanate Concentration (M)	Initial Glucose Concentration (M)	Final pH	Isomerization Ratio
0.6	1.14	7.8	92.7
0.6	1.67	7.8	80.2
0.6	2.18	7.8	73.1
0.6	2.68	7.7	69.4
0.6	3.33	7.5	66.2
1.2	2.22	7.8	93.8

From S. A. Barker et al. (1978) and Pelmore (1978).

(Tatuki, 1972). This process requires careful monitoring of the pH and concentration of the components of the solution. The process also requires the concentration of the solution to Brix 78% when the sodium chloride–glucose double-salt crystallizes out, leaving fructose in solution. The glucose can be recovered by dissolving the double-salt in water and deionizing. However, the pH required during the concentration procedure is 9.0, and it is possible that some alkaline isomerization may occur, as well as alkaline degradation, which contaminates the fructose solution.

A similar type of separation has been carried out by using calcium chloride to complex with fructose and then removing the complex using electrodialysis (Ozaki and Hagiuara, 1973). This process requires the use of ethanol to effect the separation of the fructose from the calcium chloride.

The more usual method of glucose–fructose separation is to use some form of column chromatography. The use of an alkaline earth metal salt of a cross-linked sulphonated polystyrene cation exchange resin was reported by The Colonial Sugar Refining Co. (1967). Separation of fructose and glucose from invert sugar was achieved by sequentially introducing syrup and water into a column, Dowex 50W cross-linked with 4% divinylbenzene with a particle size of 35–70 mesh proving a suitable packing.

Much of the work in this area originates from the work of Jones and Wall (1969), who discovered that monosaccharides were eluted by water from columns of Dowex 50W $\times$ 8 (200–400 mesh, Ba^{2+}) in the order D-glucose, 2 acetamido 2-deoxy D-glucose, D-sorbose, D-galactose, D-xylose, D-mannoheptulose, D-mannose, D-altroheptulose, L-arabinose, D-lyxose, D-mannitol, D-fructose, D-glycerol, D-galacto-heptitol, galactitol, D-xylitol, D-ribose, and D-galacturonic acid. The most popular cation to use would appear to be calcium, owing to its lack of toxicity.

Other types of resin have been used for the purpose of separating glucose–fructose mixtures. Strong acid cation exchange resin (Duolite C-3) in the hydrogen form in conjunction with a weak base anion exchange resin (Duolite A-6) formed part of a procedure proposed for refining enzymically produced fructose-containing solutions (Khaleeluddin et al. 1974). Takasaki (1974) treated invert sugar with a Ca^{2+}-type cation exchanger followed by a HSO_3-type anion exchanger to isolate fructose. Even more recently, Takasaki (1979) used Dowex 1 $\times$ 8 (HSO_3^-) at 55°C to separate glucose and fructose.

All of these methods of separation using columns are effectively batch production and not the more highly desired continuous production of fructose.

There is one form of continuous separation that is possible and that uses the process of P. E. Barker and Deeble (1974). It has been applied to the separation of glucose and fructose (P. E. Barker and Ching, 1980). This process utilized the "Semicontinuous Chromatographic Refiner" (SCCR), in which the columns were packed with a polystyrene resin in the calcium form. In the SCCR there were ten glass columns 2.54 cm in internal diameter and 75 cm in length that were packed with a cation exchange resin in the calcium form. The process works on the principle of the moving port system. Figure 5.1 illustrates

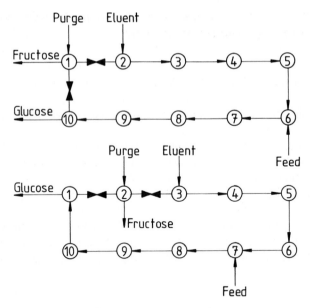

FIGURE 5.1   Schematic representation of two stages of the operation of a Semicontinuous Chromatographic Refiner.

the principle of the operation of the process. In the upper part of the diagram the glucose–fructose mixture is introduced into the system at column 6. The less strongly complexed glucose moves faster in the deionized water, entering at column 2, and is eluted at column 10. In the meantime, column 1 has been isolated from the loop, and under the influence of a high flow rate of deionized water the fructose is eluted. The cycle proceeds by one column after a predetermined time. This time interval is known as the "switch time" or "sequencing interval." Column 2 is now the isolated column that is being purged by deionized water to produce a fructose-rich eluent. Each successive move around the cycle results in simulated movement of the packing in a direction counter-current to the direction of the deionized water (the mobile phase).

Initial purities of the components eluted from the system were of the order of 93–95%. It was found that the columns were able to operate for 600 hours continuously without any significant effect on the stability of the resin (P. E. Barker and Chuah, 1981). The use of columns of a larger internal diameter (10.8 cm) resulted in good yields of high-purity fructose (98% recovered and 99.9% pure) (P. E. Barker et al., 1982). This was achieved with an input concentration of 20% w/v sugar solution. In further work using a fructose/glucose/dextran mixture, fructose was again separated in good yields and purity. The best results so far obtained using this mixture at a 70% sugar solid feed are fructose at 99.9% pure and 99.9% recovery with a fructose product concentration of 16% w/v (P. E. Barker, personal communication, 1983).

Since this process allows the use of glucose/fructose/dextran, it is especially useful for high-fructose corn syrups.

VII. FRUCTOSE DEHYDRATION REACTIONS

Future developments of fructose as industrial feedstock will center around its dehydration reactions, and efficient methods of producing hydroxymethylfurfural (Szmant and Chundury, 1981a) and chloromethylfurfural (Szmant and Chundury, 1981b) are already known. Laevulinic acid can also be obtained (Schraufnagel and Rase, 1975). Brown et al. (1982) claimed an improved method of hydroxymethylfurfuraldehyde preparation using fructose with Amberlyst 15 resin and triethylamine hydrochloride in ethanonitrile at 95–100°C for 2.5 hours under nitrogen (yield 69%). One development of this procedure is that in the presence of an alcohol as a solvent and reactant it gives a mixture of the corresponding hydroxymethylfurfural ether and laevulinic ester:

$$\text{Fructose} + \text{ROH} \xrightarrow[\text{dry resin}]{\text{Amberlyst 15}} \text{CH}_3\text{COCH}_2\text{CH}_2\text{COOR} \qquad (5.1)$$

$$+$$

$$(5.2)$$

Alcohols employed included methanol (I, 47%; II 43%) and *n*-butyl alcohol (I, 77%).

Alkaline conditions should be avoided during carbohydrate processing, since the reducing ends of polyglucosans are easily transformed and defects may accumulate during subsequent enzyme degradation. Thus Hicks et al. (1983) recently demonstrated the mild conditions under which maltose is transformed to maltulose. At high concentrations of up to 40% w/v sugar, sodium hydroxide could replace the base in the preferred treatment with boric acid and triethylamine at pH 11 at 70°. Without the boric acid and using alkali alone the pathway favors the production of acids.

REFERENCES

Adler, D., Lim, H., Cottle, D., and Emery, A. (1979) *Biotech. Bioeng. 21*, 1345–1359.
Allen, B. R., Charles, M., and Coughlin, R. W. (1979) *Biotechnol. Bioeng. 21*, 689–706.
Aunstrup, K. (1972), U.S. Patent 3,677,902.
Barker, P. E., and Ching, C. B. (1980) paper presented at Kem Tek Fifth International Congress, Copenhagen.

Barker, P. E., and Chuah, C. H. (1981) *Chem. Eng. 371–2*, 389–393.

Barker, P. E., and Deeble, R. E. (1974) British patent 1,418,503.

Barker, P. E., Gould, J. C., and Irlam, G. A. (1982) in *Proceedings of the Institute of Chemical Engineers (London) Jubilee Conference*, London April 1982.

Barker, S. A., Chopra, A. K., Hatt, B. W., and Somers, P. J. (1973) *Carb. Res. 26*, 33–40.

Barker, S. A., Somers, P. J., Woodbury, R. R., and Stafford, G. H. (1978) U. K. Patent 1,497,888.

Benson, C. P., Kelly, C. T., and Fogarty, W. M. (1982) *J. Chem. Tech. Biotechnol. 32*, 790–798.

Boersma, J. G., Vellenga, K., de Wilt, H. G. J., and Joosten, G. E. H. (1979) *Biotechnol. Bioeng. 21*, 1711–1724.

Bohnenkamp, C. D., and Reilly, P. J. (1980) *Biotech. Bioeng. 22*, 1752–1758.

Brown, D. W., Floyd, A. J., Kinsman, R. G., and Roshan-Ali, Y. (1982) *J. Chem. Tech. Biotechnol. 32*, 920–924.

The Colonial Sugar Refining Co. (1967) British Patent 1,083,500.

Gong, C. S., Chen, L. F., and Tsao, G. T. (1980) *Biotech. Bioeng. 22*, 833–845.

Hicks, K. B., Symanski, E. V., and Pfeffer, P. E. (1983) *Carb. Res. 112*, 37–50.

Hofreiter, B. T., Smiley, K. L., Boundy, J. A., Swanson, C. L., and Fecht, R. J. (1978) *Cereal Chem. 55*, 995–1006.

Huitron, C., and Limon-Lason, J. (1978) *Biotech. Bioeng. 20*, 1377–1391.

Hurst, T. L. (1970) German Patent 1,943,096.

Jones, J. K. N., and Wall, R. A. (1969) *Can. J. Chem. 38*, 2290.

Kathrein, H. R. (1963) U.S. Patent 3,108,928.

Khaleeluddin, K., Sutthoff, R. F., and Nelson, W. (1974) British patent 1,359,236.

Kikkert, A., Vellenga, K., de Wilt, H. G. J., and Joosten, G. E. H. (1981) *Biotech. Bioeng. 23*, 1087–1101.

Knutson, C. A., Khoo, U., Cluskey, J. E., and Inglett, G. E. (1982) *Cereal Chem. 59*, 512–515.

Lee, Y. Y., Fratzke, A. R., Wun, K., and Tsao, G. T. (1976a) *Biotech. Bioeng. 18*, 389–413.

Lee, Y. Y., Lee, D. D., and Tsao, G. T. (1976b) in Proc. 3rd Int. Biodegradation Symp., Sharpley, J. M., and Kaplan, A. M., eds. Applied Science, Barking, Eng. pp. 1021–1032.

Loos, P. J., Hood, L. F., and Graham, H. D. (1981) *Cereal Chem. 58*, 282–286.

Mendicino, J. F., and Muntz, J. A. (1960) *J. Amer. Chem. Soc. 82*, 4975–4979.

Ozaki, Y., and Hagiuara, H. (1973) Japan Kokai 73 77,039.

Park, S. H., Lee, S. B., and Ryu, D. D. Y. (1981) *Biotech. Bioeng. 23*, 1237–1254.

Pazur, J. H., and Ando, T. (1959) *J. Biol. Chem. 234*, 1966–1970.

Pelmore, H. (1978) Ph.D. thesis, University of Birmingham, Birmingham, England.

Ryu, D. Y., and Chung, S. H. (1977) *Biotech. Bioeng. 19*, 159–184.

Saleh, M. R. S. (1978) M. Sc. thesis, University of Birmingham, Birmingham, England.

Schraufnagel, R. A., and Rase, H. F. (1975) *Ind. Eng. Chem. Prod. Res. Dev. 14*, 40–44.

Stanley, W. L., Watters, G. G., Kelly, S. H., Chan, B. G., Garibaldi, J. A., and Schade, J. E. (1976) *Biotech. Bioeng. 18*, 439–443.

Szmant, H. H., and Chundury, D. D. (1981a) *J. Chem. Tech. Biotechnol. 31*, 135–145.

Szmant, H. H., and Chundury, D. D. (1981b) *J. Chem. Tech. Biotechnol. 31*, 205–212.

Takasaki, Y. (1966) *Agr. Biol. Chem. 30*, 1247.

Takasaki, Y. (1971) *Agr. Biol. Chem. 35*, 1371–1375.
Takasaki, Y. (1972) U.S. Patent 3,689,362.
Takasaki, Y. (1974) *Japan Kokai 74*, 07,442.
Takasaki, K. (1979) *Japan Kokai 79*, 20,577.
Tatuki, R. (1972) U.S. Patent 3,671,316.
Underkofler, L., Denault, J. and Hou, E. F. (1965) *Die Stärke 17*, 179–84.

Immobilized Plant Cells

Peter Brodelius

I. INTRODUCTION

Pharmaceuticals, flavorings, and coloring agents from higher plants have been utilized by humans for many centuries, and such compounds continue to play an important role in modern technology. Most of these substances are of complex chemical structure, and they are classified as natural products. Over 80% of the approximately 30,000 known natural products are of plant origin; consequently, higher plants play an important role as a resource for complex natural products. Some examples of plant chemicals and their uses are listed in Table 6.1. The price of plant chemicals may vary from a few cents per kilogram (e.g., foodstuffs) to several million dollars per kilogram (e.g., pharmaceuticals).

The gradual incursion into the natural environment by modern civilization has made the discovery and exploitation of new substances of such economically important commodities increasingly more difficult. The supply of certain raw plant material is today, or may in the near future be, limited. It has become increasingly important to find alternative resources.

About 30 years ago it was suggested that plant tissue culture may be exploited as a source of potentially useful compounds as an alternative to the collection or cultivation of plants. Any plant species can, in principle, be grown in

TABLE 6.1 Selected Natural Compounds from Higher Plants and Their Properties or Uses

Compound	Property/Use	Source
Alkaloids		
Caffeine	central stimulant	coffee, tea, cola nuts
Cocaine	local anesthetic	*Erythroxyloa coca*
Emetine	antiamebic	*Ceohaelis ipecacuanha*
Harmine	central stimulant	*Peganum harmala* and *Banisteriopsis caapi*
Hyoscyamine	anticholinergic	*Hyoscyamine niger, Atropa belladonna,* and *Datura* species
Morphine and codeine	narcotic analgesic	*Papaver somniferum* and *Papaver album*
Nicotine	insecticide	*Nicotiana tabacum*
Pilocarpine	cholinergic	*Pilocarpus jaborandi*
Quinine	antimalaric/flavor	*Cinchona* species
Reserpine	antihypertensive	*Rauwolfia* species
Scopolamine	anticholinergic	*Datura metal, Scopolia carniolica,* and *Duboisia myoporoides*
Tubocurarine	muscle relaxant	*Chondodendron tomentosum*
Vinblastin/ vincristine	antileukemic	*Catharanthus roseus*
Flavonoids		
Morin	dye	*Chlorophora tinctoria*
Rotenone	antiprotozoal/ insecticide	*Derris* species
Rutin	capillary strengthening	*Eucalyptus* species
Phenolics		
Chrysarobin	dermatologic	*Andira araroba*
Cotoin	antidiarrheal	*Aniba* species
Coumarines	flavors	tonka beans, lavender oil, and sweet clover
Steroids		
Digitoxin	cardiotonic	*Digitalis purpurea*
Digoxin	cardiotonic	*Digitalis lanata*
Diosginin/ solasodine	raw material for steroidal drugs	*Discorea* species *Solanum* species
Terpenoids		
Menthol	flavor compound	*Mentha* species
Stevioside	sweetening agent	*Stevia rebandiana*
Tetrahydrocan- nabinol	psychotropic	*Cannabis sativa*

submerged culture, in analogy to microorganisms. Thus it may be possible to produce any compound found in the parent plant by cultivating the cells originating from this plant species in a fermentor, with the following major advantages:

1. production under controlled conditions
2. possible continuous production
3. constant supply (no seasonal variations)

However, progress in utilizing plant cell culture for the production of biochemicals has been relatively slow. Some of the reasons for this slow development are:

1. low productivity of the target substance
2. slow growth of plant cells
3. instability of plant cells in culture
4. difficulties in large-scale cultivation

Despite these problems, the first chemical compound produced from plant tissue culture was introduced into the market recently. The compound, shikonin, is a dye and pharmaceutical used in Japan for its antibacterial and anti-inflammatory effects. It is produced by Mitsui Petrochemical Industries Ltd., using cultures of *Lithospermum erythrorhizon* in a two-stage batch process. The Mitsui process may be considered a major breakthrough in plant tissue culture technology, even though the operation is on a relatively small scale (5 kg of shikonin after 3 weeks of cultivation). At least it has been shown that a secondary metabolite can be produced by plant tissue culture techniques on a commercial basis.

Recently, an extra dimension was introduced to plant tissue culture technology, that is, the immobilization of viable plant cells (Brodelius et al., 1979; Brodelius and Nilsson, 1980), and in this review possible advantages of such immobilized plant cells are discussed. In particular, the potential of immobilized plant cells to overcome or reduce the above-listed problems of producing biochemicals in culture are considered. It may, however, be appropriate first to give a general introduction to plant tissue culture techniques.

II. PLANT TISSUE CULTURE TECHNIQUES

Plant tissue culture has been studied and used by various scientists as schematically illustrated in Fig. 6.1. Plant cells in culture have been demonstrated to express "totipotency," which means that any living nucleated parendryma cell is capable of complete genetic expression, independent of its origin. Entire plants may therefore be regenerated from cultured cells. Morphogenesis, that

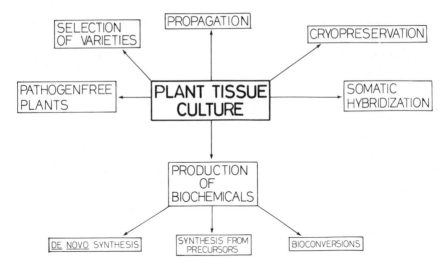

FIGURE 6.1   Schematic diagram of some selected applications of plant tissue cultures.

is, root and/or shoot formation, may be induced by altering the growth medium, in particular the hormone concentrations. This quality of cultured plant cells has been the basis for the wider use of plant cell culture by plant physiologists, geneticists, and breeders. Biochemists, and more recently biotechnologists, have found plant cell culture a powerful tool for many studies.

A. From Plant to Suspension Culture

The standard procedure to establish a plant cell suspension culture is schematically shown in Fig. 6.2. A piece of tissue (explant) is taken from the plant and, after surface sterilization, is placed on a medium solidified with

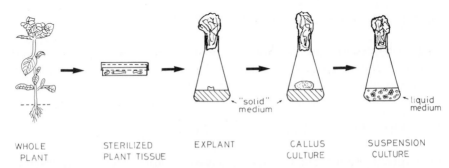

FIGURE 6.2   Schematic diagram of the principle procedure to establish a suspension culture from an intact plant. (Reprinted with permission from Mattiasson, *Immobilized Cells and Organelles*, Vol. 1, pp. 27–55. Copyright CRC Press, Inc., Boca Raton, FL.)

agar. Various parts of the plant, such as root, stem, or leaves, may serve as a source of the explant.

After the explant has been exposed to the medium for some time, a callus, which consists of meristematic cells, starts to form. Newly formed callus tissue is transferred to fresh medium, and this procedure is repeated until a more or less uniform tissue is obtained.

In the next step, friable callus tissue is transferred to a liquid medium, and the suspended callus pieces are incubated on a gyratory shaker. When required, the suspension is transferred to fresh medium to sustain growth. In order to obtain a finer suspension, larger aggregates may be filtered off during the transfer. Most cells in suspension grow in aggregates of various sizes. In general, plant cells in suspension grow more rapidly than cells on a solid medium, and suspended cells are easier to manipulate by alterations or additions to the medium.

B. The Medium

Although whole plants have relatively simple requirements for growth, plant tissue cultures require a complex medium. The composition of the medium has to be empirically established for each plant species, but in most cases a slight modification of a commonly used medium is sufficient for the successful establishment of a new culture. In Table 6.2 the basic composition of a few commonly used media are given. In most cases the same medium is used for callus and suspension cultures.

As can be seen in Table 6.2 the ingredients of the media can be grouped into three classes, macro-, micro-, and organic nutrients. These include vitamins and growth regulators (phytohormones) in addition to a carbon source (most commonly sucrose). Establishing and maintaining a culture is very much dependent on the type and concentration of the hormones used. The major classes of phytohormones are the auxins and the cytokinins, which often are used in combination. Suitable concentrations of hormones very between 0.01 and 10 mg/l depending on the plant species.

The most commonly used auxins are the naturally occurring indole-3-acetic acid (IAA) and the synthetic naphthalene acetic acid (NAA) and 2,4-dichlorophenoxy acetic acid (2,4-D). The most frequently used cytokinins are kinetin and benzyladenine (BA). These two cytokinins are synthetic, but the naturally occurring zeatin is also used. Other types of growth regulators, such as gibberellins, abscisic acid, and ethylene, are occasionally used.

C. Selection Techniques

For the production of biochemicals using cultured plant cells, a primary requirement is a good yield of the final product. As was mentioned in the introduction, the yield of a particular compound is generally low in culture. There are, however, methods for establishing high-producing cell lines; indeed,

TABLE 6.2   Composition of Three Commonly Used Plant Tissue Culture Media (All Values Expressed as mg/l)

Ingredient	MS	B5	SH
Inorganic macronutrients			
KNO_3	1900	2500	2500
NH_4NO_3	1650	–	–
NaH_2PO_4	–	150	–
KH_2PO_4	170	–	–
$NH_4H_2PO_4$	–	–	300
$CaCl_2\ 2H_2O$	440	150	200
$MgSO_4\ 7H_2O$	370	250	400
$(NH_4)_2\ SO_4$	–	134	–
Inorganic micronutrients			
$Na_2MoO_4 \cdot 2H_2O$	0.25	0.25	0.1
$CuSO_4 \cdot 5H_2O$	0.025	0.025	0.2
$MnSO_4 \cdot H_2O$	16.9	10.0	10.0
$ZnSO_4 \cdot 7H_2O$	8.6	2.0	1.0
$CoCl_2 \cdot 6H_2O$	0.025	0.025	0.01
H_3BO_3	6.2	3.0	5.0
KI	0.83	0.75	0.1
$FeSO_4 \cdot 7H_2O$	27.8	27.8	15.0
Na_2EDTA	37.3	37.3	20.0
Organic nutrients			
Thiamine-HCl	0.1	10.0	5.0
Pyridoxine-HCl	0.5	1.0	0.5
Nicotinic acid	0.5	1.0	5.0
Glycine	2.0	–	–
Inositol	100	100	100
Sucrose	30000	20000	30000
Hormones			
2,4-D	–	1.0	0.5
IAA	1.0	–	–
Kinetin	0.1	0.1	0.1
CPA	–	–	2.0
pH	6.0	5.8	6.0

a number of examples have been reported in recent years in which the yield on a dry weight basis is of the same order as, or even higher than, for the parent plant, as summarized in Table 6.3. Normally, the yield is given in % of dry weight, but from a biotechnological point of view it may be more appropriate to give yield as a function of volume (g/l). This value can be given only for suspension cultures, as shown in Table 6.3.

The most widely used procedure to obtain high-producing plant cell cultures is to prepare single-cell clones and subsequently screen for productivity. Single cells are obtained from the suspension cell aggregates by preparing protoplasts by enzymic digestion of the cell wall under hypertonic conditions. The prepared protoplasts are plated at a low density and allowed to grow into

TABLE 6.3 Some Selected Natural Products Formed in Plant Tissue Cultures in a Yield Close to or Higher Than the Parent Plant

Compound	% of Dry Weight (in Culture)	Type of Culture*	g/l Medium	Plant Species	% of Dry Weight (in Plant)	Ratio Culture/Plant	Reference
Ginsengoside	27	C	—	*Panax ginseng*	4.1	6.7	Furuya and Ishii (1972)
Anthraquinones	18	S	2.5	*Morinda citrifolia*	2.2	8.2	Zenk et al. (1975)
Anthocyanins	16	S	0.83	*Vitis* sp.	n.s.	—	Yamakawa et al. (1983)
Rosmarinic acid	15	S	3.6	*Coleus blumei*	3.0	5.0	Zenk et al. (1977a)
Shikonin	12	S	1.4	*Lithospermum erythrorhizon*	1.5	8.0	Fujita et al. (1981)
Benzylisoquino-line alkaloids	11	S	1.7	*Coptis japonica*	5–10	1–2	H. Fukui et al. (1982)
Berberine	10	S	1.2	*Coptis japonica*	2–4	2.5–5	Sato et al. (1982)
Shikimic acid	10	S	1.2	*Galium mollugo*	n.s.	—	Amrhein et al. (1980)
Cinnamoyl putrescine	10	S	1.0	*Nicotiana tabacum*	n.s.	—	Berlin et al. (1982)
yatrorrhizine	7	S	2.7	*Berberis stolonifera*	n.s.	—	Hinz and Zenk (1981)
Glycyrrhizin	7	C	—	*Glycyrrhiza urarensis*	n.s.	—	Fujita et al. (1978)
Anthraquinones	6	C	—	*Cassia tora*	0.6	10	Tabata et al. (1975)
Nicotine	3.4	C	—	*Nicotiana tabacum*	2	1.7	Ogino et al. (1978)
Biscoclaurine	2.3	C	—	*Stephania cepharantha*	0.8	2.9	Akasu et al. (1976)
Saponins	2.2	S	n.s.	*Panax ginseng*	0.53	4	Furuya et al. (1983)
Diosgenin	2	S	n.s.	*Dioscorea deltoides*	2	1	Staba and Kaul (1971)
Harmane	2	S	0.12	*Peganum harmala*	n.s.	—	Sasse et al. (1982)
Paeoniflorin	2	C	—	*Paeonia lactiflora*	n.s.	—	H. Yamamoto et al. (1982)
Saponins	1.9	S	n.s.	*Bupleurum falcatum*	1.8	1	Tomita and Uomori (1976)
Benzophenan-thridine	1.7	S	0.15	*Eschscholtzia californica*	n.s.	—	Berlin et al. (1983)

TABLE 6.3 (continued)

Compound	% of Dry Weight (in Culture)	Type of Culture*	g/l Medium	Plant Species	% of Dry Weight (in Plant)	Ratio Culture/Plant	Reference
Caffeine	1.6	C	—	*Coffea arabica*	1.6	1	Frischknecht et al. (1977)
Valtrate	1.4	S	n.s.	*Fedia cornucopiae*	0.6	2.1	Becker and Schrall (1980)
Ajmalicine	1.0	S	0.26	*Catharanthus roseus*	0.3	3.3	Zenk et al. (1977b)
Serpentine	0.8	S	0.16	*Catharanthus roseus*	0.5	1.6	Zenk et al. (1977b)
Quillaic acid	0.66	S	0.08	*Saponaria officinalis*	0.5–1.3	1	Henry and Guignard (1982)
Ubiquinone-10	0.5	S	n.s.	*Nicotiana tabacum*	0.003	173	Matsumato et al. (1982)
Protopine	0.4	C	—	*Macleaya microcarpa*	0.32	1.2	Koblitz et al. (1975)
Visnagin	0.3	C	—	*Amni visnaga*	0.1	3	Kaul and Staba (1967)
Scopolamine	0.3	C	—	*Hyoscyamus niger*	n.s.	—	Hashimoto et al. (1982)
Codeine	0.15	S	0.004	*Papaver somniferum*	n.s.	—	Tam et al. (1980)
Hyoscyamine	0.09	C	—	*Hyoscyamus niger*	n.s.	—	Hashimoto et al. (1982)
Tripdiolide	0.05	S	0.004	*Tripterygium wilfordii*	0.001	50	Kutney et al. (1983)

* C = callus culture; S = suspension culture.
n.s. = not stated.

callus cultures, which subsequently are screened for productivity. Various screening methods may be applied, but a semiautomatic method based on radioimmunoassay appears to be particularly suitable (Weiler and Zenk, 1979).

It has been shown that plants having higher than normal content of a particular compound yield cultures producing this substance in elevated quantities (Deus and Zenk, 1982). High-producing cell lines can thus be established by careful selection of the original plant material.

Still another approach to obtain a high-producing cell line has been developed in Japan (Y. Yamamoto et al., 1982). The procedure involves selection of a callus culture with a comparably high content of the target substance. This callus is divided into a number of pieces and subcultured. The selection and subculture procedures are repeated until satisfactory production is reached.

An important requirement for an "industrial" plant cell culture is that the cells be stable, in order to give a constant yield of product. However, the yield often declines with age of culture, and a reselection of a high-producing cell line is necessary. This may be achieved relatively easily once a high-producing line has been established. Reasons for this instability have not been established, but it appears that various mechanisms are involved (e.g., changes of genetic expression).

It may be appropriate to point out that for product formation in certain cases, differentiation of the cells is required. The formation of some products appears to be strictly regulated and linked to dedifferentiation of the cultured cells.

III. IMMOBILIZATION OF PLANT CELLS

Plant cells in suspension cultures are relatively large (up to 100 μm in diameter), and they grow, as was mentioned above, in aggregates of various sizes. These characteristics of cultivated plant cells have to be considered when the method of immobilization is chosen. We have found that entrapment within gels is convenient for these heterogeneous cell preparations. Other immobilization techniques, such as covalent attachment and adsorption, have also been used; these latter two methods of immobilization will be reviewed only briefly, whereas entrapment in various gels will be discussed in more detail.

A. Entrapment

The entrapment of microorganisms in gels has been studied extensively (Birnbaum et al., 1983; Brodelius and Vandamme, 1984), and some of these techniques can also be used without significant alterations for the immobilization of sensitive plant cells. Immobilization techniques resulting in viable cell preparations are of particular interest, and the entrapment within gel-forming polysaccharides appears to be most suitable, but other polymers may also be employed. Table 6.4 lists some polymers used for the entrapment of cells.

TABLE 6.4   Some Polymers Used for Entrapment of Cells

Polymer	Principle of Gel Formation	Gel-forming Procedure
Synthetic		
Polyacrylamide	polymerization	addition of initiators to a solution of monomers
Epoxy resin	polycondensation	epoxy and polyamine precursors are cured
Polyurethane	polycondensation	reaction of polyisocyanates with water
Carbohydrates		
Cellulose	precipitation	cellulose is dissolved in an organic solvent and precipitated in water
Agar, agarose	thermal gelation	cooling of heated polymer solution
x-Carrageenan	thermal gelation + ionotropic stabilization	cooling of heated polymer solution
Alginate	ionotropic	dripping into a calcium chloride solution
Proteins		
Collagen	neutralization	neutralization of an acidic solution
Gelatin	thermal gelation	cooling of heated polymer solution
Fibrin	enzymatic	enzymatic conversion of fibrinogen to fibrin

1. Alginate.   In our initial investigations of immobilized plant cells, entrapment in calcium alginate was used (Brodelius et al., 1979). These experiments were initiated to study the behavior of immobilized plant cells and to investigate possible advantages of using such preparations for the biosynthesis and bioconversion of valuable plant products. Alginate was chosen as being a very mild method of immobilization that can be carried out under sterile conditions.

Alginate is a block copolymer of D-mannuronate and L-guluronate joined by 1,4-linkages. An aqueous solution of this polysaccharide forms a relatively stable gel in the presence of multivalent cations such as Ca^{2+}, Ba^{2+}, or Al^{3+}. Spherical beads of alginate containing entrapped cells are obtained by dripping a cell/alginate suspension into a medium containing calcium chloride (50–200 mM).

The final concentration of alginate in the gel varies depending on the preparation used, but a normal concentration is 2–5% (w/w). The cell concentration within the gel may be as high as 50% (w/w), but it is normally 5–20% (w/w).

Large quantities of alginate-entrapped cells are readily prepared under sterile conditions. A disadvantage of this gel type is the requirement for multivalent cations for gel stability. The most commonly used cation is calcium, which must be present at a concentration of 5–10 mM in the medium. This may impose problems when the medium also contains phosphate. However, this calcium dependence makes it possible to reverse the immobilization under relatively mild conditions. The gel may be dissolved by adding a calcium-chelating agent (e.g., EDTA or citrate), and the entrapped cells are released and can be studied.

2. Agar and agarose. Agar and agarose (a purified preparation of agar) are polysaccharides composed of a repeating agarobiose unit consisting of alternating 1,3-linked D-galactopyranose and 1,4-linked 3,6-anhydro-1-galactopyranose. These polysaccharides form gels upon cooling of a heated aqueous solution. The gelling temperature can be altered by chemical modification of the structure (introduction of hydroxyethyl groups). There are various agarose preparations available with different gelling temperatures. For the immobilization of most plant cells a gelling temperature around 30°C is convenient.

The plant cells (normally a final concentration of 5–20% (w/w)) are mixed with an agarose solution (final agarose concentration 2–5% (w/w)) at 35–40°C, and the mixture is then cooled to induce gel formation. Gel particles (beads) may be obtained in various ways. The gel can be made as a block with subsequent mechanical disintegration into particles. More homogeneous gel particles can be made directly by molding in a template form. A more convenient method to make spherical beads has recently been developed (Nilsson et al., 1983). This general procedure for the immobilization of cells with preserved viability is based on dispersion of the cell/polymer suspension in a nontoxic hydrophobic phase with stirring, as schematically shown in Fig. 6.3. When droplets of appropriate size have been obtained, the whole mixture is cooled to allow gel formation. The size of the beads can be controlled by the stirring speed. Plant cells entrapped in agarose by this procedure are shown in Fig. 6.4. Some hydrophobic phases which may be used are listed in Table 6.5.

3. $\varkappa$-Carrageenan. $\varkappa$-Carrageenan contains a repeating disaccharide unit made up of D-galactose-4-sulfate and 3,4-anhydro-D-galactose. This strong gelling, anionic hydrocolloid, can be used for the entrapment of cells under relatively mild conditions in a manner similar to that described for alginate, except that the polymer solution must be heated in order to be maintained in a liquid state. Beads are formed by dripping the cell/carrageenan suspension into a medium containing potassium ions (0.3 M). Gel particles (beads) may also be prepared in analogy with agarose beads. In this case the gel particles are treated with a potassium solution to improve the mechanical stability. Most plant cell media contain sufficient amounts of potassium (cf. Table 6.2) and therefore no extra additions of potassium are required.

The temperature required for keeping the carrageenan in a liquid state is highly dependent upon the preparation of the polymer. A low-temperature

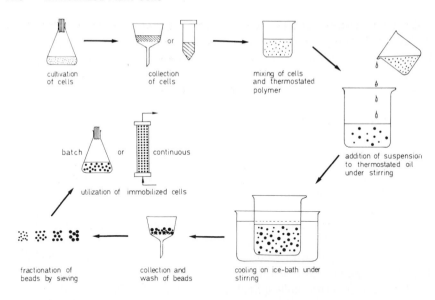

cultivation
of cells

collection
of cells

mixing of cells
and thermostated
polymer

addition of suspension
to thermostated oil
under stirring

batch   or   continuous

utilization of immobilized cells

fractionation of
beads by sieving

collection and
wash of beads

cooling on ice-bath under
stirring

FIGURE 6.3   Schematic diagram of a general method to entrap cells in gel-forming polymers with preserved viability. (From Nilsson et al., 1983.)

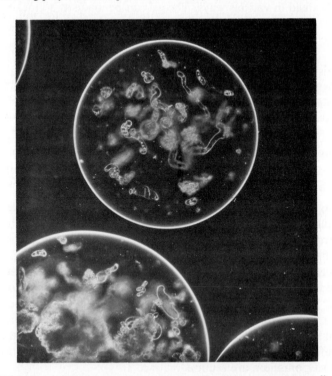

FIGURE 6.4   Cells of *Daucus carota* entrapped in agarose according to the procedure schematically shown in Fig. 6.3.

TABLE 6.5   Relative Respiration of Entrapped Plant Cells after Immobilization in Various Hydrophobic Phases

Hydrophobic Phase	Relative Respiration (%)*
Soy oil	100
Paraffin oil	91
Silicon oil	100
Tri-*n*-butylphosphate	100
Dibutylphthalate	82
Toluene:chloroform (73:27)	0

* Soybean cells (10% w/w) in agar (3% w/v).

gelling carrageenan can be prepared by removing potassium ions from the polymer. There are preparations of carrageenan commercially available that will stay liquid at room temperature at a polymer concentration of up to 5% (w/w). Carrageenan is typically used at a final concentration ranging from 2% to 5% (w/w).

4. Polyacrylamide.   The very first cell immobilization was carried out in polyacrylamide (Mosbach and Mosbach, 1966); since then, this technique has been widely used for the immobilization of microbial cells (Birnbaum et al., 1983; Brodelius and Vandamme, 1984). Attempts to entrap plant cells in polyacrylamide resulted in nonviable cell preparations (Brodelius and Nilsson, 1980) because of the toxicity of the chemicals used to prepare the polyacrylamide gel.

Plant cells have, however, been immobilized in polyacrylamide with preserved viability according to a newly developed and elegant technique (Freeman and Aharonowitz, 1981; Galun et al., 1983). In this method the plant cells are suspended in an aqueous solution of prepolymerized linear polyacrylamide, partially substituted with acylhydrazide functional groups. Subsequently, the suspension is cross-linked with a controlled amount of dialdehyde (e.g., glyoxal) under physiological conditions. The gel obtained is mechanically disintegrated into smaller particles.

Another approach to avoid the toxicity of the chemicals used in the procedure to entrap cells in polyacrylamide was taken by Rosevear and Lambe (1982). The plant cells were in this case mixed with a viscous solution (alginate or xanthan gum) before they were entrapped in polyacrylamide. The cells were protected by the viscous solution, and cells within sheets of this composite materials were shown to be viable.

5. Polyurethane.   Polyurethane foam particles were used to immobilize plant cells (Lindsey et al., 1983). Freely suspended cells were found to invade such porous particles and were strongly retained over a 21-day culture period. The mechanism responsible for the immobilization is not quite clear, but it is suggested that the cells are entrapped within the foam particles.

Plant cells have also been entrapped by cross-linking of urethane pre-polymers according to a method used for the immobilization of microbial cells (S. Fukui and Tanaka, 1984; Tanaka et al., 1983).

6. Membranes.   Various membrane configurations may also be used for the entrapment of plant cells. Hollow-fiber reactors were the first examples of this type of immobilization (Shuler, 1981; Jose et al., 1983). The membranes form a tubular structure and are available commercially in different designs. The fibers are readily arranged as a parallel bundle inside a containing sheath. The plant cells are introduced on the shell side of the reactor, and an oxygenated medium is supplied at a relatively high flow rate through the fiber lumen. It has been shown in other type of reactors that the pore size of the membrane is of great importance for cell behavior (Shuler and Hallsby, 1983).

B. Adsorption

Adsorption to a solid support offers a simple and mild method to immobilize cells. However, the relatively weak binding forces involved will most likely lead to release of cells from the support, especially if the cells are allowed to grow and divide in situ. One example of adsorbed plant cells in the form of protoplasts has been reported (Bornman and Zachrisson, 1982). The protoplasts were adsorbed on microcarriers (Cytodex), previously used for growth of anchorage-dependent animal cells.

C. Covalent Linkage

Covalent linkage involves, as the name implies, covalent bonds between the polymer and the biocatalyst. The introduction of such bonds requires reactive chemicals and may result in loss of cell viability. Therefore this method is less attractive for the immobilization of sensitive plant cells. An advantage of the method, however, is the strength of the bond between matrix and cells.

One example can be found in the literature (Jirku et al., 1981). Cells of *Solanum aviculare* were covalently linked to glutaraldehyde-activated poly-phenylene oxide beads. The immobilized cells were biosynthetically active, but no information on cell viability was given.

IV. CHARACTERISTICS OF IMMOBILIZED PLANT CELLS

The immobilization methods described above have been used for the immobilization of various plant cells as summarized in Table 6.6. Some characteristics of these immobilized preparations are described below.

TABLE 6.6   Viable Preparations of Plant Cells Entrapped in Various Matrices

Matrix	Plant Species	Reference
Alginate	*Catharanthus roseus*	Brodelius and Nilsson (1980)
	Morinda citrifolia	Brodelius et al. (1979)
	Digitalis lanata	Brodelius et al. (1979); Alfermann et al. (1980)
	Daucus carota	Jones and Veliky (1981a,b)
	Cannabis sativa	Jones and Veliky (1981b)
	Ipomoea sp.	Jones and Veliky (1981b)
	Nicotiana tabacum	Morris and Fowler (1981)
	Silybum marianum	Cabral et al. (1984)
Agarose	*Catharanthus roseus*	Brodelius and Nilsson (1980)
	Daucus carota	Brodelius and Nilsson (1983)
	Datura innoxia	Brodelius and Nilsson (1983)
Agar	*Catharanthus roseus*	Brodelius and Nilsson (1980)
	Glycine max	Nilsson et al. (1983)
	Daucus carota	Linsefors and Brodelius (unpublished, 1983)
	Silybum marianum	Cabral et al. (1984)
Carrageenan	*Catharanthus roseus*	Brodelius and Nilsson (1980)
	Daucus carota	Linsefors and Brodelius (unpublished, 1983)
Polyacrylamide	*Catharanthus roseus*	Rosevear (1981)
	Nicotiana tabacum	Rosevear (1981)
	Mentha species	Galun et al. (1983)
Polyurethane	*Daucus carota*	Lindsey et al. (1983)
	Capsicum frutescens	Lindsey et al. (1983)
Membrane	*Glycine max*	Shuler (1981)
	Daucus carota	Jose et al. (1983)

A. Viability

It is of great importance that the plant cells remain viable after immobilization, since major parts of the cell metabolism are required for the conversion of a simple carbon source to complex molecules. However, viable cells have the potential to grow and divide; but in the immobilized state it is desirable to prevent the cells from multiplying. In suspension culture the slow growth of cells is a major limitation (doubling time is normally 25–60 hours), whereas with immobilized systems the growth of cells is actually a problem. Extensive cell growth will in the latter case result in cell release and even destruction of beads. Therefore at present, attempts are being made to formulate media that keep the cells in a viable state over an extended period of time and at the same time do not allow multiplication of the cells. An additional advantage of having viable cells in a resting state is the possibility of inducing cell growth in situ by administration of a growth medium if the biosynthetic capacity of the immobilized preparation is deteriorating.

Various techniques, both direct and indirect, may be employed to investigate the viability of immobilized plant cells. Plasmolysis and some staining techniques may be used to study the integrity of the plasma membrane. For example, staining with fluorescein diacetate is widely utilized as a qualitative method to determine cell viability.

Respiration is another indirect method to study cell viability. Respiration is a vital part of cell metabolism and therefore may be used as a good indication of cell viability in combination with plasmolysis.

Cell growth and cell division are absolute proofs of cell viability. These events may be studied by following the increase of cell number (cell density) within the beads or by studying the mitotic index, which reflects the number of cells in mitosis. The increase in dry weight as well as the mitotic index as a function of incubation time for freely suspended and agarose-entrapped cells are shown in Fig. 6.5 as an illustration of these methods.

All of the above-described methods have been used to show viability of immobilized plant cells, as indicated in Table 6.6.

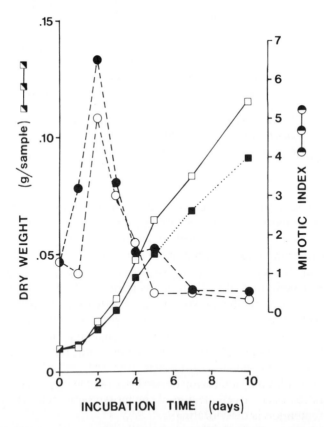

FIGURE 6.5 Dry weight and mitotic index for freely suspended (open symbols) and agarose-entrapped (solid symbols) cells of *Catharanthus roseus* as a function of incubation time. (From Brodelius et al., 1982.)

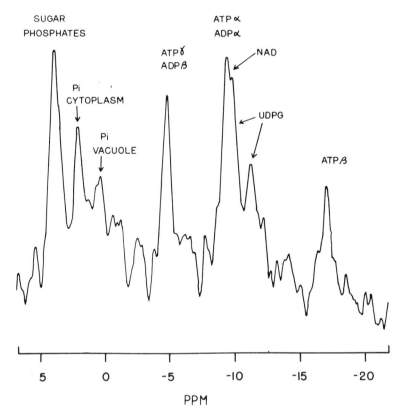

FIGURE 6.6   A typical ^{31}P NMR spectrum (145.7 MHz) of cells of *Catharanthus roseus*. (From Brodelius and Vogel, 1984b.)

We recently initiated studies on the metabolism of immobilized plant cells by ^{31}P NMR (Brodelius and Vogel 1984a, b; Vogel and Brodelius, 1984). This technique does not, however, have as broad applicability as the techniques described above, since it requires very special and expensive equipment. However, it is a powerful tool to study cell metabolism under noninvasive conditions. The uptake, storage, and metabolism of inorganic phosphate can be followed. In Fig. 6.6 a typical ^{31}P NMR spectrum of *Catharanthus roseus* is depicted. Many phosphorylated metabolites can be identified, and the change in concentration may be followed by taking spectra at various points in the growth curve. Figure 6.7 illustrates the conversion of inorganic phosphate stored in the vacuoles of agarose-entrapped *C. roseus* cells to various phosphorylated compounds. The chemical shift of the inorganic phosphate resonance is pH-dependent, and from this phenomenon it can be concluded that the phosphate is stored within the vacuoles of the cells before it is utilized for the synthesis of phosphorylated compounds such as ATP, ADP, NAD(P) (H), and sugar phosphates (Brodelius and Vogel, 1984b). An additional ad-

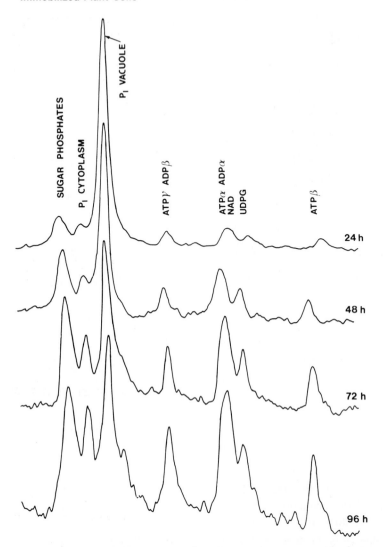

FIGURE 6.7 ^{31}P NMR spectra (103.2 MHz) of agarose-entrapped cells of *Catharanthus roseus* at various times after inoculation into fresh medium. The spectra were obtained by accumulating 50,000 scans. (Vogel and Brodelius, unpublished.)

vantage of ^{31}P NMR is the possibility of studying compartments within the cells. The ^{31}P NMR spectra obtained with immobilized cells are very similar to the corresponding spectra of freely suspended cells, as shown in Fig. 6.8.

The methods described above for the determination of cell viability have shown that the immobilization of plant cells does not affect the viability to any great extent.

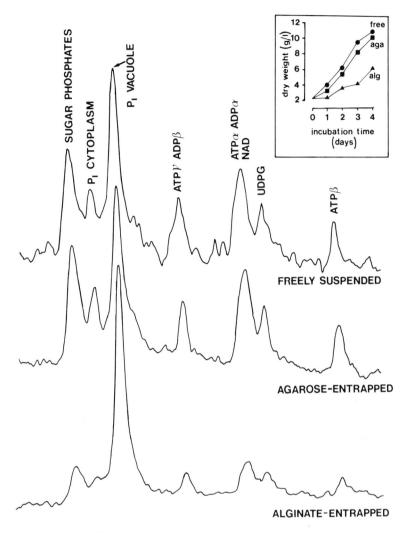

FIGURE 6.8 ^{31}P NMR spectra (103.2 MHz) of various preparations of *Catharanthus roseus* cells after 4 days of incubation. The dry weight increase of the corresponding preparations is shown in the inserted diagram. (Vogel and Brodelius, unpublished.)

B. Biosynthetic Capacity

Most studies on immobilized plant cells have been carried out to study the biosynthetic capacity of the cells and to compare it to corresponding freely suspended cells. The immobilized cells have been utilized for a wide range of biosynthetic activities ranging from bioconversions to de novo synthesis from a simple carbon source.

It may be appropriate to point out a unique feature of cultivated plant cells; the cells have the potential to synthesize completely new compounds. This can be achieved by feeding the cultures analogs of natural substrates that may be converted or incorporated into new compounds. This approach, which is most likely only applicable toward the end of a biosynthetic pathway, has not yet been explored to any great extent.

The biosynthetic studies carried out with immobilized plant cells are grouped into three categories (i.e., bioconversions, synthesis from added distant precursors, and de novo synthesis) in the discussion below.

1. Bioconversions.   Plant cells in culture have been utilized for a wide range of bioconversions (Reinhard and Alfermann, 1980). These include hydroxylations, methylations, acetylations, and glycosylations. Immobilized plant cells have also been utilized to a limited extent for bioconversions, as summarized in Table 6.7.

The most extensively studied bioconversion with immobilized plant cells is the 12-β-hydroxylation of digitoxin derivatives to the corresponding digoxin derivatives (Fig. 6.9). Cells of *Digitalis lanata* entrapped in alginate have been employed for this conversion. Both digitoxin (Brodelius et al., 1979, 1981) and β-methyldigitoxin (Moritz et al., 1982; Alfermann et al., 1983) have been used as substrates.

A packed-bed reactor containing 140 g of beads, corresponding to about 32 g of cells (fresh weight), was used continuously for 70 days to convert digitoxin to digoxin (Brodelius et al., 1981). Approximately 70% of the maximal hydroxylation activity was still present after this period of time. The flow rate of the substrate was 22 ml/h, and the concentration of digitoxin was 50 μM. The maximum yield of digoxin was 77%.

The conversion of β-methyldigitoxin has been studied extensively in batch-change and bubble-column reactors (Moritz et al., 1982, Alfermann et al., 1983). In the former case the same preparation of alginate-entrapped *Digitalis* cells could be used for the hydroxylation for up to 180 days (Fig. 6.10). Fifty beads (diameter 4–5 mm) were used in 25 ml of medium; the medium was changed every third day.

For a practical application of this bioconversion a continuous or at least semicontinuous hydroxylation would be preferred; therefore the same preparation of alginate-entrapped *Digitalis* cells have been tested in a bubble-column reactor. There was no significant difference of productivity between the bubble-column and the batch-change reactor configurations (Fig. 6.11). The experiment was discontinued after 29 days because of extensive growth of the immobilized cells. Attempts are now being made to limit the growth and to optimize the bubble-column reactor.

The 5-β-hydroxylation of the aglycones digitoxigenin and gitoxigenin by alginate-entrapped cells of *Daucus carota* (Fig. 6.12) has been studied extensively (Jones and Veliky, 1981a, b; Veliky and Jones, 1981). The effect of

TABLE 6.7 Use of Immobilized Plant Cells for the Production of Biochemicals

Species	Immobilization Method	Substrate	Product	Operation	Duration (days)	Reference
Conversions						
Digitalis lanata	entrapment in alginate	digitoxin	digoxin	batch	33	Brodelius et al. (1979)
				continuous	70	Brodelius et al. (1981)
Daucus carota	entrapment in alginate	methyldigitoxin	methyldigoxin	batch change	180	Alfermann et al. (1983)
		digitoxigenin	periplogenin	batch change	12	Jones and Veliky (1981a)
		gitoxigenin	5-hydroxygitoxigenin	continuous	21	Veliky and Jones (1981)
Catharanthus roseus	entrapment in agarose	cathenamine	ajmalicine isomers	batch	1	Felix et al. (1981)
Mentha sp.	entrapment in prepolymerized acrylamide	(−) menthone	(+) neomenthol	batch	1	Galun et al. (1983)
		(+) pulegone	(+) isomenthone	batch	1	Galun et al. (1983)
Synthesis from precursors						
Catharanthus roseus	entrapment in alginate agarose agar carrageenan	tryptamine + secologanin	ajmalicine isomers	batch	5	Brodelius and Nilsson (1983)
	entrapment in alginate agarose	tryptamine + secologanin	ajmalicine isomers	batch change	14	Brodelius and Nilsson (1983)
De novo synthesis						
Catharanthus roseus	entrapment in alginate agarose agar carrageenan	sucrose	ajmalicine	batch	14	Brodelius and Nilsson (1980)
	entrapment in alginate	sucrose	serpentine	batch	40	Brodelius et al. (1981)
	entrapment in alginate/ polyacrylamide	sucrose	ajmalicine	batch change	220	Lambe and Rosevear (1983)
	entrapment in xanthan gum/ polyacrylamide	sucrose	serpentine	batch change	180	Lambe and Rosevear (1983)
Morinda citrifolia	entrapment in alginate	sucrose	anthraquinones	batch	21	Brodelius et al. (1980)

TABLE 6.7 (continued)

Species	Immobilization Method	Substrate	Product	Operation	Duration (days)	Reference
Capsicum frutescens	entrapment in macroporous polyurethane foam	sucrose	capsaicin	batch	10	Lindsey et al. (1983)
Lavandula vera	entrapment in urethane prepolymers	sucrose	pigments	—	—	Tanaka et al. (1984)
Glycine max	entrapment in hollow fibre reactors	sucrose	phenolics	continuous	30	Shuler (1981)
Daucus carota	entrapment in hollow fibre reactors	sucrose	phenolics	continuous	21˙	Jose et al. (1983)
Solanum aviculare	covalent attachment to poly-phenyleneoxide beads	sucrose	steroid glycosides	continuous	11	Jirku et al. (1981)

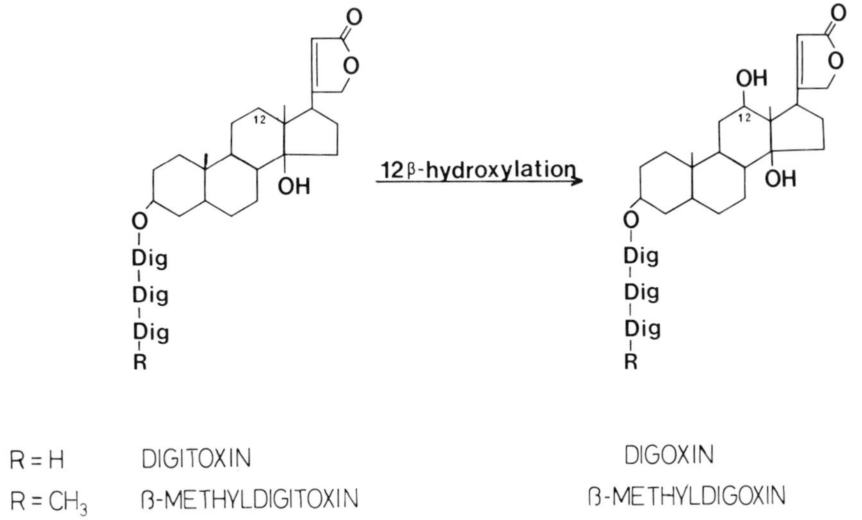

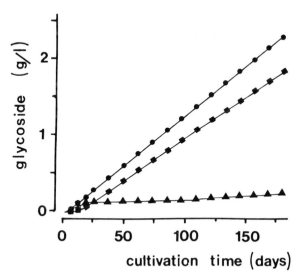

R = H   DIGITOXIN   DIGOXIN
R = CH₃   β-METHYLDIGITOXIN   β-METHYLDIGOXIN

FIGURE 6.9 Bioconversion of heart glycosides carried out with immobilized cells of *Digitalis lanata*.

FIGURE 6.10 Bioconversion of methyldigitoxin to methyldigoxin by alginate-entrapped cells of *Digitalis lanata* in a batch procedure. The medium was changed every third day. (●———●) methyldigitoxin added; (■———■) methyldigoxin formed; (▲———▲) methyldigitoxin unconverted. (Adapted from Alfermann et al., 1983.)

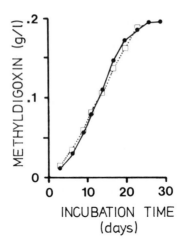

FIGURE 6.11 Bioconversion of methyldigitoxin to methyldigoxin by alginate-entrapped cells of *Digitalis lanata* in shaker flasks and in a bubble column. The medium was changed every third day, and new substrate (40 mg/l) was added. (□————□) shaker flasks; (●————●) bubble column. (Adapted from Moritz et al., 1982.)

various medium constituents on the viability of the immobilized cells was investigated (Jones and Veliky, 1981a). A MES-buffer was designed that could sustain viability of the immobilized cells for an extended period of time. However, the viability could be improved by intermittently incubating the immobilized cells in a growth medium for 2 days as illustrated in Fig. 6.13.

Conversion of digitoxigenin to periplogenin was carried out in a repeated batch procedure using the MES-buffer (Jones and Veliky, 1981b). Alginate beads containing cells corresponding to a dry weight of 6–7 mg were incubated

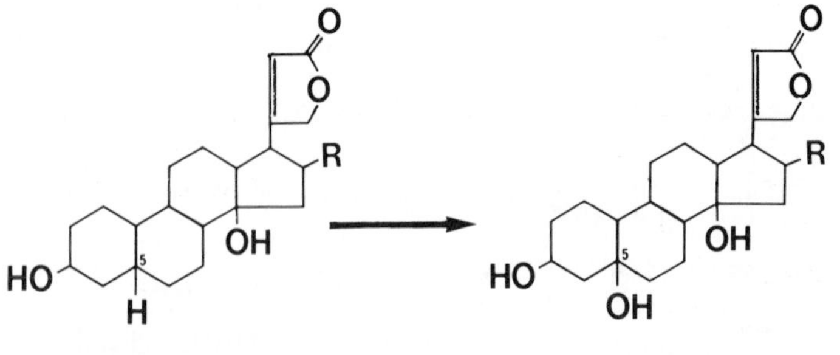

R = H	DIGITOXIGENIN
R = OH	GITOXIGENIN

PERIPLOGENIN

5β-HYDROXYGITOXIGENIN

FIGURE 6.12 Bioconversion of aglycones carried out with immobilized cells of *Daucus carota*.

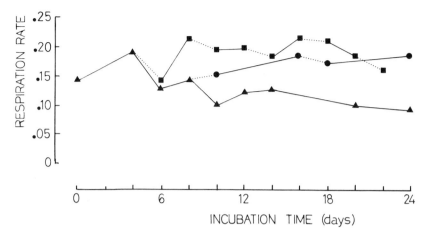

FIGURE 6.13   Effect of activation with growth medium on long-term survival of alginate-entrapped cells of *Daucus carota*. (▲——▲) incubation in 10 mM of MES-buffer without activation; (■——■) at 4-day intervals the immobilized cells were activated for 2 days with growth medium (dotted line); (●——●) at 8-day intervals the immobilized cells were activated for 2 days with growth medium (dotted line). (Adapted from Jones and Veliky, 1981b.)

in 25 ml of medium containing 10 mg digitoxigenin/l; after 48 h, 60% of the substrate had been converted. After five batches (= 10 days total reaction time) the activity of the immobilized cells was rapidly declining, as illustrated in Fig. 6.14. This was assumed to be caused by the accumulation of untransformed toxic substrate within the beads and/or cells.

The same prepartion of immobilized *Daucus carota* cells was also used to hydroxylate the less toxic substrate gitoxigenin in a column bioreactor (Veliky and Jones, 1981). Air supply to the entrapped cells within the reactor was achieved in two different ways: (1) by an air-lift pump with a continuous flow of medium buffer solution or (2) by direct upward aeration by closing the air-lift pump. In both cases the entrapped cells appeared to be intact after 30 days of incubation. It took 5–6 days to establish a steady state rate of hydroxylation. With the upward aeration a conversion of 75–80% was maintained for 23 days; with the air-lift pump a rate of 60–65% was observed. The higher conversion observed in the reactor with direct aeration can be explained by a better mixing in this reactor design (cf. fluidized-bed reactor). A periodical activation of the immobilized cells with growth medium was also shown to be beneficial.

Alginate-entrapped cells of *Ipomoea* sp. and *Cannabis sativa* were shown to convert digitoxigenin to 3-epidigitoxigenin and 3-dehydrodigitoxigenin, respectively (Jones and Veliky, 1981b).

Agarose-entrapped cells of *C. roseus* have been employed for the reduction of a double bond in cathenamine to yield ajmalicine and its isomers

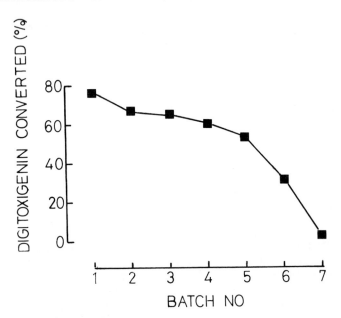

FIGURE 6.14   Formation of periplogenin by alginate-entrapped cells of *Daucus carota* in a repeated batch experiment. (Adapted from Jones and Veliky, 1981a.)

19-epi-ajmalicine and tetrahydroalstonine (Felix et al., 1981). The immobilized cells were essentially as efficient as the corresponding freely suspended cells in converting the amine to ajmalicine isomers.

A final example of bioconversion is illustrated by *Mentha* cells entrapped in polyacrylamide by cross-linking of prepolymerized polyacrylamide-hydrazide with glyoxal (Galun et al., 1983). Two different cell lines have been investigated. One cell line has been employed for the bioconversion of ($-$) menthone to ($+$) neomenthol, and another has been used for the conversion of ($+$) pulegone to ($+$) isomenthone. The conversion efficiency of the cells entrapped in this gel is in both cases as high as that of freely suspended cells. In Fig. 6.15 the kinetics of ($+$) neomenthol formation by freely suspended and entrapped cells incubated with ($-$) menthone is shown. The rate of product formation is essentially the same until a maximum yield is reached after 12 h. However, the product disappears completely after 24 h from an extract of the medium of the freely suspended cells but is completely retained in the medium of the entrapped cells. This can be explained by the fact that both neomenthol and menthone are readily glycosylated by enzymes in *Mentha* tissue. It seems that less glycosylation occurs in entrapped cells as compared with freely suspended cells. The reason for this difference between freely suspended and immobilized cells is not known. In a repeated batch experiment it was shown that the immobilized cells were able to perform three consecutive conversions without any significant loss of capacity.

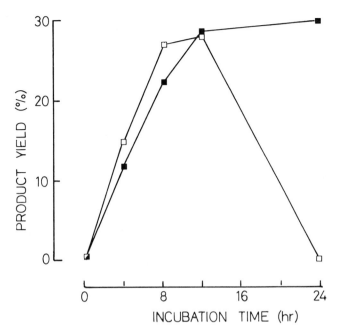

FIGURE 6.15   Time course of the conversion of (−) menthone to (+)
neomenthol by freely suspended (□————□) and immobilized (■————■)
cells of *Mentha* sp. (Adapted from Galun et al., 1983.)

2. Synthesis from precursors.   The yield of a particular compound in plant
tissue culture may be increased considerably by feeding appropriate precursors. However, knowledge about the biosynthetic pathway is a requirement for
this approach; furthermore, the precursors should be readily available.
Although this approach to the synthesis of complex biochemicals appears to be
attractive, relatively few examples can be found in the literature.

The synthesis of various indole alkaloids from the common precursors
tryptamine and secologanin by cell cultures of *C. roseus* has been studied extensively in various laboratories (Stockigt and Zenk, 1977; Mizukami et al.,
1979). The synthesis of ajmalicine isomers from these two precursors by immobilized cells of *C. roseus* has been studied in our laboratory as summarized
in Table 6.7 (Brodelius et al., 1979; Brodelius and Nilsson 1980, 1983). The
plant cells were entrapped in various gels, and the biosynthetic capacity of the
viable cell preparations were investigated (Brodelius and Nilsson, 1980). Cells
entrapped in agar, agarose, or carrageenan produced ajmalicine isomers at approximately the same rate as the corresponding freely suspended cells, and
cells entrapped within alginate gels showed an increased synthesis (up to 160%
of that observed for freely suspended cells) as illustrated in Fig. 6.16. The
reason for this increased synthesis is believed to be related to the restricted
growth of the cells observed with this polymer. On a quantitative basis, approximately 12 times as much product is formed from the added precursors

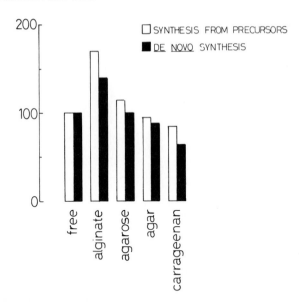

FIGURE 6.16 Relative production of ajmalicine isomers by various preparations of *Catharanthus roseus* by synthesis from added precursors and by de novo synthesis. The productivity of freely suspended cells is defined as 100% in both cases.

after 5 days incubation as is formed by de novo synthesis after 2 weeks of incubation.

The comparative studies described above were carried out in a batch procedure in shaker flasks. Alginate-entrapped cells have also been used in a column bioreactor with recirculation of medium (Brodelius et al., 1979). The effluent from the column was bubbled through an organic solvent layer (chloroform) in order to continuously extract lipophilic compounds before it was recirculated to the bioreactor. Lipophilic compounds were extracted, and a portion of these were ajmalicine isomers, illustrated in Fig. 6.17. The yield of ajmalicine isomers was about 10%, but it should be pointed out that no optimization of the yield was attempted.

The model studies described in this section have clearly demonstrated that synthesis of complex compounds from added precursors is also possible with immobilized plant cells and that in certain instances this approach may improve the yield of the target compound.

3. De novo synthesis. As was discussed in the introduction, a wide variety of compounds have been isolated from plant tissue cultures. Recently, there has been considerable interest in the production of such compounds by immobilized plant cells. A number of model studies have been carried out, as summarized in Table 6.7, in order to investigate whether immobilized cells behave differently

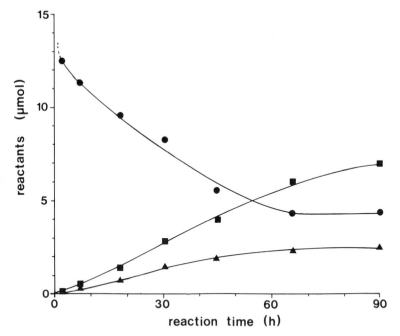

FIGURE 6.17 Synthesis of ajmalicine isomers from added precursors by alginate-entrapped cells of *Catharanthus roseus* in a recirculated batch reactor. (●———●) tryptamine in medium; (■———■) chloroform-extractable compounds in medium; (▲———▲) ajmalicine isomers in the chloroform phase. (From Brodelius et al., 1979.)

from freely suspended cells in this context. The normal carbon source is sucrose, which is metabolized and incorporated into the products.

Indole alkaloids—in particular, ajmalicine and serpentine—have been the target substances in a number of investigations on immobilized cells of *C. roseus* (Brodelius and Nilsson, 1980; Brodelius et al., 1981; Rosevear and Lambe, 1982; Lambe and Rosevear, 1983).

Cells entrapped in agarose, agar, and carrageenan showed approximately the same productivity as freely suspended cells under growth limiting conditions (no hormones added to the medium) (Brodelius and Nilsson, 1980). On the other hand, alginate-entrapped cells showed an increased synthesis (140% of that observed for freely suspended cells). The behavior of these four viable preparations of immobilized *C. roseus* cells is similar for synthesis from precursors and for synthesis de novo as illustrated in Fig. 6.16. Cells immobilized by methods resulting in nonviable preparations did not show any synthesis of ajmalicine isomers. As was expected, viable cell preparations are a requirement for the synthesis of complex natural products.

Alginate-entrapped cells of *C. roseus* have also been used to synthesize the related indole alkaloid serpentine in a batch procedure (Brodelius et al., 1981).

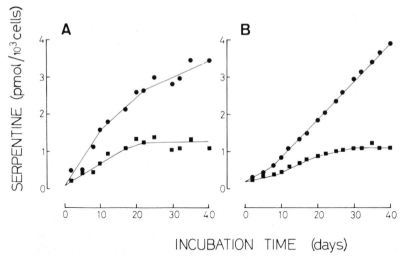

FIGURE 6.18   Kinetics of the de novo synthesis of the indole alkaloid serpentine by *Catharanthus roseus* cells (a) under phosphate limiting conditions or (b) under osmotic stress. (■——■) freely suspended cells; (●——●) alginate-entrapped cells. (From Brodelius et al., 1981.)

In this study an increased synthesis was also observed when immobilized cells were used, as shown in Fig. 6.18. In this case, growth was restricted by limiting the phosphate in the medium (Fig. 6.18a), or by exposing the cells to osmotic stress (Fig. 6.18b). In both cases the freely suspended cells ceased to produce the alkaloid after about 20 days, whereas the immobilized cells carried on the synthesis for at least 40 days. At this point the product yield was three to four times that obtained in freely suspended cells.

Still another example of the synthesis of ajmalicine and serpentine by immobilized cells of *C. roseus* is shown in Fig. 6.19 (Rosevear and Lambe, 1982). The cells were immobilized by entrapment in polyacrylamide. The toxicity of the reagents was reduced by mixing the cells with a viscous polysaccharide solution before immobilization. At the beginning of the incubation a relatively slow synthesis of alkaloids was observed, but this increased after 40 days and was maintained for another 110 days. In comparison to freely suspended cells, the production by the immobilized cells was considerably lower at the beginning of the experiment, but the total yield after 150 days was much higher than the maximum yield observed with suspended cells. Recently, it was reported that the same preparation of immobilized cells had been utilized for up to 220 days for the production of ajmalicine and serpentine (Lambe and Rosevear, 1983). This prolonged production phase may play an important role in the future development of plant tissue culture as a source of biochemicals.

Alginate-entrapped cells of *Morinda citrifolia* were used to synthesize anthraquinones from sucrose (Brodelius et al., 1980). After 21 days of incubation the immobilized cells had produced about ten times as much anthra-

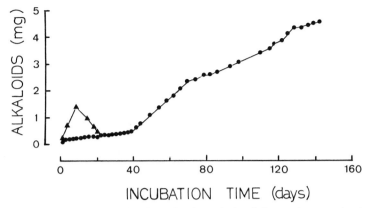

FIGURE 6.19   Accumulated production of ajmalicine and serpentine by cells of *Catharanthus roseus* as a function of incubation time. (▲——▲) freely suspended cells; (●——●) polyacrylamide-entrapped cells. (Adapted from Rosevear and Lambe, 1982.)

quinone as did freely suspended cells under growth-limiting conditions (no hormones added to the medium).

The synthesis of phenolics by cells of *Glycine max* and *Daucus carota* entrapped in hollow fiber reactors was chosen as a model to study the behavior of plant cells immobilized by this method (Shuler, 1981; Jose et al., 1983). The entrapped cells could synthesize phenolics continuously for 30 days, as shown in Fig. 6.20. Recently, it was reported that by immobilizing the *G. max* cells in

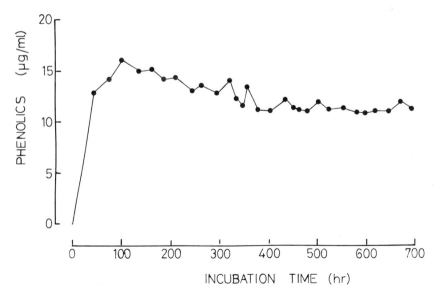

FIGURE 6.20   Phenol production of *Glycine max* cells entrapped in a hollow-fiber reactor. (Adapted from Shuler, 1981.)

a membrane reactor with relatively large pores (125 μm) the cells could be employed for at least 110 days for synthesis of phenolics (Shuler and Hallsby, 1983). The productivity is about four times higher, on a per cell dry weight basis, for the membrane-entrapped cells compared to the suspended cells.

Cells of *Capsicum frutescens* adsorbed to macroporous polyurethane foam also showed an increased productivity of a secondary metabolite (cap-saicin) compared to the corresponding freely suspended cells (Lindsey et al., 1983). On the average, the adsorbed cells produced around 120 times more capsaicin after 10 days of incubation (Table 6.8). Addition of the precursor isocapric acid (5 mM) to the immobilized cells enhanced the yield five-fold.

Finally, cells of *Solanum aviculare* covalently coupled to polyphenylene-oxide beads were able to synthesize steroid glycosides from sucrose during 11 days of incubation (Jirku et al., 1981).

The repeatedly observed increase in productivity by immobilized plant cells is at present difficult to explain. It is, however, of great importance for the future development of this technology. One may speculate that the immobilized cells are closer, metabolically, to cells in the whole plant (similar to differen-tiated cells) than the cells in suspension and therefore are more productive.

C. Product Release

In plant tissue culture the secondary products are often stored within the cells and then in most cases within vacuoles. A major advantage of immobilized biocatalysts is the possible continuous operation of a bioreactor. However, this distinct advantage cannot be realized if the product is retained within the immobilized cells. Attempts are therefore being made to induce product release from immobilized plant cells. It appears that the immobilization as such can, in certain instances, induce such a product release (spontaneous release), but in many cases the cells must be induced to release product by ad-ditions to the medium (permeabilization).

1. Spontaneous release.   There are some reports of spontaneous release from immobilized cells of products that are normally stored within cells in suspen-

TABLE 6.8   Capaicin Production by Freely Suspended and Immobilized Cells of Papsicum frutescens. (Adapted from Lindsey, et al., 1983)

Replicate	Suspension Culture		Immobilized Cells	
	Cells (ng/gDW)	Medium (ng/ml)	Cells (ng/gDW)	Medium (ng/ml)
1	<0.1	10.4	<0.1	1500
2	<0.1	8.3	<0.1	900
3	<0.1	1.2	<0.1	1700
4	<0.1	20.4	<0.1	1000
Average	<0.1	12.1	<0.1	1275

sion (Brodelius et al., 1981; Rosevear and Lambe, 1982). Indole alkaloids were released into the medium from entrapped cells of *C. roseus* (cf. Fig. 6.19). The mechanism behind this spontaneous release is not known, but it has been concluded that the release is not caused by lysis of the cells.

2. Permeabilization. The spontaneous release described above is not a general phenomenon; therefore we have been involved in developing methods to manipulate the immobilized plant cells to release the intracellularly stored products into the surrounding medium (Felix et al., 1981; Brodelius and Nilsson, 1983). This induced product release preferably should not affect the viability of the cells.

In order to release products from the vacuoles of immobilized cells, two membrane barriers have to be overcome (the plasma membrane and the tonoplast surrounding the vacuoles). The permeability of these membranes may be monitored in the following manner.

Plasma membrane. Various enzyme activities within the cells can be employed for monitoring the permeability of the plasma membrane. Enzymes requiring nucleotide coenzymes such as NADP, ATP, or CoA are convenient to use for this purpose. These coenzymes cannot penetrate an intact plasma membrane; thus no enzyme activity is expressed unless the plasma membrane has been made permeable (Felix et al., 1981). Isocitrate dehydrogenase, for example, may be employed to monitor the permeability of the plasma membrane. Fig. 6.21 illustrates the measured activity of this enzyme in immobilized untreated and 5% DMSO-treated cells of *C. roseus*. The DMSO treatment rendered the cells permeable to isocitrate and NADP(H), allowing the direct assay of the intracellular enzyme.

Tonoplast. To monitor the permeability of the tonoplast, we have used ^{31}P NMR spectroscopy (Vogel and Brodelius, unpublished). Cells of *C. roseus* accumulate inorganic phosphate in the vacuoles as indicated in Fig. 6.7. The position of this peak in the spectrum (2 ppm) corresponds to the vacuolar pH (6.0). After treatment with the uncoupler CCCP (destroys proton gradients in the cell) the phosphate peak is shifted to a position corresponding to the external pH (7.0), as illustrated in Fig. 6.22a. No dilution of the phosphate is observed; therefore the permeability of the tonoplast for inorganic phosphate is not changed by this treatment. If the cells are treated with DMSO, on the other hand, both a shift of the peak and a dilution of phosphate (peak-to-noise ratio) is observed (Fig. 6.22b). Obviously, the tonoplast is made permeable by DMSO treatment.

Permeabilization for product release. DMSO was chosen as a permeabilization agent because treatment with this compound resulted in high activities of intracellular enzymes (Felix et al., 1981) and because it is widely used as a

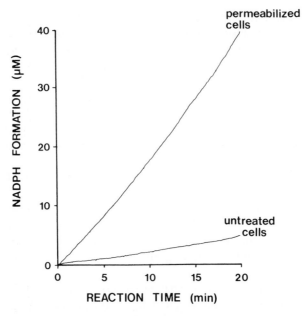

FIGURE 6.21 Time course of the isocitrate dehydrogenase reaction in agarose-entrapped cells of *Catharanthus roseus*. Permeabilization treatment was carried out for 30 min. (From Brodelius and Nilsson, 1983.)

cryoprotective agent and therefore it was expected that the viability of the cells would be unaffected by a moderate treatment with this agent.

It was shown that treatment with 5% DMSO for 30 min was sufficient to make both the plasma membrane and the tonoplast permeable (Brodelius and Nilsson, 1983; Vogel and Brodelius, unpublished) and also that the cells remained viable after this treatment (Brodelius and Nilsson, 1983). A process involving intermittent permeabilization for product release according to Fig. 6.23 was suggested and was also tested in model studies (Brodelius and Nilsson, 1983). Indeed, a cyclic process with intermittent product release was possible, as is illustrated in Fig. 6.24 for agarose-entrapped cells of *C. roseus*. An increased yield of product was observed for each cycle due to increase of biomass during the experiment. As compared to extraction with organic solvents (reflux) of the immobilized cells, the DMSO treatment released about 90% of the intracellularly stored products.

The experiments described above indicate that it may be possible to utilize immobilized plant cells in a repeated fashion for the production of secondary products. A process of this kind would enable the reuse of the otherwise prohibitively expensive plant cell biomass, thereby making such a process economically feasible. However, it should be pointed out that the technique is probably applicable only to water-soluble products.

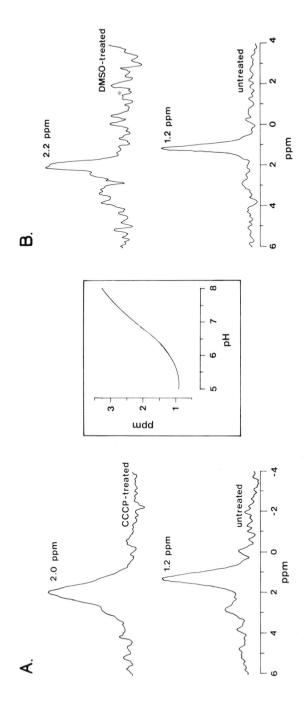

FIGURE 6.22 ^{31}P NMR spectra (103.2 MHz) of agarose-entrapped cells of *Catharanthus roseus* before and after treatment with (a) CCCP and (b) DMSO. The inserted diagram shows the pH-dependence of the chemical shift of the inorganic phosphate resonance. (Adapted from Vogel and Brodelius, unpublished.)

143

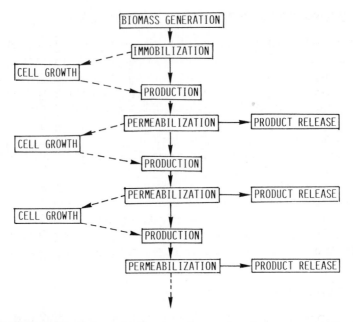

FIGURE 6.23 Schematic diagram of a semicontinuous process for the production of intracellularly stored products.

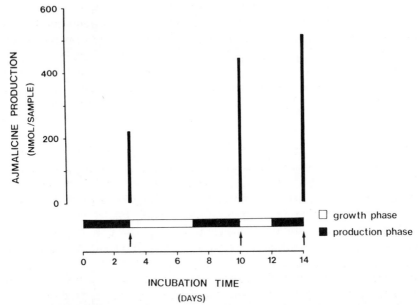

FIGURE 6.24 Semicontinuous production of ajmalicine isomers by agarose-entrapped cells of *Catharanthus roseus* according to the process outlined in Fig. 6.23. The cells were intermittently permeabilized (arrows). (From Brodelius and Nilsson, 1983.)

V. REACTORS FOR IMMOBILIZED PLANT CELLS

It can be expected that immobilized plant cells may be utilized in relatively un-complicated reactors in contrast to freely suspended cells, which in most cases require sophisticated reactor design owing to the low stress tolerance of the cells. However, relatively little research has so far been carried out on the design of reactors for immobilized plant cell preparations. Most experiments have been performed in simple batch configurations (i.e., shaker flasks), but there are also examples of packed-bed reactors (Brodelius et al., 1979, 1981), bubble-column reactors (Veliky and Jones, 1981; Moritz et al., 1982), and a fluidized-bed reactor (Morris et al., 1983).

Recently, membrane reactors were introduced as potential immobilized plant cell reactors (Shuler, 1981; Jose et al., 1983). In this type of reactor the cells are maintained on one side of the membrane, and medium is circulated on the other side of the membrane. This appears to be a convenient way to im-mobilize plant cells and simultaneously solve the problem of reactor design.

VI. CONCLUDING REMARKS

The first report on the immobilization of plant cells appeared in 1979 (Brodelius et al., 1979); since then, there have been an increasing number of re-ports. It is believed that by immobilizing plant cells, some major problems in utilizing this type of cells on a large scale may be overcome.

The immobilization allows growth of cells to a certain extent, but the pro-duction of compounds that are directly associated to the growth rate would not be advisable with entrapped cells. It is therefore likely that immobilized plant cells will find applications mainly for products that are nongrowth associated, that is, products of the cells in stationary phase. The extended sta-tionary phase observed for immobilized plant cells may circumvent some of the problems resulting from the apparent slow growth of cells (biomass generation).

Cell aggregation occurs in culture; this leads to a wide distribution in ag-gregate size, resulting in problems with agitation of, and mass transfer to, the cells. These types of problems may be reduced by immobilization, since "ag-gregation" becomes a design parameter. A relatively homogeneous prepara-tion of beads containing the cell aggregates is readily made.

Plant cell culture often shows significant changes upon long-term serial passages. For instance, cell lines selected for high productivity of a particular compound frequently lose this capability. The extended stationary phase observed with immobilized cells may also, to a limited extent, reduce this prob-lem, since the cells are utilized in a nondividing state, which should reduce the appearance of genetic changes.

The problem of low shear tolerance of plant cells is eliminated by the en-trapment of the cells in a protective polymeric matrix. This results in a simpler reactor design.

The possibility of releasing intracellularly stored products by intermittent permeabilization of the immobilized cells is still another advantage that may prove valuable in the future development of plant tissue culture technology. This allows the reutilization of the biomass over an extended period of time.

A number of possible advantages of immobilized plant cells have been listed and briefly discussed. The results so far obtained are promising, and immobilization of plant cells may prove a valuable tool to overcome or reduce some of the problems encountered in the utilization of plant tissue cultures for the production of complex natural products.

ACKNOWLEDGMENT

The author wishes to express his gratitude to Dr. Stephen Cliffe for linguistic advice.

REFERENCES

Akasu, M., Itokawa, H., and Fujita, M. (1976) *Phytochemistry 15*, 471–473.

Alfermann, A. W., Schuller, I., and Reinhard, E. (1980) *Planta Med. 40*, 218–223.

Alfermann, A. W., Bergmann, W., Figur, C., Helmbold, U., Schwantag, D., Shuller, I., and Reinhard, E. (1983) in *Plant Biotechnology*, (Mantell, S. H. and Smith, H., eds.) Society for Experimental Biology Seminar Series, Vol. XVIII, pp. 67–74, Cambridge University Press, Cambridge, England.

Amrhein, N., Deus, B., Gehrke, P., and Steinrocken, H.-C. (1980) *Plant Physiol. 66*, 830–834.

Becker, H., and Schrall, R. (1980) *J. Nat. Prod. 43*, 721–723.

Berlin, J., Knobloch, K.-H., Hofle, G., and Witte, L. (1982) *J. Nat. Prod. 45*, 83–87.

Berlin, J., Forche, E., Wray, V., Hammer, J., and Hosel, W. (1983) *Z. Naturforsch. 38C*, 346–352.

Birnbaum, S., Larsson, P.-O., and Mosbach, K. (1983) in *Solid Phase Biochemistry: Analytical and Synthetic Aspects* (Scouten, W. H., ed.), pp. 679–762, John Wiley, New York.

Bornman, C. H., and Zachrisson, A. (1982) *Plant Cell Rep. 1*, 151–153.

Brodelius, P., and Nilsson, K. (1980) *FEBS Lett. 122*, 312–316.

Brodelius, P., and Nilsson, K. (1983) *Eur. J. Appl. Microbiol. Biotechnol. 17*, 275–280.

Brodelius, P., and Vandamme, E. J. (1984) in: *Biotechnology*, Vol. VII (Kennedy, J. E., ed.) Verlag Chemie, Weinheim, FRG. (in press).

Brodelius, P., and Vogel, H. J. (1984a) in: *Enzyme Engineering*, Vol. VII (Laskin, A. I., Tsao, G. T., and Wingard, L. B., Jr., eds.) pp. 496–500, New York Academy of Sciences.

Brodelius, P., and Vogel, H. J. (1984b) *J. Biol. Chem.* (in press).

Brodelius, P., Deus, B., Mosbach, K., and Zenk, M. H. (1979) *FEBS Lett. 103*, 93–97.

Brodelius, P., Deus, B., Mosbach, K., and Zenk, M. H. (1980) in *Enzyme Engineering*, Vol. V (Weetall, H. H., and Royer, G. P., eds.), pp. 373–381, Plenum Press, New York.

Brodelius, P., Deus, B., Mosbach, K., and Zenk, M. H. (1981) European Patent Application 80850105.0.

Brodelius, P., Constabel, F., and Kurz, W. G. W. (1982) in: *Enzyme Engineering*, Vol. VI (Chibata, I., Fukui, S., and Wingard, L. B., Jr., eds.), pp. 203-204, Plenum Press, New York.

Cabral, J. M. S., Fevereiro, P., Novais, J. M., and Pais, M. S. S. (1984) in *Enzyme Engineering*, Vol. VII (Laskin, A. I., Tsao, G. T., and Wingard, L. B., Jr., eds.) pp. 501-503, New York Academy of Sciences.

Deus, B., and Zenk, M. H. (1982) *Biotechnol. Bioeng. 24*, 1965-1974.

Felix, H., Brodelius, P., and Mosbach, K. (1981) *Anal. Biochem. 116*, 462-470.

Freeman, A., and Aharonowitz, Y. (1981) *Biotechnol. Bioeng. 23*, 2747-2759.

Frischknecht, P. M., Baumann, T. W., and Wanner, H. (1977) *Planta Med. 31*, 344-350.

Fujita, Y., Teranishi, K., and Furukawa, T. (1978) Japanese Patent 78-91188.

Fujita, Y., Hara, Y., Suga, C., and Morimoto, T. (1981) *Plant Cell Rep. 1*, 61-63.

Fukui, H., Nakagawa, K., Tsuda, S., and Tabata, M. (1982) in *Proceedings of the Fifth International Congress on Plant Tissue and Cell Culture* (Fujiwara, A., ed.), pp. 313-314, Japanese Association for Plant Tissue Culture, Tokyo.

Fukui, S., and Tanaka, A. (1984) in *Advances in Biochemical Engineering Biotechnology*, Vol. XXIX (Fiechter, A., ed.), pp. 1-33, Springer, New York.

Furuya, T., and Ishii, T. (1972) German Patent 2143946.

Furuya, T., Yoshikawa, T., Orihara, Y., and Oda, H. (1983) *Planta Med. 48*, 83-87.

Galun, E., Aviv, D., Dantes, A., and Freeman, A. (1983) *Planta Med. 49*, 9-13.

Hashimoto, T., Sato, F., Mino, M., and Yamada, Y. (1982) in *Proceedings of the Fifth International Congress on Plant Tissue and Cell Culture* (Fujiwara, A., ed.), pp. 305-306, Japanese Association for Plant Tissue Culture, Tokyo.

Hinz, H., and Zenk, M. H. (1981) *Naturwiss. 67*, 620.

Henry, M., and Guignard, J.-L. (1982) in *Proceedings of the Fifth International Congress on Plant Tissue and Cell Culture* (Fujiwara, A., ed.), p. 299, Japanese Association for Plant Tissue Culture, Tokyo.

Jirku, V., Macek, T., Vanek, T., Krumphanzl, V., and Kubanek, V. (1981) *Biotechnol. Lett. 3*, 447-450.

Jones, A., and Veliky, I. A. (1981a) *Eur. J. Appl. Microbiol. Biotechnol. 13*, 84-89.

Jones, A., and Veliky, I. A. (1981b) *Can. J. Bot. 59*, 2095-2101.

Jose, W., Pedersen, H., and Chin, C. K. (1983) *Ann. N.Y. Acad. Sci. 413*, 409-412.

Kaul, B., and Staba, E. J. (1967) *Planta Med. 15*, 145-156.

Koblitz, H., Schumann, U., Bohm, H., and Franke, J. (1975) *Experientia 31*, 768-769.

Kutney, J. P., Choi, L. S. L., Duffin, R., Hewitt, G., Kawamura, N., Kurihara, T., Salisbury, P., Sindelar, R., Stuart, K. L., Townsley, P. M., Chalmers, W. T., Webster, F., and Jacoli, G. G. (1983) *Planta Med. 48*, 158-163.

Lambe, C. A., and Rosevear, A. (1983) *Proceedings of Biotech83, London, May 4-6, 1983*, Online Publications Ltd., Northwood, U. K., pp. 565-576.

Lambe, C. A., Reading, A., Roe, S., Rosevear, A., and Thomson, A. R. (1982) in *Enzyme Engineering*, Vol. VI (Chibata, I., Fukui, S., and Wingard, L. B. Jr., eds.), Plenum Press, New York.

Lindsey, K., Yeoman, M. M., Black, G. M., and Mavituna, F. (1983) *FEBS Lett. 155*, 143-149.

Matsumoto, T., Ikeda, T., Okimura, C., Obi, Y., Kisaki, T., and Noguchi, M. (1982) in *Proceedings of the Fifth International Congress on Plant Tissue and Cell Culture* (Fujiwara, A., ed.), pp. 275-276, Japanese Association for Plant Tissue Culture, Tokyo.

Mizukami, H., Nordlov, H., Lee, S., and Scott, A. I. (1979) *Biochemistry 18*, 3760-3763.

Moritz, S., Schuller, I., Figur, C., Alfermann, A. W., and Reinhard, E. (1982) in *Pro-

ceedings of the Fifth International Congress on Plant Tissue and Cell Culture (Fujiwara, A., ed.), pp. 401–402, Japanese Association for Plant Tissue Culture, Tokyo.

Morris, P., and Fowler, M. W. (1981) *Plant Cell Tiss. Org. Cult. 1*, 15–24.

Morris, P., Smart, N. J., and Fowler, M. W (1983) *Plant Cell Tiss. Org. Cult. 2*, 207–216.

Mosbach, K., and Mosbach, R. (1966) *Acta Chem. Scand. 20*, 2807–2810.

Nilsson, K., Birnbaum, S., Flygare, S., Linse, L., Schroder, U., Jeppsson, U., Larsson, P.-O., Mosbach, K., and Brodelius, P. (1983) *Eur. J. Appl. Microbiol. Biotechnol. 17*, 319–326.

Ogino, T., Hiraoka, N., and Tabata, M. (1978) *Phytochemistry 17*, 1907–1910.

Reinhard, E., and Alfermann, A. W. (1980) in *Advances in Biochemical Engineering*, Vol. XVI (Fiechter, A., ed.), pp. 49–83, Springer, New York.

Rosevear, A. (1981) European Patent Application 81304001.1.

Rosevear, A., and Lambe, C. A. (1982) European Patent Application 82301571.4.

Sasse, F., Heckenberg, U., and Berlin, J. (1982) *Z. Pflanzenphysiol. 105*, 315–322.

Sato, F., Endo, T., Hashimoto, T., and Yamada, Y. (1982) in *Proceedings of the Fifth International Congress on Plant Tissue and Cell Culture* (Fujiwara, A., ed.), pp. 319–320, Japanese Association for Plant Tissue Culture, Tokyo.

Shuler, M. (1981) *Ann. N.Y. Acad. Sci. 369*, 65–79.

Shuler, M., and Hallsby, A. G. (1983) Paper #76b at AIChE 1983 Summer National Meeting, Denver, Colorado, August 28–31.

Staba, E. J., and Kaul, B. (1971) U. S. Patent 3628287.

Stockigt, J., and Zenk, M. H. (1977) *J.C.S. Chem. Comm. 1977*, 646–648.

Tabata, M., Hiraoka, N., Ikenoue, M., Sano, Y., and Konoshima, M. (1975) *Lloydia 38*, 131–134.

Tam, W. H. J., Constabel, F., and Kurz, W. G. W. (1980) *Phytochemistry 19*, 486–487.

Tanaka, A., Sonomoto, K., and Fukui, S. (1984) in *Enzyme Engineering*, Vol. VII (Laskin, A. I., Tsao, G. T., and Wingard, L. B. Jr., eds.) pp. 479–482, New York Academy of Sciences.

Tomita, Y., and Uomori, A. (1976) Japanese Patent 76-12988.

Veliky, I. A., and Jones, A. (1981) *Biotechnol. Lett. 3*, 551–554.

Vogel, H. J., and Brodelius, P. (1984) *J. Biotechnol. 1*, 159–170.

Weiler, E. W., and Zenk, M. H. (1979) *Anal. Biochem. 92*, 147–155.

Yamakawa, T., Kato, S., Ishida, K., Kodama, T., and Minoda, Y. (1983) *Agr. Biol. Chem. 47*, 2185–2191.

Yamamoto, H., Machida, A., and Tomimori, T. (1982) in *Proceedings of the Fifth International Congress on Plant Tissue and Cell Culture* (Fujiwara, A., ed.), pp. 351–352, Japanese Association for Plant Tissue Culture, Tokyo.

Yamamoto, Y., Mizuguchi, R., and Yamada, Y. (1982) *Theor. Appl. Genet. 61*, 113–116.

Zenk, M. H., El-Shagi, H., and Schulte, U. (1975) *Planta Med. Suppl.* 79–101.

Zenk, M. H., El-Shagi, H., and Ulbrich, B. (1977a) *Naturwiss. 64*, 585.

Zenk, M. H., El-Shagi, H., Arens, H., Stockigt, J., Weiler, E. W., and Deus, B. (1977b) in *Plant Tissue Culture and Its Biotechnological Application* (Barz, W., Reinhard, E., and Zenk, M. H., eds.), pp. 27–43, Springer, New York.

7

Bioconversion of Lipophilic Compounds by Immobilized Biocatalysts in the Presence of Organic Solvents

Atsuo Tanaka
Saburo Fukui

I. INTRODUCTION

Biocatalysts — enzymes, microbial cells, plant cells, animal cells, and cellular organelles — have been mainly used in aqueous reaction systems including water-soluble reactants, unlike most ordinary chemical catalysts. With the recent development of enzyme technology, much wider applications of biocatalysts are expected, including bioconversions of biological and xenobiotic compounds having a hydrophobic character. In such bioconversion systems it is often necessary to introduce organic solvents to enhance the solubility of reactants in water. However, biocatalysts are often liable to denature in the presence of organic solvents, resulting in the loss of their catalytic abilities. Modification or loss of substrate specificities, stereo-, and/or regio-specificities is also observed under these conditions. To render biocatalysts resistant to organic solvents, proper immobilization on or in suitable supports seems to be most promising, although attempts have also been made using chemical modification of the biocatalysts (Klibanov, 1979).

Introduction of appropriate organic solvents is very useful to construct homogeneous reaction systems containing lipophilic and water-insoluble substrates and to carry out such reactions continuously. Furthermore, for the syntheses of useful compounds using hydrolytic enzymes, such as proteases, lipases, and esterases, it is essential to reduce the water ratio in the systems by replacing water with appropriate organic solvents.

Although only limited information is available at present about the bioconversions of such lipophilic and water-insoluble compounds by immobilized biocatalysts in the presence of organic solvents, we would like to summarize the recent results mainly obtained in the authors' laboratory (Fukui et al., 1980b; Fukui and Tanaka, 1981, 1982a, b).

II. EFFECTS OF ORGANIC SOLVENTS ON ENZYMES

To carry out bioconversions of lipophilic compounds effectively, it is essential to introduce organic solvents into the reaction systems. For these purposes, one must have knowledge about the effects of organic solvents on properties and functions of enzymes (Butler, 1979; Singer, 1962).

Enzymes as well as other proteins maintain their structural conformation through intramolecular interactions among side chains of the component amino acids. Hydrophobic side chains also contribute to such interactions. In general, enzyme molecules in aqueous solutions have both hydrophilic domains in contact with water and hydrophobic domains folded inside the molecules. When the polarity of the medium surrounding the enzyme molecules is reduced by adding organic solvents, the hydrophobic domains are liable to disperse, resulting in the unfolding of the molecules. Furthermore, hydrophobic interactions between the enzyme and substrate molecules are also disrupted. The facts mentioned above indicate that suitably hydrated states of enzyme molecules should be maintained to keep the enzymes active and stable even when organic solvents are utilized in enzymatic reaction systems. It is also useful to suppress the unfolding of the molecules by binding or interacting enzymes with appropriate supporting materials at multiple points (Martinek and Berezin, 1977). Use of two-phase systems composed of an aqueous phase and a water-insoluble organic solvent phase will also be useful in some cases as long as the enzymes are kept in the aqueous phase (Antonini et al., 1981).

Organic solvents produce various physicochemical effects on enzyme molecules, and the effects differ depending upon the kinds of organic solvents and the enzymes. As a result, enzymes are stabilized or instabilized owing to the change not only in the tertiary structure but also in the secondary structure, such as α-helix and β-structure. Change in the content of α-helix in the presence of alcohols or glycols differs from enzyme to enzyme (Herskovits et al., 1970).

In many cases, low concentrations of organic solvents — alcohols and water-soluble solvents — do not affect the catalytic activities of enzymes but indeed may stabilize them. This phenomenon has often been applied for the stabilization of enzymes during purification and storage. In some cases, activities of enzymes are even enhanced by organic solvents. For example, detergent-resistant phospholipase A from *Escherichia coli* is active only when methanol is present at appropriate concentrations (Doi et al., 1972). These facts suggest that some enzymes work in vivo under conditions that are similar to the in vitro reaction mixtures containing organic solvents. Conformational changes of enzyme molecules interacting with organic solvents result in the alteration of substrate specificity of enzymes and in the affinity of substrates toward enzymes (Minato and Hirai, 1979). Thus organic solvents exhibit contradictory actions as stimulators and inhibitors for enzymatic reactions, depending on the kinds of enzymes and on the types and concentrations of organic solvents. Enzymes are sometimes stabilized in reversed micelles composed of surfactant in organic solvents (Grandi et al., 1981; Martinek et al., 1981a).

Hydrolytic enzymes, which require water as one of the substrates in hydrolytic reactions, can catalyze synthetic reactions and group exchange reactions when the concentration of water in the system is low. If the water fraction in the reaction mixture is reduced by introducing organic solvents, it is possible to synthesize useful hydrophobic compounds more efficiently by changing the ionic equilibrium (Martinek and Semenov, 1981) or the partition equilibrium (Martinek et al., 1981b) in the desired direction. In the latter case, hydrophobic products are extracted into the organic solvent phase, the reaction equilibrium being shifted to the favorable direction.

III. ENZYME REACTIONS IN THE PRESENCE OF ORGANIC SOLVENTS

Enzyme reactions, even in those cases in which water-insoluble, lipophilic compounds are substrates, used to be carried out in aqueous systems because biocatalysts were believed to be unstable in organic solvents. However, it is desirable to perform enzymatic reactions with lipophilic compounds in mixtures of water and suitable organic cosolvents or in appropriate organic solvent systems, if the catalytic activities of biocatalysts can be maintained under such reaction conditions. The use of organic solvents can improve the poor solubility in water of substrates or other reaction components of a hydrophobic nature. In fact, bioconversions of steroids by free enzymes (Cremonesi et al., 1973, 1974, 1975; Lugaro et al., 1973) or by free microbial cells (Buckland et al., 1975; Duarte and Lilly, 1980; Lilly, 1982) were carried out in two-phase systems composed of aqueous solutions and water-immiscible organic solvents. Epoxidation of 1,7-octadiene by bacterial cells was also achieved in a two-phase system (Schwartz and McCoy, 1977).

As was mentioned above, hydrolytic enzymes, such as proteases, lipases, and esterases, can be utilized in the formation of peptide and ester bonds when the water fraction in the reaction system is reduced by the introduction of appropriate organic solvents. One of the most important applications is the plastein reaction to prepare high-molecular-weight proteins having a high solubility (Yamashita et al., 1975) or a high nutritive value (Yamashita et al., 1976) from partially digested peptides and appropriate amino acids. In addition to the plastein reaction, several examples, such as the syntheses of low-molecular-weight peptides (Homandberg et al., 1978; Morihara and Oka, 1977; Semenov et al., 1981), amino acid esters (Ingalls et al., 1975; Klibanov et al., 1977), and urea (Butler and Reithel, 1977), peptidation of porcine insulin to human insulin (Morihara et al., 1979, 1980; Jonczyk and Gattner, 1981), and transfer of a phosphate group (Wan and Horvath, 1975), have been reported. These examples indicate that alteration of the reaction equilibrium to the desired direction by introducing organic solvents into reaction mixtures containing hydrolytic enzymes is becoming a very important technique in biochemical syntheses.

Organic solvents to be utilized in such reaction systems should be selected according to the following criteria: (1) solubility of reactants, (2) stability of biocatalysts, (3) toxicity, and (4) flammability (Lilly, 1982).

IV. BIOCONVERSION OF LIPOPHILIC COMPOUNDS BY IMMOBILIZED BIOCATALYSTS

Immobilization often imparts a markedly enhanced stability on biocatalysts against denaturation by organic solvents (Klibanov, 1979; Martinek and Berezin, 1977). Thus bioconversions of water-insoluble compounds such as steroids have been achieved with immobilized biocatalysts. For the conversion of lipophilic compounds, several solvent systems are applicable (Antonini et al., 1981; Butler, 1979) — water–water-miscible organic solvent homogeneous systems, water–water-immiscible organic solvent two-phase systems, and organic solvent systems containing a small amount of water. However, two-phase systems, which have been applied for bioconversions with free or immobilized biocatalysts (Carrea et al., 1979; Duarte and Lilly, 1980; Kimura et al., 1983; Lilly, 1982), seem to be inconvenient in some cases to carry out the reactions continuously.

V. ENTRAPMENT OF BIOCATALYSTS BY PREPOLYMER METHODS

Of various immobilization methods proposed hitherto, entrapment of biocatalysts in appropriate gels is most promising because this technique can be applicable for immobilization not only of single enzymes but also of multiple enzymes, cellular organelles, microbial cells, plant cells, and animal cells.

For the development of immobilized biocatalyst techniques, gel materials that have a variety of physicochemical properties, such as hydrophilicity–hydrophobicity balance, are required, depending on the nature of the reactions. However, it is difficult to satisfy such demands by using natural polymer gels because of the difficulty in preparing their derivatives. Synthetic polymer gels that have different characteristics may be prepared by simple procedures from appropriate monomers or prepolymers that have different properties, although the conventional method using acrylamide sometimes inactivates biocatalysts.

We have developed simple and convenient methods to use photo-cross-linkable resin prepolymers (Fukui et al., 1976, 1980c) and urethane prepolymers (Fukushima et al., 1978; A. Tanaka et al., 1979). These prepolymer methods have the following merits.

1. Polymer matrices can be formed under extremely mild conditions with simple procedures. Photo-cross-linkable resin prepolymers are polymerized in the presence of an appropriate photosensitizer by illumination with near-UV light for several minutes. Urethane prepolymers can be polymerized simply by mixing the prepolymer solution with an aqueous solution of enzymes or an aqueous suspension of organelles or microbial cells. Thus the polymerization can be achieved without heating, shifting pH to extreme values, or the use of chemicals that modify the structures of biocatalysts.
2. Prepolymers containing multiple photo-sensitive functional groups or isocyanate groups at a desired location can be prepared first in the absence of biocatalysts. This makes it possible to select the size of the network of gel matrices. Furthermore, a suitably hydrophobic, hydrophilic, or ionic property can be introduced at this stage.

These prepolymers have been applied widely to the immobilization not only of enzymes but also of cellular organelles and microbial cells (Fukui and Tanaka, 1984). Water-insoluble, hydrophobic prepolymers were also utilized successfully to entrap enzymes and microbial cells (Omata et al., 1979a). Advantages of hydrophobic gels in the transformations of lipophilic compounds by immobilized biocatalysts under hydrophobic conditions are described below. Structures of several photo-cross-linkable resin prepolymers and urethane prepolymers are shown in Figs. 7.1 and 7.2, respectively. Some properties of these prepolymers are also summarized in Tables 7.1 and 7.2.

VI. BIOCONVERSION IN WATER–ORGANIC COSOLVENT SYSTEMS

Water–water-miscible organic solvent systems have been widely employed to dissolve water-insoluble, lipophilic compounds to prepare homogeneous reaction systems and to shift reaction equilibrium to a desired direction, especially

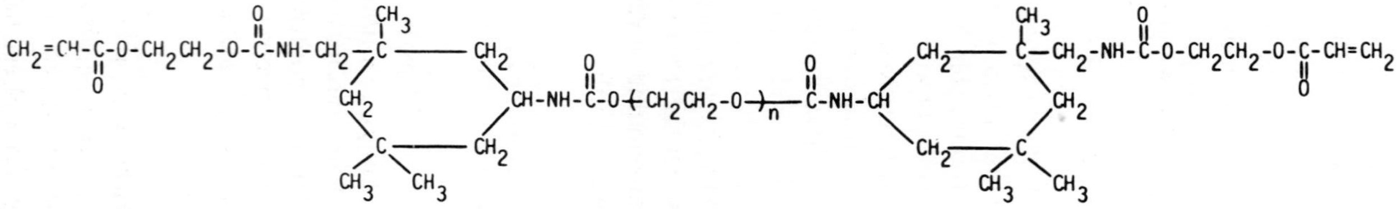

FIGURE 7.1 Structures of typical photo-cross-linkable resin prepolymers.

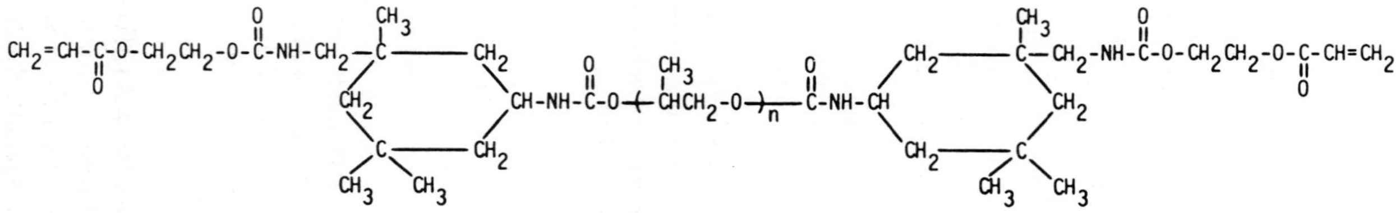

154

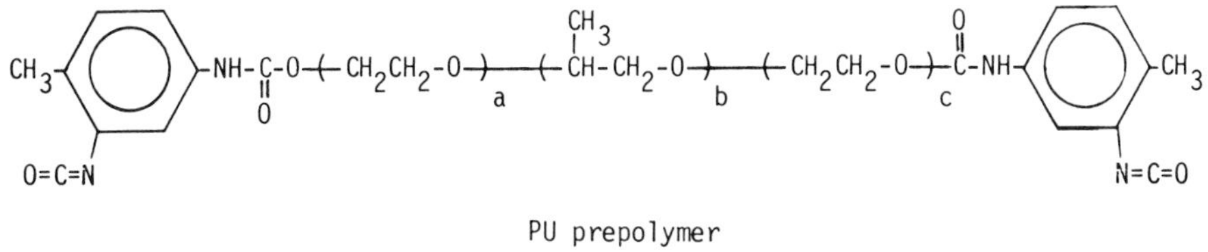

CH$_3$—〈benzene ring〉—NH-C-O-(-CH$_2$CH$_2$-O-)$_a$—(-CH-CH$_2$-O-)$_b$—(-CH$_2$CH$_2$-O-)$_c$-C-NH-〈benzene ring〉—CH$_3$

O=C=N   N=C=O

PU prepolymer

FIGURE 7.2   General formula of urethane prepolymers.

TABLE 7.1   Properties of Typical Photo-cross-linkable Resin Prepolymers

Prepolymer	Main Chain	Molecular Weight of Main Chain	Property
ENT-1000	Poly(ethylene glycol)	~1000	Hydrophilic
-2000		2000	
-4000		4000	
-6000		6000	
ENTP-1000	Poly(propylene glycol)	~1000	Hydrophobic
-2000		2000	
-3000		3000	
-4000		4000	

TABLE 7.2   Properties of Typical Urethane Prepolymers

Prepolymer	Molecular Weight of Polyether Diol	NCO Content in Prepolymer (%)	Poly(ethylene glycol) Content in Polyether Diol (%)	Property of Gel Formed
PU-3	2592	4.2	57	Hydrophobic
PU-6	2627	4.0	91	Hydrophilic
PU-9	2616	4.0	100	Hydrophilic

to the synthetic direction with hydrolytic enzymes. For the bioconversion of steroids with enzymes, treated cells, or living cells, relatively low concentrations of organic solvents such as methanol, ethanol, N,N-dimethylformamide, dimethyl sulfoxide, and the like have been used to dissolve substrates and/or products (Atrat et al., 1980; Chun et al., 1981; Maddox et al., 1981; Ohlson et al., 1978, 1980; Yang and Studebaker, 1978). To direct proteases to the synthesis of peptide bonds, high concentrations of organic solvents have been used. Such reactions were applied to plastein synthesis, to the peptidation reaction to yield human insulin from porcine insulin, and so on, as described previously. Biosynthesis of α-hydroxynitriles was also facilitated in the presence of organic cosolvents (Becker and Pfeil, 1966). Hydrolytic activity for peptides by immobilized trypsin was examined in the presence of high concentrations of organic cosolvents (Weetall and Vann, 1976).

TABLE 7.3   Bioconversions in Water–Organic Cosolvent Systems by Microbial Cells Entrapped with Prepolymers

Microorganism (Condition)	Organic Cosolvent	Application
Arthrobacter simplex (acetone-dried)	10% Methanol	Δ^1-Dehydrogenation of hydrocortisone
Curvularia lunata (living)	2.5% Dimethyl sulfoxide	11β-Hydroxylation of cortexolone
Rhizopus stolonifer (living)	2.5% Dimethyl sulfoxide	11α-Hydroxylation of progesterone
Sepedonium ampullosporum (living)	0.65% *N,N*-Dimethylformamide	16α-Hydroxylation of estrone
Corynebacterium sp. (living)	15% Dimethyl sulfoxide	9α-Hydroxylation of 4-androstene-3,17-dione
Enterobacter aerogenes (thawed)	40% Dimethyl sulfoxide	Synthesis of adenine arabinoside

Microbial cells entrapped with various prepolymers were also applied to bioconversions of lipophilic or water-insoluble compounds in water–organic cosolvent systems (Table 7.3)

Acetone-dried cells of *Arthrobacter simplex* entrapped with photo-cross-linkable resin prepolymers (Sonomoto et al., 1979) or urethane prepolymers (Sonomoto et al., 1980) catalyzed Δ^1-dehydrogenation of hydrocortisone to form prednisolone (Fig. 7.3). Introduction of organic cosolvents (10% by volume), such as methanol, ethylene glycol, propylene glycol, trimethylene glycol, and glycerol, stimulated significantly the reaction mediated by the entrapped cells, presumably because of the increased solubilities of substrate and product. Entrapment markedly enhanced the stability of the cells in organic cosolvents of high concentrations in long-term operations.

Hydroxylation of steroids by entrapped resting or living cells was also carried out in the presence of organic cosolvents (Fig. 7.4) (Sonomoto et al., 1981, 1982, 1983a, b; A. Tanaka et al., 1982). Cortexolone (Reichstein's Compound

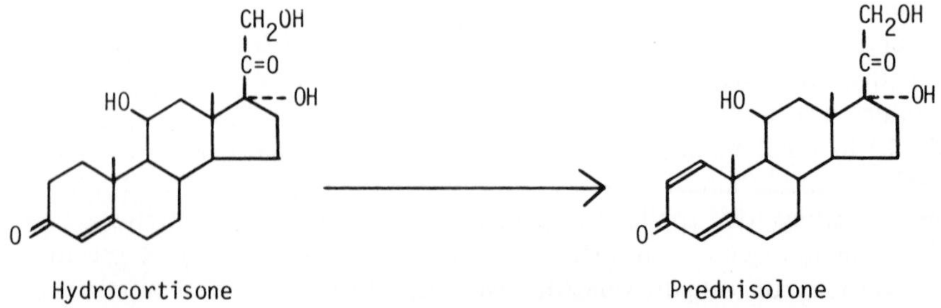

Hydrocortisone   Prednisolone

FIGURE 7.3   Hydrocortisone dehydrogenation catalyzed by *Arthrobacter simplex*.

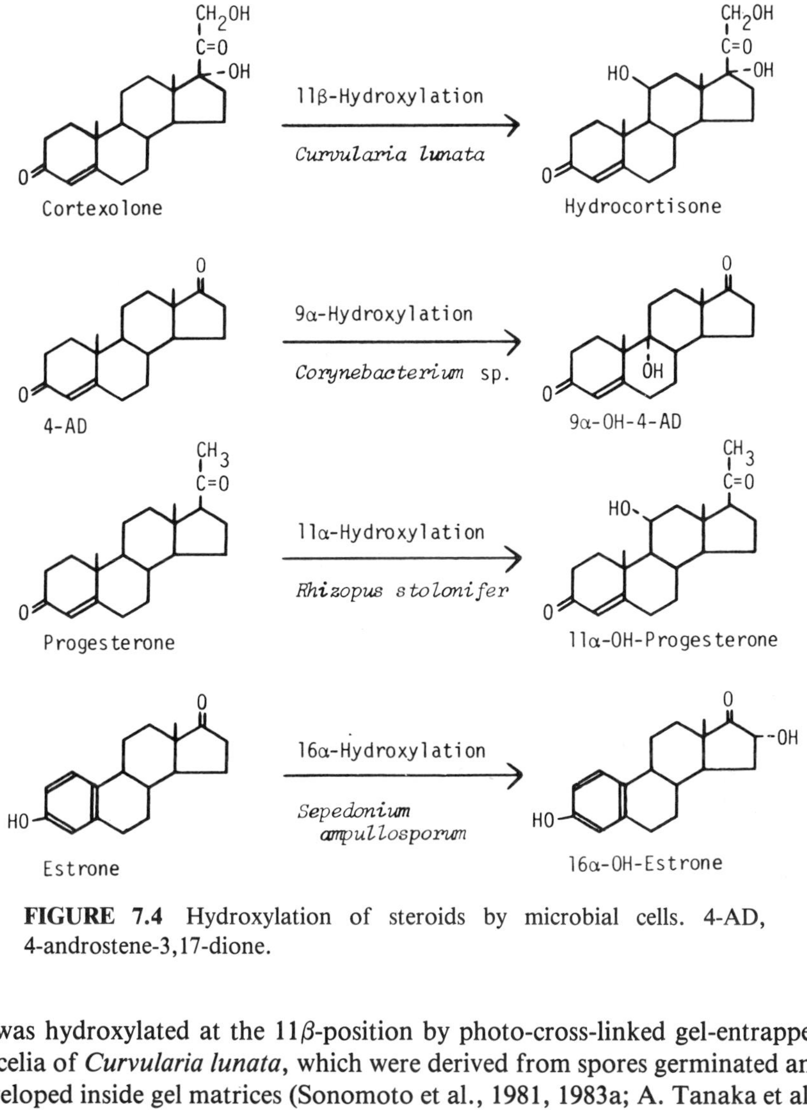

FIGURE 7.4 Hydroxylation of steroids by microbial cells. 4-AD, 4-androstene-3,17-dione.

S) was hydroxylated at the 11β-position by photo-cross-linked gel-entrapped mycelia of *Curvularia lunata*, which were derived from spores germinated and developed inside gel matrices (Sonomoto et al., 1981, 1983a; A. Tanaka et al., 1982); this was similar to the work reported by Ohlson et al. (1980). Dimethyl sulfoxide or methanol at 2.5% by volume was insufficient to dissolve the substrate (cortexolone) but was sufficient to dissolve the product (hydrocortisone). The hydroxylation system in the entrapped mycelia could be reactivated by incubating the mycelium-entrapping gels in a nutrient medium containing cortexolone as an inducer. The hydroxylation activity was correlated to the growth of mycelia inside gel matrices, both of which were affected by the structure of the gel network as determined by the chain length of prepolymers used to entrap the spores. The entrapped mycelia were far more stable than their free counterparts and could be utilized repeatedly for at least 50 batches (total

operational period, 100 days) (Fig. 7.5). This reaction system could be connected sequentially with the Δ^1-dehydrogenation system of *A. simplex* cells to
produce prednisolone from cortexolone (S. Fukui, T. K. Mazumder, K.
Sonomoto, and A. Tanaka, unpublished results). Spores of *Rhizopus stolonifer*
were also entrapped with photo-cross-linkable resin prepolymers and allowed
to germinate and develop inside gel matrices (Sonomoto et al., 1982). The entrapped fungal mycelia thus obtained hydroxylated progesterone at the
11α-position to form 11α-hydroxyprogesterone in the presence of 2.5% (by
volume) methanol, ethanol, or dimethyl sulfoxide. In this case also, the enzyme system in the entrapped mycelia was reactivated by incubation in a
nutrient broth. Photo-cross-linkable resin prepolymers were also used for the
entrapment of spores of *Sepedonium ampullosporum* (S. Fukui, J. M. Kim,
K. Sonomoto, and A. Tanaka, unpublished results). Cells that developed inside the gel matrices hydroxylated estrone at the 16α-position to yield
16α-hydroxyestrone in the presence of a low concentration of *N,N*-dimethyl-
formamide. Entrapped living cells of *Corynebacterium* sp. were more tolerant
to organic solvents than were the entrapped fungal mycelia. Thus 9α-hydroxyla-
tion of 4-androstene-3,17-dione to 9α-hydroxy-4-androstene-3,17-dione was
performed in a nutrient medium containing 15% (by volume) dimethyl sulfoxide (Sonomoto et al., 1983b). The reaction in a buffer solution instead of in
the medium resulted in a rapid decrease in the hydroxylation activity.

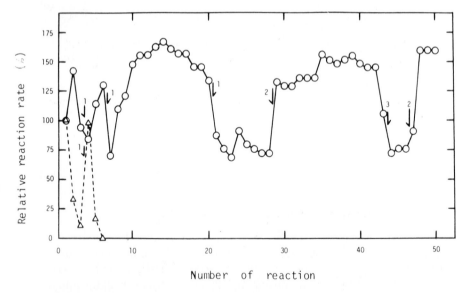

FIGURE 7.5   Repeated use of *Curvularia lunata* mycelia in hydrocortisone formation. Each reaction was carried out for 48 h in the presence of
2.5% dimethyl sulfoxide or methanol. Arrows indicate the reactivation of
mycelia under proper conditions. (○), ENT-4000-entrapped mycelia; (△),
free mycelia. (From Sonomoto et al., 1983a.)

The synthesis of adenine arabinoside, an anti-viral antibiotic, from uracil arabinoside and adenine (Fig. 7.6) was achieved successfully using entrapped cells of *Enterobacter aerogenes* in a system containing 40% (by volume) dimethyl sulfoxide (Yokozeki et al., 1982c). Although the reaction could be carried out in an aqueous system (Utagawa et al., 1980), the productivity was low because of the low solubility of adenine and adenine arabinoside in water. Dimethyl sulfoxide (40%) was selected as an organic cosolvent on the basis of the criteria of stability of the enzyme system and solubility of adenine arabinoside (Fig. 7.7). High concentrations of the substrates in the optimized reaction system resulted in the formation of a high concentration of the product, which was easily recovered from the reaction solution simply by cooling and filtering. Entrapment of *E. aerogenes* cells with photo-cross-linkable resin prepolymers or urethane prepolymers markedly enhanced the operational stability of the enzyme system during repeated reactions. Thus the entrapped cells could be used for at least 35 days in the presence of 40% dimethyl sulfoxide at 60°C without any loss of enzyme activity (Fig. 7.8)

As described above, water–organic cosolvent systems can be applied to bioconversions catalyzed by immobilized biocatalysts. Immobilization often

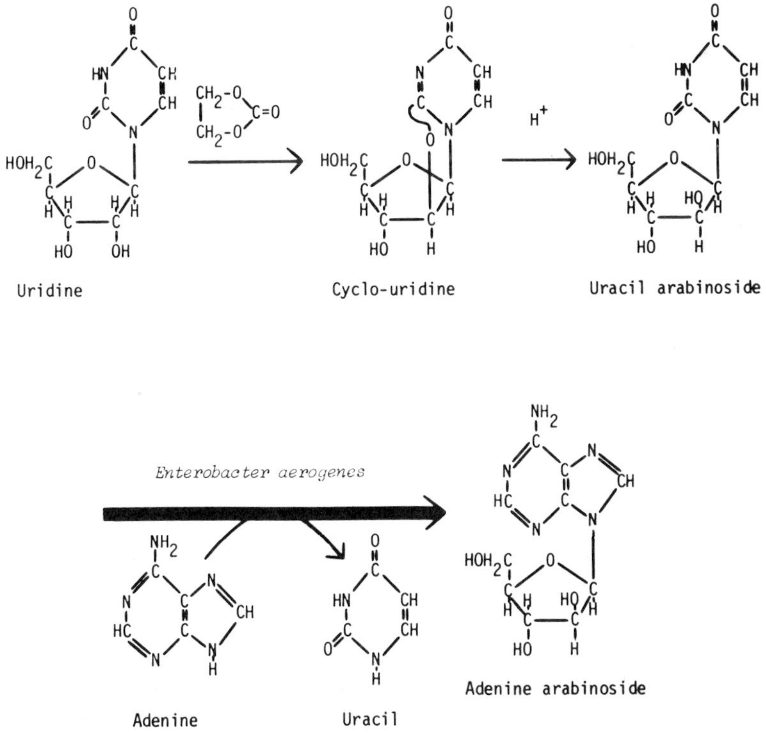

FIGURE 7.6 Synthesis of adenine arabinoside by a combination of chemical and enzymatic reactions. (From Yokozeki et al., 1982c.)

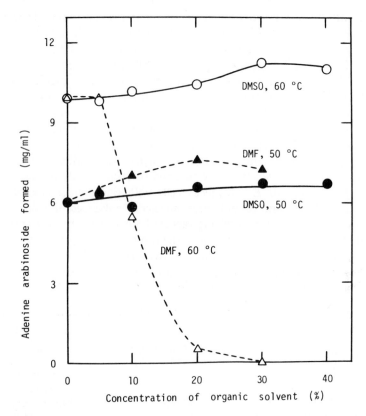

FIGURE 7.7   Effect of organic cosolvents on adenine arabinoside synthesis by free cells of *Enterobacter aerogenes*. (○), Reaction at 60°C with dimethyl sulfoxide; (●), reaction at 50°C with dimethyl sulfoxide; (△), reaction at 60°C with *N,N*-dimethylformamide; (▲), reaction at 50°C with *N,N*-dimethylformamide. (From Yokozeki et al., 1982c.)

gives a markedly enhanced stability to biocatalysts. This effect is probably ascribable to maintenance of the active conformation of enzyme molecules by efficient interactions between biocatalysts and support gels, which prevents inactivation by organic solvents, and to effective coating of microbial cells, which protects leakage of enzymes.

VII. BIOCONVERSION IN ORGANIC SOLVENT SYSTEMS

Although water-immiscible organic solvents have been employed for bioreactions to increase the solubility of substrates and products of a hydrophobic nature, most reports deal with water-organic solvent two-phase systems

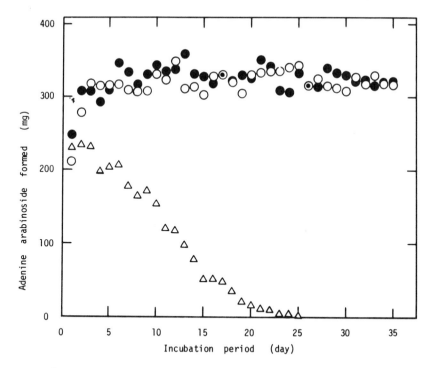

FIGURE 7.8 Repeated use of *Enterobacter aerogenes* cells for adenine arabinoside production. Each reaction was carried out for 24 h at 60°C in the presence of 40% dimethyl sulfoxide. (○), ENT-4000-entrapped cells; (●), PU-6-entrapped cells; (△), free cells. (From Yokozeki et al., 1982c.)

(Antonini et al., 1981). Only a limited number of papers have described the single use of organic solvents (mostly water-saturated organic solvents) in bioreactions.

Klibanov et al. (1977) reported a pioneering work on the use of organic solvents in the synthesis of *N*-acetyl-L-tryptophan ethyl ester from *N*-acetyl-L-tryptophan and ethanol by chymotrypsin covalently bound to porous glass. In this case, chloroform, benzene, or ether was used as the reaction solvent. Toluene or carbon tetrachloride was also employed in the dehydrogenation of cholesterol to cholestenone by DEAE-cellulose-adsorbed cells of *Nocardia erythropolis* (Atrat el al., 1980).

We have extensively investigated bioconversions of lipophilic compounds by immobilized biocatalysts in organic solvent systems (Table 7.4). The activities of immobilized biocatalysts were found to be affected by the hydrophilicity–hydrophobicity balance of the gels, hydrophobicity of substrates, and polarity of reaction solvents (Fukui et al., 1980b; Fukui and Tanaka, 1981, 1982a).

TABLE 7.4   **Bioconversions in Organic Solvent Systems by Biocatalysts Entrapped with Prepolymers**

Biocatalyst	Organic Solvent*	Application
Nocardia rhodochrous cells	Benzene-n-Heptane (1:1 by volume)	Δ^1-Dehydrogenation of 4-andro-stene-3,17-dione
	Benzene-n-Heptane (4:1 by volume)	Δ^1-Dehydrogenation of testosterone
	Benzene-n-Heptane (1:1 by volume)	3β-Hydroxysteroid dehydrogenation
	Benzene-n-Heptane (4:1 by volume)	17β-Hydroxysteroid dehydrogena-tion
	Chloroform-n-Heptane (1:1 by volume)	3β-Hydroxysteroid dehydrogenation
Rhodotorula minuta cells	n-Heptane	Resolution of dl-menthol
Rhizopus delemar lipase	n-Hexane	Interesterification of triglyceride

*Water-saturated.

As shown in Fig. 7.9, *Nocardia rhodochrous* cells mediate different types of bioconversions of a variety of steroids: Δ^1-dehydrogenation, 3β-hydroxy-steroid dehydrogenation, and 17β-hydroxysteroid dehydrogenation (Fukui et al., 1980a, b; Fukui and Tanaka, 1981; Omata et al., 1979a, b, 1980; Yamane et al., 1979). In order to study the effect of gel hydrophobicity— that is, the influence of affinity between hydrophobic substrates and gels entrapping biocatalysts—a very simple parameter, the partition coefficient (a ratio of substrate concentration between gels and external solvent) was employed. The partition coefficient (*P*) was estimated by using the following equation:

$$P = \left(\frac{C_0 - C}{C}\right)\left(\frac{V_0}{V - V_0}\right)$$

where C_0 is the initial substrate concentration in the solvent, C is the final substrate concentration in the solvent, V_0 is the initial volume of the system without gels, and V is the final volume of the system with gels.

The effect of gel hydrophobicity was investigated by using the conversion of 3β-hydroxy-Δ^5-steroids to 3-keto-Δ^4-steroids (Fig. 7.10) in a water-saturated mixture of benzene and n-heptane (1:1 by volume) (Omata et al., 1979b). The solvent system was chosen on the basis of the following criteria: Substrates and products are adequately soluble; the enzyme system is not damaged; and the solvent does not cause the gels to swell. As shown in Table 7.5, *N. rhodochrous* cells entrapped in hydrophobic gels, such as ENTP-2000 and PU-3, converted dehydroepiandrosterone (DHEA) to 4-androstene-3,17-dione (4-AD) with a high reaction rate comparable to that of the free cells. On the other hand, the cells entrapped in hydrophilic gels, ENT-4000 and PU-6, were less active.

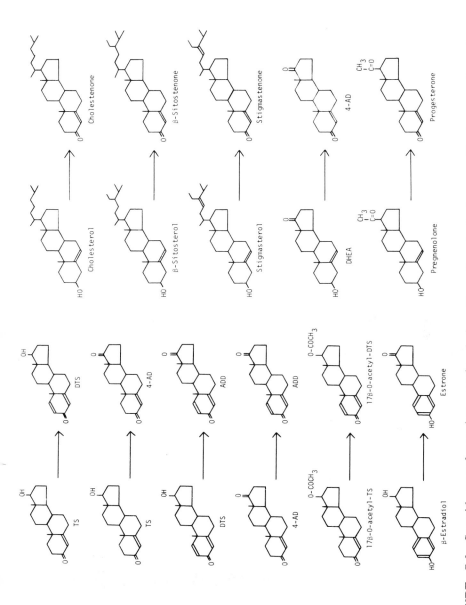

FIGURE 7.9 Steroid transformation catalyzed by *Nocardia rhodochrous* cells in organic solvents. 4-AD, 4-androstene-3,17-dione; ADD, 1,4-androstadiene-3,17-dione; DHEA, dehydroepiandrosterone; DTS, Δ^1-dehydrotestosterone; TS, testosterone.

163

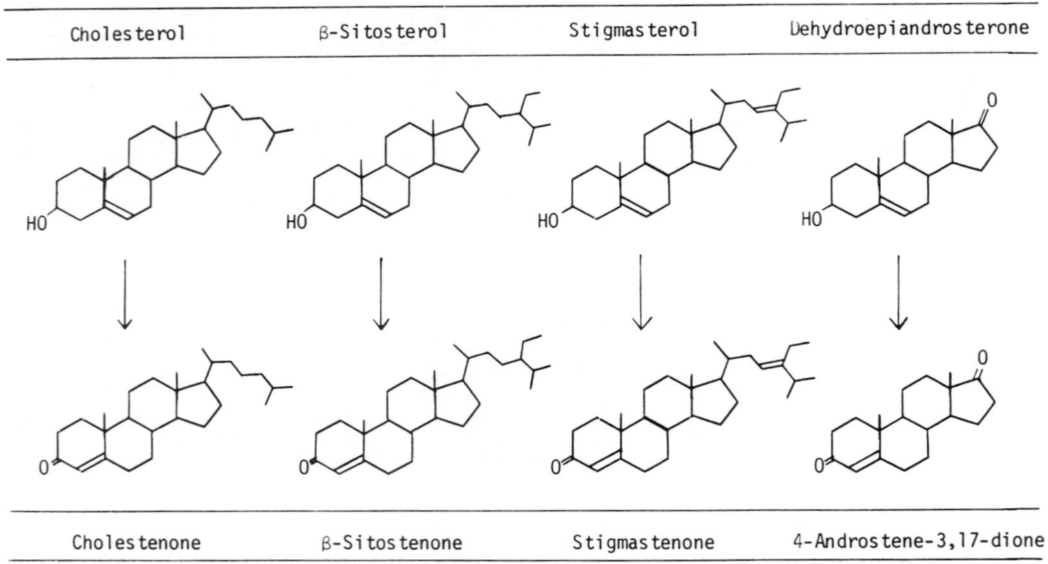

FIGURE 7.10 Bioconversion of 3β-hydroxy-Δ⁵-steroids to the corresponding 3-keto-Δ⁴-steroids by *Nocardia rhodochrous* cells. (From Omata et al., 1979b.)

TABLE 7.5   Steroid Transformation Activities of Free and Entrapped *Nocardia rhodochrous* **Cells**

	Specific (Relative) Activity on			
Cell	Cholesterol	β-Sitosterol	Stigmasterol	Dehydroepi-androsterone
	$\mu mol \cdot h^{-1} \cdot g$ wet cell^{-1}(%)			
Free	82 (100)	92 (100)	79 (100)	86 (100)
ENTP-2000-entrapped	84 (102)	76 (83)	88 (111)	92 (107)
ENT-4000-entrapped	0 (0)	0 (0)	0 (0)	66 (78)
PU-3-entrapped	62 (77)	50 (54)	67 (85)	90 (105)
PU-6-entrapped	0 (0)	NT	0 (0)	62 (72)
PU-9-entrapped	0 (0)	0 (0)	NT	NT

NT = not tested.
From Omata et al. (1979b).

Figure 7.11b shows the relationship between the relative activity of the entrapped cells and the partition coefficient of DHEA, both of which changed depending on the hydrophobicity of gels. The abscissa shows the mixing ratio of hydrophobic prepolymer PU-3 and hydrophilic prepolymer PU-6. A close relationship between the relative activity of the gel-entrapped cells and the partition coefficient of the substrate can be observed. In the case of transformations of cholesterol, β-sitosterol, and stigmasterol, more lipophilic substrates than DHEA, to the corresponding 3-keto-Δ^4-steroids, a much more clear-cut interrelationship was observed between gel hydrophobicity and activity of gel-entrapped cells (Table 7.5). Only the cells entrapped with hydrophobic prepolymers (ENTP-2000 and PU-3) exhibited the catalytic activity. This phenomenon was also confirmed when the hydrophobicity of gels was changed by mixing PU-3 and PU-6 (Fig. 7.11a). No activity for cholesterol conversion, in accordance with the low partition coefficient of cholesterol, was observed when PU-3, a hydrophobic prepolymer, was present only in low amounts.

In addition to gel hydrophobicity and substrate hydrophobicity the polarity of the solvents also had a marked effect on the conversion of steroids (Omata et al., 1980). *N. rhodochrous* cells entrapped in hydrophilic gels could not transform cholesterol in a nonpolar solvent mixture, benzene-*n*-heptane (1:1 by volume). However, substitution of benzene with chloroform made the hydrophilic gel-entrapped cells active in keeping with the increased partition coefficient (Fig. 7.12). Increase in solvent polarity lowered the activity and stability of the free cells (Table 7.6) and, subsequently, those of the entrapped cells. In a water-saturated mixture of chloroform and *n*-heptane (1:1 by volume), little effect of gel hydrophobicity was observed in the conversion of cholesterol and DHEA, differing from the results obtained in the nonpolar solvent system shown in Fig. 7.11. The hydrophobic gel-entrapped cells rather had a low activity in the conversion of pregnenolone, a less hydrophobic substrate which was not soluble in benzene-*n*-heptane. In spite of these facts, the

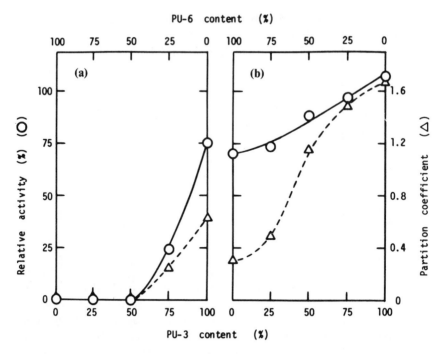

FIGURE 7.11   Effect of hydrophobicity of polyurethane resins on relative activity of steroid transformation and partition coefficient. (a) cholesterol; (b) dehydroepiandrosterone. (○), Relative activity; (△), partition coefficient. (From Omata et al., 1979b.)

TABLE 7.6   Effect of Solvents on Cholesterol Transformation Activity of Free and ENT-4000-entrapped *Nocardia rhodochrous* **Cells**

			Transformation Activity $(\mu mol \cdot h^{-1} \cdot g \ wet \ cell^{-1})$	
Solvent (1:1 by volume)	(D_m^a, D_c^b)	P^c	*Free Cells*	*Immobilized Cells*
Carbon tetrachloride-*n*-Heptane	(0.0, 2.2)	0.02	68	0
Benzene-*n*-Heptane	(0.0, 2.3)	0.02	57	0
Toluene-*n*-Heptane	(0.4, 2.4)	0.06	68	0
Chloroform-*n*-Heptane	(1.1, 4.7)	0.82	42	29 (69% [d])
Methylene chloride-*n*-Heptane	(1.5, 8.9)	0.81	19	15 (80%)
Ethyl acetate-*n*-Heptane	(1.9, 6.0)	0.25	27	0
Acetone-*n*-Heptane	(2.7, 20.7)	—	trace	0
Ethanol-*n*-Heptane	(1.7, 24.3)	—	0	0
Methanol-*n*-Heptane	(1.7, 32.6)	—	0	0

[a] Dipolar moment of organic solvent excluding *n*-heptane.
[b] Dielectric constant of organic solvent excluding *n*-heptane.
[c] Partition coefficient of cholesterol between ENT-4000 gel and external solvents.
[d] Activity of free cells in each solvent was expressed as 100%.
From Omata et al. (1980).

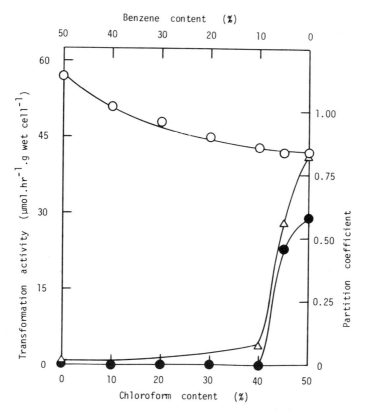

FIGURE 7.12 Effect of solvent polarity on cholesterol transformation by free and ENT-4000-entrapped *Nocardia rhodochrous* cells. Solvents employed were prepared by mixing benzene and chloroform in different ratios keeping the content of *n*-heptane constant (50% by volume). (○), Activity of free cells; (●), activity of ENT-4000-entrapped cells; (△), partition coefficient. (From Omata et al., 1980.)

usefulness of hydrophobic gels is clear in the transformation of highly lipophilic compounds because the transformation activity of the gel-entrapped cells is usually high and stable in less polar solvents. In conclusion, it would be very important for successful bioconversions of lipophilic compounds to select gel materials with suitable hydrophobicity and reaction solvents with proper polarities, both of which should be selected depending on substrate hydrophobicity. For the entrapment of biocatalysts, the prepolymer methods are very useful because these methods offer a very easy choice of gel hydrophobicity.

It has been observed that gel hydrophobicity affected the conversion routes from testosterone (TS) to 1,4-androstadiene-3,17-dione (ADD) (Fukui et al., 1980a). Figure 7.13 illustrates the transformation pathway of TS into

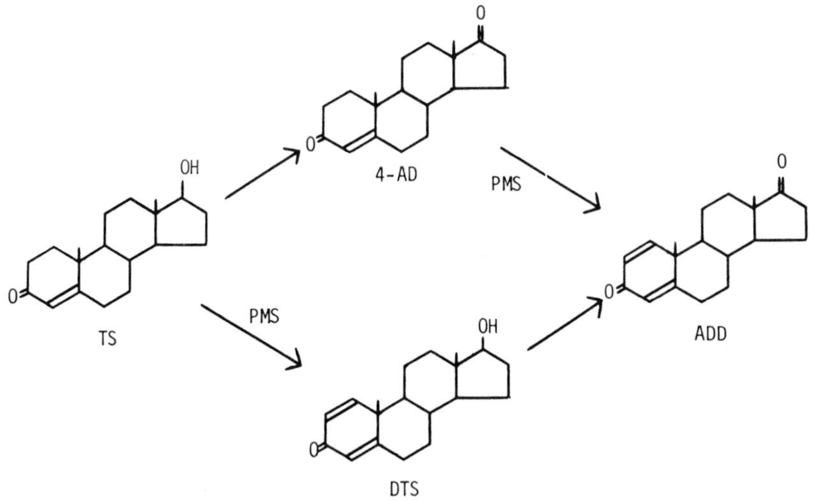

FIGURE 7.13 Transformation of testosterone catalyzed by *Nocardia rhodochrous* cells in organic solvent. 4-AD, 4-androstene-3,17-dione; ADD, 1,4-androstadiene-3,17-dione; DTS, Δ^1-dehydrotestosterone; PMS, phenazine methosulfate; TS, testosterone. (From Fukui et al., 1980a.)

ADD mediated by *N. rhodochrous* in water-saturated benzene-*n*-heptane (4:1 by volume). In the presence of an electron acceptor, such as phenazine methosulfate (PMS), the free bacterial cells converted TS to ADD via two diverse routes. As shown in Fig. 7.13, the 17β-dehydrogenation product, 4-androstene-3,17-dione (4-AD), and the Δ^1-dehydrogenation product, Δ^1-dehydrotestosterone (DTS), appeared as intermediates in ADD formation. In these reactions, Δ^1-dehydrogenation absolutely required PMS, whereas 17β-dehydrogenation could proceed without the exogenous electron acceptor, although PMS stimulated the reaction. When the cells were entrapped in gels of different hydrophilicity or hydrophobicity, the property of the gels gave striking effects on the conversion routes. With hydrophobic (PU-3-rich) gel-entrapped cells, 4-AD was formed as major reaction product (Fig. 7.14). On the other hand, DTS was the main product with hydrophilic (PU-6-rich) gel-entrapped cells. This different profile in dehydrogenation products can be explained by a marked difference in the affinity of PMS, a hydrophilic compound, to the hydrophobic and hydrophilic gels. With hydrophilic gel-entrapped cells, Δ^1-dehydrogenation of TS to yield DTS is stimulated by PMS taken up inside the gels, and the DTS so accumulated inhibits 17β-hydroxysteroid dehydrogenase converting DTS to ADD even at a low concentration. On the contrary, PMS was hardly taken up by the cells entrapped within hydrophobic gels, and hence 17β-dehydrogenation of TS becomes the main route.

These results indicate that the dehydrogenation reactions at two distinct positions of TS—Δ^1-dehydrogenation and 17β-hydroxy group dehydrogena-

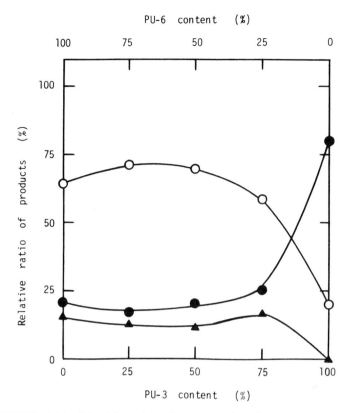

FIGURE 7.14 Selectivity of product formation from testosterone by *Nocardia rhodochrous* cells immobilized in different combination of PU-3 and PU-6. (○), Δ¹-Dehydrotestosterone; (●), 4-androstene-3,17-dione; (▲), 1,4-androstadiene-3,17-dione. (From Fukui et al., 1980a.)

tion — can be controlled by controlling the hydrophilic or hydrophobic nature of the gels entrapping *N. rhodochrous* cells. Thus selective formation of a desired product among diverse products from a single substrate has been achieved by appropriate use of hydrophilic or hydrophobic gels. This concept could be applicable to the bioconversions of many organic compounds.

The stereoselective hydrolysis of *dl*-menthyl succinate by gel-entrapped cells of *Rhodotorula minuta* var. *texensis* (Fig. 7.15) was carried out successfully in water-saturated *n*-heptane (Omata et al., 1981). *l*-Menthol, the compound having a peppermint flavor and useful in the food and pharmaceutical industries, was obtained by stereospecific hydrolysis of an appropriate ester of chemically synthesized *dl*-menthol. The ammonium salt of *dl*-menthyl succinate is water-soluble, and its stereoselective hydrolysis could be achieved in an aqueous system by using free cells of the yeast having the esterase activity. However, owing to its poor solubility in aqueous buffers, the

dl-Menthyl succinate   l-Menthol   d-Menthyl succinate

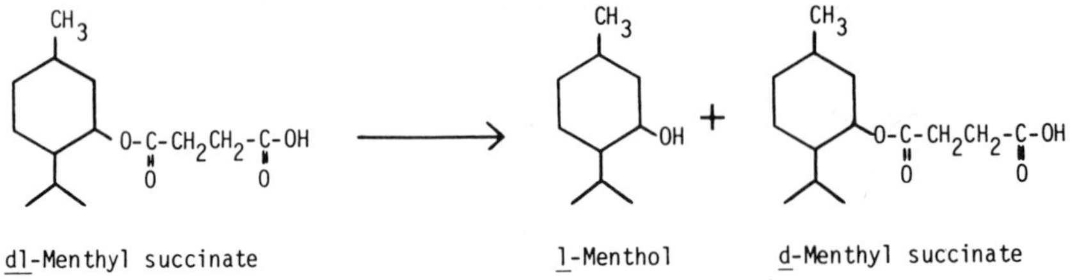

FIGURE 7.15   Stereoselective hydrolysis of *dl*-menthyl succinate catalyzed by *Rhodotorula minuta* var. *texensis*.

l-menthol formed accumulated on the surface of the yeast cells, thus decreasing the activity. To prevent the accumulation of *l*-menthol on the cell surface, various kinds of water-miscible organic solvents were tested as cosolvents. However, the hydrolytic activity of the yeast cells was reduced in the presence of organic cosolvents. After many trials using free cells in appropriate organic solvents and appropriate two-phase systems consisting of potassium phosphate buffer and organic solvents, a combination of gel-entrapped cells and water-saturated *n*-heptane was finally employed to obtain a homogeneous reaction system. Although the effect of gel hydrophobicity was not so remarkable as in the case of steroid conversion, the activity of the gel-entrapped cells increased with increased gel hydrophobicity. Figure 7.16 shows the comparison of operational stability between free cells and gel-entrapped cells over repeated reactions. The half-life of the free cells was 2 days, while that of the PU-3-entrapped cells was estimated to be 63 days. Thus immobilization greatly improved the operational stability of the hydrolytic enzyme in the yeast cells. The optical purity of the product was also constantly maintained at 100% after long-term operation. In a semipilot-scale production using immobilized *R. minuta* cells the overall yield of *l*-menthol from the starting material, *dl*-menthol, was 86%.

   Another example is the interesterification of triacylglyceride by lipase. For reformation of olive oil to cacao butter–like fat, the exchange of oleoyl moieties at positions 1 and 3 of triglyceride by saturated fatty acyl moieties, such as a stearoyl group, was successfully achieved by the use of 1- and 3-position-specific lipase from *Rhizopus delemar* (Fig. 7.17) (T. Tanaka et al., 1981). In this reaction it is very important to control the water content in the reaction system, because the ester exchange reaction cannot be initiated without water, but hydrolysis of the ester is preferred at a high concentration of water. Therefore water-saturated *n*-hexane was selected as the reaction solvent, considering the activity and stability of the enzyme and also the solubility of the reactants. To provide water in the vicinity of the enzyme, lipase was adsorbed on an appropriate porous support, such as Celite, with a controlled amount of water. When Celite-adsorbed lipase was entrapped with different prepolymers (Fig. 7.18) (Yokozeki et al., 1982a, b), the enzyme entrapped with

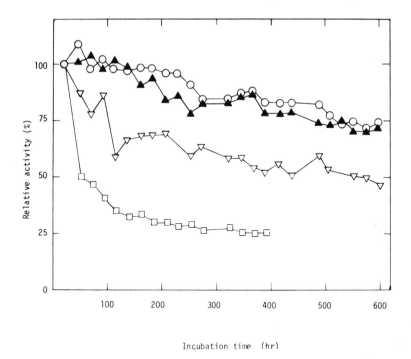

FIGURE 7.16 Repeated use of free and entrapped *Rhodotorula minuta* cells in hydrolysis of *dl*-menthyl succinate. Each reaction was carried out for 24 h. (□), Free cells; (○), PU-3-entrapped cells; (▲), PU-6-entrapped cells; (▽), ENT-4000-entrapped cells. (From Omata et al., 1981.)

a hydrophobic photo-cross-linkable resin prepolymer, ENTP-2000, showed the highest activity of interesterification, about 75% of that of lipase simply adsorbed onto Celite (Table 7.7). On the contrary, the hydrophilic gel-entrapped enzyme exhibited only a low activity. This result again indicates the usefulness of hydrophobic gels in bioconversion of lipophilic compounds in a suitable nonpolar solvent. Entrapment markedly enhanced the operational stability of lipase, the enzyme losing only 10% of the original activity after 12 batches (operational period, 12 days) (Fig. 7.19).

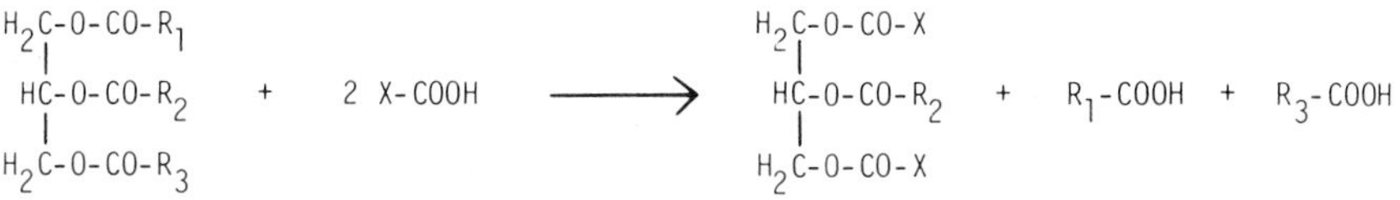

$$H_2C\text{-}O\text{-}CO\text{-}R_1$$
$$|$$
$$HC\text{-}O\text{-}CO\text{-}R_2 \quad + \quad 2\ X\text{-}COOH \quad \longrightarrow \quad HC\text{-}O\text{-}CO\text{-}R_2 \quad + \quad R_1\text{-}COOH \quad + \quad R_3\text{-}COOH$$
$$|$$
$$H_2C\text{-}O\text{-}CO\text{-}R_3 \qquad\qquad\qquad\qquad\qquad H_2C\text{-}O\text{-}CO\text{-}X$$

with $H_2C\text{-}O\text{-}CO\text{-}X$ at the top of the product.

FIGURE 7.17 1- and 3-position specific ester exchange of triglyceride catalyzed by *Rhizopus delemar* lipase.

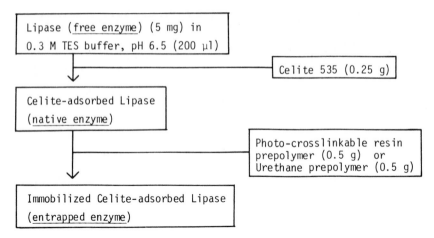

FIGURE 7.18   Preparation of Celite-adsorbed and gel-entrapped lipase.

The results mentioned above, together with those using immobilized chymotrypsin reported by Klibanov et al. (1977), indicate that not only microbial cells but also enzymes can be employed as biocatalysts in organic solvent systems after appropriate immobilization.

TABLE 7.7   Interesterification Activity of Lipase Entrapped with Different Prepolymers

Prepolymer	Adsorption on Celite	Activity Yield (%)
None	+	100
	−	0
ENT-1000	+	14
	−	13
ENT-2000	+	20
	−	18
ENT-4000	+	22
	−	20
ENT-6000	+	15
	−	14
ENTP-2000	+	75
	−	29–82
PU-3	+	21
	−	19
PU-6	+	17
	−	15
PU-9	+	14
	−	3

From Yokozeki et al. (1982b).

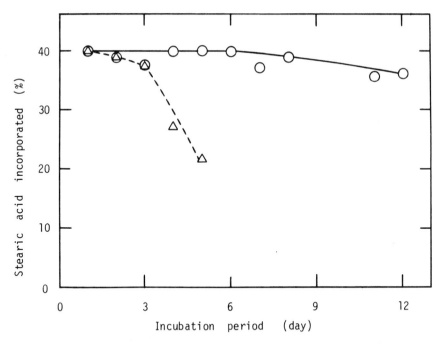

FIGURE 7.19 Repeated use of lipase preparations for interesterification of triglyceride. Each reaction was carried out for 24 h. ($\triangle$), Celite-adsorbed lipase; ($\bigcirc$), ENTP-2000-entrapped Celite-adsorbed lipase. (From Yokozeki et al., 1982b.)

VIII. FUTURE PROSPECTS

At present, the introduction of biochemical processes into such industries as the chemical, food, and pharmaceutical industries is attracting worldwide attention. For this purpose the synthesis and conversion of a variety of lipophilic compounds or compounds hardly soluble in water must be carried out in the presence of organic solvents. In the case of the use of biochemical processes for producing commodity petrochemicals, the substrates themselves are usually organic solvents. Thus biocatalysts must maintain their activities toward such unconventional substrates and unfavorable conditions. In these cases the properties of gels used for entrapping biocatalysts will seriously affect the efficiency of the reactions and the stability of the biocatalysts. The prepolymer methods that we have developed have proved to be very effective in carrying out bioconversions in the presence of organic solvents because of the ease of selection of hydrophilic or hydrophobic gels. However, new types of gel materials, which are easy to handle and for which it is easy to control the gel properties, should be developed for extensive applications of immobilized biocatalysts in bioindustries.

REFERENCES

Antonini, E., Carrea, G., and Cremonesi, P. (1981) *Enzyme Microb. Technol. 3*, 291-296.

Atrat, P., Hüller, E., and Hörhold, C. (1980) *Z. Allg. Mikrobiol. 20*, 79-84.

Becker, W., and Pfeil, E. (1966) *J. Amer. Chem. Soc. 88*, 4299-4300.

Buckland, B. C., Dunnill, P., and Lilly, M. D. (1975) *Biotechnol. Bioeng. 17*, 815-826.

Butler, L. G. (1979) *Enzyme Microb. Technol. 1*, 253-259.

Butler, L. G., and Reithel, F. J. (1977) *Arch. Biochem. Biophys. 178*, 43-50.

Carrea, G., Colombi, F., Mazzola, G., Cremonesi, P., and Antonini, E. (1979) *Biotechnol. Bioeng. 21*, 39-48.

Chun, Y. Y., Iida, M., and Iizuka, H. (1981) *J. Gen. Appl. Microbiol. 27*, 505-509.

Cremonesi, P., Carrea, G., Sportoletti, G., and Antonini, E. (1973) *Arch. Biochem. Biophys. 159*, 7-10.

Cremonesi, P., Carrea, G., Ferrara, L., and Antonini, E. (1974) *Eur. J. Biochem. 44*, 401-405.

Cremonesi, P., Carrea, G., Ferrara, L., and Antonini, E. (1975) *Biotechnol. Bioeng. 17*, 1101-1108.

Doi, O., Ohki, M., and Nojima, S. (1972) *Biochim. Biophys. Acta 260*, 244-258.

Duarte, J. M. C., and Lilly, M. D. (1980) *Enzyme Eng. 5*, 363-367.

Fukui, S., and Tanaka, A. (1981) *Acta Biotechnol. 1*, 339-350.

Fukui, S., and Tanaka, A. (1982a) *Enzyme Eng. 6*, 191-200.

Fukui, S., and Tanaka, A. (1982b) *Ann. Rev. Microbiol. 36*, 145-172.

Fukui, S., and Tanaka, A. (1984) *Adv. Biochem. Eng. Biotechnol. 29*, 1-33.

Fukui, S., Tanaka, A., Iida, T., and Hasegawa, E. (1976) *FEBS Lett. 66*, 179-182.

Fukui, S., Ahmed, S. A., Omata, T., and Tanaka, A. (1980a) *Eur. J. Appl. Microbiol. Biotechnol. 10*, 289-301.

Fukui, S., Omata, T., Yamane, T., and Tanaka, A. (1980b) *Enzyme Eng. 5*, 347-353.

Fukui, S., Sonomoto, K., Itoh, N., and Tanaka, A. (1980c) *Biochimie 62*, 381-386.

Fukushima, S., Nagai, T., Fujita, K., Tanaka, A., and Fukui, S. (1978) *Biotechnol. Bioeng. 20*, 1465-1469.

Grandi, C., Smith, R. E., and Luisi, P. L. (1981) *J. Biol. Chem. 256*, 837-843.

Herskovits, T. T., Gadegbeku, B., and Jaillet, H. (1970) *J. Biol. Chem. 245*, 2588-2598.

Homandberg, G. A., Mattis, J. A., and Laskowski, M., Jr. (1978) *Biochemistry 17*, 5220-5227.

Ingalls, R. G., Squires, R. G., and Butler, L. G. (1975) *Biotechnol. Bioeng. 17*, 1627-1637.

Jonczyk, A., and Gattner, H.-G. (1981) *Hoppe-Seyler's Z. Physiol. Chem. 362*, 1591-1598.

Kimura, Y., Tanaka, A., Sonomoto, K., Nihira, T., and Fukui, S. (1983) *Eur. J. Appl. Microbiol. Biotechnol. 17*, 107-112.

Klibanov, A. M. (1979) *Anal. Biochem. 93*, 1-25.

Klibanov, A. M., Samokhin, G. P., Martinek, K., and Berezin, I. V. (1977) *Biotechnol. Bioeng. 19*, 1351-1361.

Lilly, M. D. (1982) *J. Chem. Technol. Biotechnol. 32*, 162-169.

Lugaro, G., Carrea, G., Cremonesi, P., Casellato, M. M., and Antonini, E. (1973) *Arch. Biochem. Biophys. 159*, 1-6.

Maddox, I. S., Dunnill, P., and Lilly, M. D. (1981) *Biotechnol. Bioeng. 23*, 345-354.

Martinek, K., and Berezin, I. V. (1977) *J. Solid-Phase Biochem. 2*, 343–385.

Martinek, K., and Semenov, A. N. (1981) *Biochim. Biophys. Acta 658*, 90–101.

Martinek, K., Levashov, A. V., Klyachko, N. L., Pantin, V. I., and Berezin, I. V. (1981a) *Biochim. Biophys. Acta 657*, 277–294.

Martinek, K., Semenov, A. N., and Berezin, I. V. (1981b) *Biochim. Biophys. Acta 658*, 76–89.

Minato, S., and Hirai, A. (1979) *J. Biochem. 85*, 327–334.

Morihara, K., and Oka, T. (1977) *Biochem. J. 163*, 531–542.

Morihara, K., Oka, T., and Tsuzuki, H. (1979) *Nature 280*, 412–413.

Morihara, K., Oka, T., Tsuzuki, H., Tochino, Y., and Kanaya, T. (1980) *Biochem. Biophys. Res. Comm. 92*, 396–402.

Ohlson, S., Larsson, P. O., and Mosbach, K. (1978) *Biotechnol. Bioeng. 20*, 1267–1284.

Ohlson, S., Flygare, S., Larsson, P. O., and Mosbach, K. (1980) *Eur. J. Appl. Microbiol. Biotechnol. 10*, 1–9.

Omata, T., Tanaka, A., Yamane, T., and Fukui, S. (1979a) *Eur. J. Appl. Microbiol. Biotechnol. 6*, 207–215.

Omata, T., Iida, T., Tanaka, A., and Fukui, S. (1979b) *Eur. J. Appl. Microbiol. Biotechnol. 8*, 143–155.

Omata, T., Tanaka, A., and Fukui, S. (1980) *J. Ferment. Technol. 58*, 339–343.

Omata, T., Iwamoto, N., Kimura, T., Tanaka, A., and Fukui, S. (1981) *Eur. J. Appl. Microbiol. Biotechnol. 11*, 199–204.

Schwartz, R. D., and McCoy, C. J. (1977) *Appl. Environ. Microbiol. 34*, 47–49.

Semenov, A. N., Berezin, I. V., and Martinek, K. (1981) *Biotechnol. Bioeng. 23*, 355–360.

Singer, S. J. (1962) *Adv. Protein Chem. 17*, 1–68.

Sonomoto, K., Tanaka, A., Omata, T., Yamane, T., and Fukui, S. (1979) *Eur. J. Appl. Microbiol. Biotechnol. 6*, 325–334.

Sonomoto, K., Jin, I.-N., Tanaka, A., and Fukui, S. (1980) *Agr. Biol. Chem. 44*, 1119–1126.

Sonomoto, K., Hoq, M. M., Tanaka, A., and Fukui, S. (1981) *J. Ferment. Technol. 59*, 465–469.

Sonomoto, K., Nomura, K., Tanaka, A., and Fukui, S. (1982) *Eur. J. Appl. Microbiol. Biotechnol. 16*, 57–62.

Sonomoto, K., Hoq, M. M., Tanaka, A., and Fukui, S. (1983a) *Appl. Environ. Microbiol. 45*, 436–443.

Sonomoto, K., Usui, N., Tanaka, A., and Fukui, S. (1983b) *Eur. J. Appl. Microbiol. Biotechnol. 17*, 203–210.

Tanaka, A., Jin, I.-N., Kawamoto, S., and Fukui, S. (1979) *Eur. J. Appl. Microbiol. Biotechnol. 7*, 351–354.

Tanaka, A., Sonomoto, K., Hoq, M. M., Usui, N., Nomura, K., and Fukui, S. (1982) *Enzyme Eng. 6*, 131–133.

Tanaka, T., Ono, E., Ishihara, M., Yamanaka, S., and Takinami, K. (1981) *Agr. Biol. Chem. 45*, 2387–2389.

Utagawa, T., Morisawa, H., Miyoshi, T., Yoshinaga, F., Yamazaki, A., and Mitsugi, K. (1980) *FEBS Lett. 109*, 261–263.

Wan, H., and Horvath, C. (1975) *Biochim. Biophys. Acta 410*, 135–144.

Weetall, H. H., and Vann, W. P. (1976) *Biotechnol. Bioeng. 18*, 105–118.

Yamane, T., Nakatani, H., Sada, E., Omata, T., Tanaka, A., and Fukui, S. (1979) *Biotechnol. Bioeng. 21*, 2133–2145.

Yamashita, M., Arai, S., Kokubo, S., Aso, K., and Fujimaki, M. (1975) *J. Agr. Food Chem. 23*, 27-30.

Yamashita, M., Arai, S., and Fujimaki, M. (1976) *J. Food Sci. 41*, 1029-1033.

Yang, H. S., and Studebaker, J. F. (1978) *Biotechnol. Bioeng. 20*, 17-25.

Yokozeki, K., Tanaka, T., Yamanaka, S., Takinami, K., Hirose, Y., Sonomoto, K., Tanaka, A., and Fukui, S. (1982a) *Enzyme Eng. 6*, 151-152.

Yokozeki, K., Yamanaka, S., Takinami, K., Hirose, Y., Tanaka, A., Sonomoto, K., and Fukui, S. (1982b) *Eur. J. Appl. Microbiol. Biotechnol. 14*, 1-5.

Yokozeki, K., Yamanaka, S., Utagawa, T., Takinami, K., Hirose, Y., Tanaka, A., Sonomoto, K., and Fukui, S. (1982c) *Eur. J. Appl. Microbiol. Biotechnol. 14*, 225-231.

Coenzyme Regeneration in Membrane Reactors

Christian Wandrey
Rolf Wichmann

I. INTRODUCTION

Enzyme technology has experienced a considerable stimulus in the past few years via the reuse of enzymes and the associated reduction of catalyst costs. However, coenzyme-independent systems have been almost exclusively employed on a commercial scale up to now. Only in cases in which coenzymes occur in the form of prosthetic groups are there no difficulties in applying carrier-fixed systems. In contrast, systems requiring coenzymes as transport metabolites experience problems, since on the one hand the catalytically active components have to be retained in the reactor and on the other hand mobility must be guaranteed. This problem is solved in the microbial cell by the selective and frequently activated transport of substrates and products via the cell membrane. Technical membranes do not display any comparable selectivities. There is, however, the possibility of achieving practically significant volume flows by means of forced convection over the membrane. Technical membrane reactors seem to be of special interest for coenzyme-dependent systems.

In addition to the problem of coenzyme retention, the problem of coenzyme regeneration must also be solved. The essential factor is to employ a cheap regeneration substrate where, in an ideal case, there should be no by-products or at most a by-product that neither interferes with, nor causes problems during, separation.

II. POSSIBLE SYSTEMS

As long ago as 1972, Wykes et al. (1972) pointed out that coenzymes bound to soluble polymers could be employed in membrane reactors. Experiments for producing such "enlarged" coenzymes have been reported for NAD/NADH, NADP/NADPH, ADP/ATP, and coenzyme A (Mosbach et al., 1976; Wykes et al., 1975; Weibel et al., 1974; Zapelli et al., 1975, 1976; Yamazaki and Maeda, 1981; Yamazaki et al., 1976; Coughlin et al., 1975; Chambers et al., 1974; Furukawa et al., 1980; Bueckmann et al., 1981; Wichmann et al., 1981; Pace et al., 1976; Miyawaki et al., 1982a; Asada et al., 1978; Le Goffic et al., 1980; Gardner, 1978).

Further conceivable examples could be ubiquinone (coenzyme Q) for hydrogen transfer and S-adenosyl methionine for methyl group transfer (Wang and King, 1979).

Redox reactions have achieved the greatest significance in coenzyme regeneration to date. In stereospecific reduction with the aid of coenzymes, electroenzymatic coenzyme regeneration is especially elegant (Simon et al., 1981; Shaked et al., 1981; DiCosimo et al., 1981), the hydrogen required being provided via the water and the necessary electrons via a cathode. The purely electrochemical regeneration of NAD to NADH causes difficulties due to undesirable secondary reactions (Aizawa et al., 1976; Schmarkel et al., 1974).

Enzymatic regeneration is suitable not only for the transfer of redox equivalents but also for other functional groups. Both the position of equilibrium of the regeneration step and the actual desired reaction step are important. With a favorable equilibrium position during the regeneration reaction, even thermodynamically unfavorable reaction steps can be overcome by coupling the equilibria (Wykes et al., 1975; Wang and King, 1979; Chambers et al., 1975; Stinson and Holbrook, 1973; Wratten and Cleland, 1963; Wandrey et al., 1983). If the desired regeneration reaction is per se not advantageous for coenzyme regeneration owing to its equilibrium position, then the regeneration coproduct can be captured by a reaction connected in series—although this makes the system more complicated—in order to thus increase exploitation of the regeneration substrate. This principle can be applied in employing ethanol as the regeneration substrate by capturing the resulting acetaldehyde with semicarbacide (Juilliard and Le Petit, 1982).

A chemical regeneration of coenzymes is conceivable and has been successfully practiced, for example, with dithionite in the case of regenerating NADH from NAD (Jones et al., 1972; Jones and Beck, 1976; Vandecasteele, 1980). However, problems can result in the presence of enzymes or by slight secondary reactions that are not of significance with a single regeneration but that lead to considerable coenzyme losses during continuous regeneration.

III. POSSIBLE REACTORS

The natural reactor for coenzyme regeneration is the microbial cell. For this reason the utilization of whole cells previously gained almost universal acceptance in practice in dealing with the fabrication of biotechnological products for which coenzyme regeneration was necessary. However, the utilization of whole cells has some drawbacks, as can be seen from Table 8.1, in which the advantages and disadvantages of cell systems and multienzyme systems for coenzyme regeneration are compared.

The isolation of biocatalysts for bioconversion is generally cheaper if whole cells are employed. A deactivation in using whole cells can be compensated for by applying additional cells or by controlled growth in the reactor. Retention of the coenzymes does not present any additional problems for the continuous operation usually desired if the cells themselves are retained in a reactor with continuous flow-through. If a whole chain of reactions with numerous individual steps has to proceed in order to achieve the desired product, then whole cells are probably superior in principle, since multienzyme systems become too complicated for more than one or two regeneration steps in adjusting activity and stability. The main disadvantage in using whole cells is the danger that secondary reactions will occur. Owing to mass transport problems, the achievable space–time yield tends in principle to be more limited when whole cells are used than when multienzyme systems are used. Processing usually proves to be more difficult in cell systems, since in addition to by-

TABLE 8.1 Advantages and Disadvantages of Microorganisms Versus Isolated Enzymes for Biotransformations

Cells		Isolated Enzymes	
Advantages	*Disadvantages*	*Advantages*	*Disadvantages*
cheap	secondary reactions	no secondary reactions	expensive
self-reproducing	mass transport problems	no mass transport problems	not directly capable of reproduction
coenzyme regeneration solved	limited space–time yield	high space–time yields	coenzyme regeneration difficult
also suitable for reaction chains	difficult downstream processing (pyrogenes possible)	simple downstream processing (free of pyrogenes)	suitable for only a few reaction steps
	limited reproducibility	good reproducibility	
	regeneration substrate can also be consumed in subsidiary reactions (e.g., during cell growth)	stoichiometric utilization of the regeneration substrates	

products, lysis products can also occur. The reproducibility of results is generally better with multienzyme systems, since they are easier to keep sterile and no mutations can occur. In multienzyme systems the regeneration substrate is used stoichiometrically, whereas a consumption of the regeneration substrate in undesirable secondary reactions must also be considered when whole cells are used.

A series of possible reactors for coenzyme regeneration are shown schematically in Fig. 8.1.

The potential of microbial cells can be exploited in a simple manner in a batch reactor with whole cells (a). Both growing cells and resting cells can be used. Product isolation can be achieved by centrifugation or cross-flow filtration (Herbert, 1961; Bhagat and Wilke, 1966; Margaritis and Wilke, 1972; Gerhardt and Gallup, 1963; Gerhardt and Schultz, 1966; Membrana, 1982).

Analogously, isolated enzymes and native coenzymes can be employed in a batch reactor (c). However, product separation while retaining the catalytic system in the reactor is difficult here (Chambers et al., 1974).

Carrier-fixed cells can be used in tube reactors (b) for continuous processes (Yamamoto et al., 1980; Nishida et al., 1979; Oda et al., 1983; Samejima et al., 1978).

There is no advantage in a joint carrier fixation of enzymes together with the necessary coenzyme in a multienzyme system because of the required mobility of the coenzyme between the actual production enzyme and the regeneration enzyme. On the other hand, joint microencapsulation of enzymes and coenzymes has been attempted in order thus to be able to realize a tube reactor for multienzyme systems by the physical enclosure of the catalytic system (d) (Grunwald and Chang, 1979, 1981).

In practice, attempts are being made to replace the natural cell by an artificial cell in the tube reactor. Since the pores of the microcapsules have to be very narrow in order to prevent bleeding off of the coenzyme, mass transport problems arise here so that only low space–time yields can be achieved. While activity losses due to cell growth can be compensated for in the cell reactor (c), an adjustment of the activity after deactivation is basically impossible in this enzyme reactor (d).

Different types of tube reactor have been suggested because of transport problems in joint carrier fixation or in joint physical enclosure. In principle, two methods can be conceived of: either the enzymes (e) or the coenzyme (f) is retained in the tube reactor by fixing to suitable carriers, while the other partial catalytic system has in each case to be separated in a subsequent separator and then fed in again if completely continuous operation is to be achieved. Separation of the soluble coenzyme from the product flow proves to be difficult because of the slight difference in size. Separation via adsorption or ion exchange is possible (Gardner et al., 1974; Fink and Rodwell, 1975).

Separation of the enzymes in a subsequent membrane reactor is more easily realizable, but nonspecific adsorption of the soluble enzyme on the carrier for coenzyme fixation must be expected with this type of reactor. The steric prob-

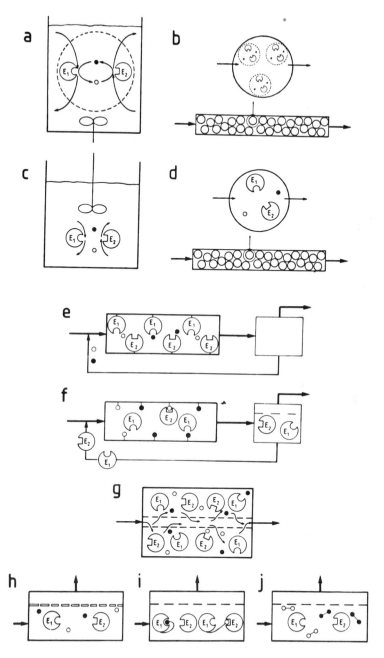

FIGURE 8.1   Reactors for coenzyme regeneration.

lems are smaller than with reactor type (d) but can still have a limiting effect in the case of reactor type (f).

With the membrane reactor (g) there are no longer any steric problems for interaction between enzymes and coenzymes; however, a mass transport problem occurs here, since only a diffusive exchange via a separating membrane is possible for the substrate and the product, and this must be designed in such a way with respect to its pore size that the coenzyme does not bleed out (Miyawaki et al., 1982a, b). For this reason, attempts are also being made to replace diffusion transport via the membrane by a forced convection–membrane reactor (h). The transport problem can thus be solved in principle; however, no commercial membranes with good retaining properties for native coenzymes and with a simultaneous low retention of comparably large product molecules are known.

By binding the coenzyme to the production enzyme or regeneration enzyme via a spacer a coenzyme loss can already be avoided in a membrane reactor (i) by utilizing ultrafiltration membranes (Manson et al., 1982). However, by combining the enzyme and coenzyme the catalytic system becomes inactive if only one of the partners is deactivated.

This problem can be avoided in a membrane reactor (j) by binding the coenzyme to water-soluble polymers. Mobility of the coenzyme is retained. Retentive capability is guaranteed by employing suitable polymers (Mosbach et al., 1976; Wykes et al., 1975; Weibel et al., 1974; Zapelli et al., 1975, 1976; Yamazaki and Maeda, 1981; Yamazaki et al., 1976; Coughlin et al., 1975; Chambers et al., 1974; Furukawa et al., 1980; Bueckmann et al., 1981; Wichmann et al., 1981; Pace et al., 1976; Miyawaki et al., 1982a; Asada et al., 1978; Le Goffic et al., 1980; Gardner, 1978; Wandrey et al., 1981, 1982).

The mass transport problems occurring during diffuse exchange across the membrane can be avoided by membrane reactors with forced convection across the membrane; but owing to a flow of liquid toward the membrane, an increased concentration of the enzymes or polymer-bound coenzymes results in front of the membrane (concentration polarization). This concentration can progress to such an extent that the solubility limit is reached. A filter cake (secondary membrane) is then in principle formed from the components that are usually soluble.

In order to avoid concentration polarization a different component must be superimposed on the flow component toward the membrane so that the effect of concentration polarization remains controllable (Fig. 8.2).

In principle, a relative velocity between the catalyst and the membrane must be generated. Moreover, with the aid of a circulating pump the catalyst can be moved past the membrane at high velocity—Fig. 8.2a. Alternatively, the membrane can be rotated so fast in the reactor—Fig. 8.2b—that the concentration polarization is limited (Bhagat and Wilke, 1966; Margaritis and Wilke, 1972). The substrates—and, if required because of deactivation, the enzymes or the coenzyme—are fed into the reaction chamber via a sterile filter with the aid of dosing pumps. The ultrafiltered product solution leaves the system con-

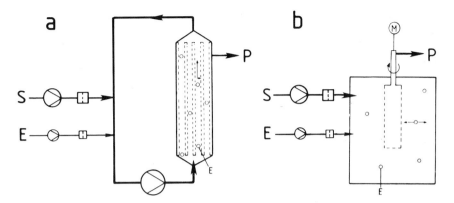

FIGURE 8.2   How to avoid concentration polarization. S.: substrate, E.: enzyme, P.: product.

tinuously. As far as the reaction technique is concerned, both cases are systems with extensive backmixing.

IV. CHARACTERIZATION OF THE COENZYME DERIVATIVES

The following representation is limited to systems with coenzymes bound to water-soluble polymers, since high cycle numbers can thus be achieved in the membrane reactor with a simultaneous high space–time yield.

Dextran, polyethylene glycol, and others were used as water-soluble polymers (Mosbach et al., 1976; Wykes et al., 1975; Weibel et al., 1974; Zapelli et al., 1975, 1976; Yamazaki and Maeda, 1981; Yamazaki et al., 1976; Coughlin et al., 1975; Chambers et al., 1974; Furukawa et al., 1980; Bueckmann et al., 1981; Wichmann et al., 1981; Pace et al., 1976; Miyawaki et al., 1982a; Asada et al., 1978; Le Goffic et al., 1980; Gardner, 1978). Polyethylene glycol has proved to be particularly effective in the following systems, since it is cheap and physiologically harmless. NAD, NADP via phosphorylation, and ATP have previously been bound to PEG. Batches of up to 200 g of NAD together with up to 18,000 g of polyethylene glycol 20,000 were used. A coenzyme activity of 75% relative to the native coenzyme employed resulted for all conversion steps (Bueckmann, 1979; Bueckmann et al., 1983). Such a batch of coenzyme derivative is sufficient, for example, to produce 1 ton of L-leucine from α-ketoisocaproate (see below).

Besides the type of polymer and the yield thus attainable during derivation, the size of the polymer is also decisive for coenzymatic activity and retentive capability in the membrane reactor.

In biotransformations with the aid of the coenzyme system NAD/ NADH, reductions are of special interest since chirality centers can be generated in this way. The task is thus to regenerate NADH from NAD. The formate–formate dehydrogenase system is especially suitable for this. The regeneration substrate is cheap and physiologically harmless and works as a mild disinfectant. The carbon dioxide originating in the regeneration step does not cause any interference; furthermore, the equilibrium of the regeneration reaction is by far on the product side, so that even with an unfavorable equilibrium position of the actual production reaction, high conversions can be achieved (Schuette et al., 1976; Mathews and Vennesland, 1950). This regeneration system is clearly superior to the ethanol–acetaldehyde system from the point of view of regeneration substrate costs and the equilibrium position. Formate dehydrogenase can be obtained today in large volumes with specific activities up to 2 U/mg and with isolation costs of about 0.1 cent/U (Kroner et al., 1982).

Because of the slight hydrogen solubility in aqueous systems the regeneration system with formate is also to be preferred to the direct regeneration of NADH with hydrogen; NAD/NADH bound to polyethylene glycol 10,000 (PEG 10,000) displays activity comparable to native NAD/NADH, as can be seen from Fig. 8.3. With polymer-bound NAD, FDH even displays an activity twice as high as when using native NAD. The K_m value is the same within the standard deviation. A significantly different picture results with the amino acid dehydrogenases, leucine dehydrogenase (LeuDH) and alanine dehydrogenase (AlaDH). Whereas the V_{max} value is over 80% in comparison with the utilization of native NADH, clearly increased K_m values result — LeuDH: factor 2.5, AlaDH: factor 16.3. In both cases the slight substrate excess inhibition observed when using native NADH disappears. However, in practice the increase in K_m values with the amino acid dehydrogenases is not problematic, since a substrate saturation with respect to the coenzyme can be achieved in the membrane reactor without difficulty.

For a more precise kinetic analysis, formate must be considered as a substrate with FDH and the α-keto acids and ammonia as substrates with the amino acid dehydrogenases. Furthermore, the influences of the reaction products of each subsystem on the respective enzyme system must be considered (Wichmann et al., 1981). As will be shown in Section V, not all the individual kinetic effects have to be measured in isolation in each case. It is also possible to record the kinetics of the coupled reaction. In addition to the activity of the coenzyme, its stability is of equal significance. Figure 8.4 shows a comparison of the results of an incubation test of native NAD+ and PEG–NAD+. It can be seen that the coenzyme derivative is significantly more stable. In this way the lower activity of the coenzyme with the amino acid dehydrogenases is more than compensated for by increased stability.

In addition to activity and stability the retentive capability of the coenzyme derivative must also be examined. Very high demands are made on the retention, since in using soluble enzymes high-volume activities can be employed, so that only short residence times are required to achieve economically interesting conversions. The experiments on coenzyme retention are

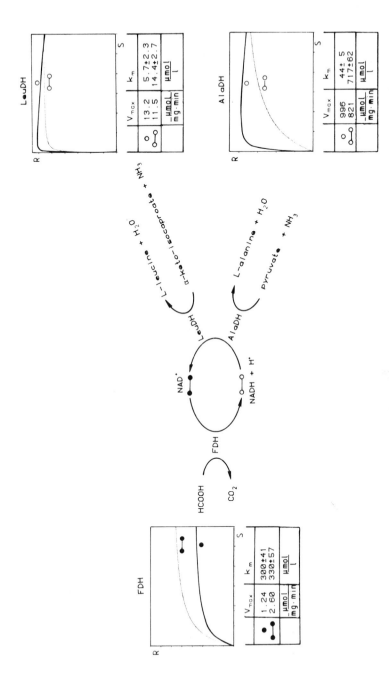

FIGURE 8.3 Activity of native NAD/NADH (●) in comparison to polymer bound NAD/NADH (●———●).

185

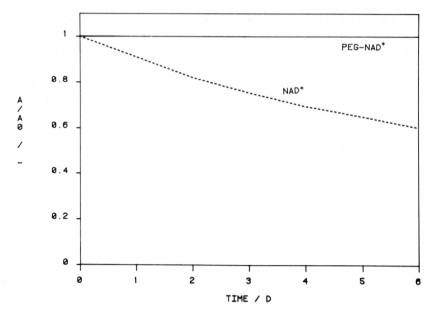

FIGURE 8.4   Stability of NAD + and PEG–NAD + .

therefore carried out in the membrane reactor at a residence time of 1 hour using an Amicon YM5 membrane (nominal exclusion limit 5000).

The retention is defined via the retentate concentration C_R and the filtrate concentration C_F:

$$R = \frac{C_R - C_F}{C_R} \tag{8.1}$$

In the flushing experiment the relative decrease in retentate concentration is described as a function of time by Eq. (8.2):

$$\frac{C_R}{C_{R_0}} = e^{-\frac{1-R}{\tau} \cdot t} \tag{8.2}$$

An apparent deactivation constant can thus be defined by means of elution losses K_{Elu}:

$$K_{Elu} = \frac{1 - R}{\tau} \tag{8.3}$$

It can be seen that the elution loss depends on both the retention and the residence time. This circumstance differs clearly from the application of ultrafiltration membranes, for example, in enzyme concentration, which in principle are only operated for one residence time. A retention of 99% is in this case completely sufficient. On the other hand, a retention of 99% in the enzyme membrane reactor, when operating with a residence time of one hour, is completely unsatisfactory, as shown in Fig. 8.5. A retention of 99% is equivalent to an elution loss of 24% per day. In contrast, PEG 10,000 NAD(H) displays a retention of 99.82%, which corresponds to a retention loss of 4.3%/d. Even better retentions can be achieved with PEG 20,000 NAD(H); 99.93% is achieved here. In this way the elution loss is reduced to 1.7%/d.

V. KINETICS OF THE COUPLED REACTION

After it had been shown that the coenzyme derivatives were active, stable, and could be retained in the membrane reactor, it appeared appropriate to investigate not only the kinetics of the production reaction but also the kinetics of the coupled reaction for a later reactor design. A membrane reactor as loop reactor was used for this purpose.

Figure 8.6 shows the corresponding arrangement. The filtrate flow, free of enzymes and coenzymes, is fed through a flow polarimeter cuvette for continuous analysis. This flow is fed in again at the pressure side of the membrane with the aid of a circulating pump. The amino acid concentration can be polarimetrically observed during reductive amination. The concentration of PEG–NADH can be continuously photometrically measured at 338 nm.

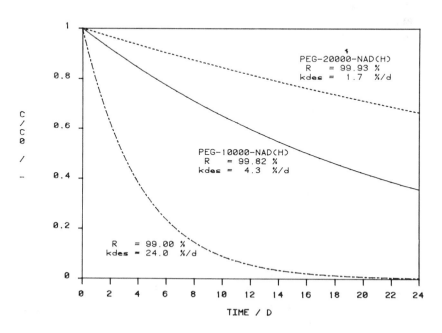

FIGURE 8.5   Retention of PEG–NAD(H).

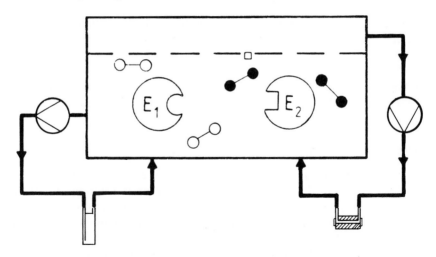

FIGURE 8.6   Membrane-loop reactor.

The results of a typical experiment are shown in Fig. 8.7. The increasing substrate conversion is first plotted against the duration of the experiment as well as the concentration of NADH relative to the total coenzyme concentration. The derivations with respect to time are shown alongside to the right or left.

If the experiment is started with PEG–NAD, then the NADH concentration first rises steeply until an approximately constant ratio of the reduced and oxidized forms of the coenzyme is achieved. The concentration of the amino acid then increases approximately linearly with time. Only shortly before reaching a practically quantitative conversion does the corresponding curve break off owing to substrate limitation. During this experimental period the NADH concentration increases rapidly, since with an excess of formate the NADH-consuming reaction gradually recedes more and more owing to exhaustion of the keto acid reservoir. In the mean conversion range the NADH concentration drops slightly, since the slight substrate excess inhibition caused by the α-keto acid recedes more and more with increasing conversion. In this way NADH-consuming reaction becomes temporarily relatively faster so that the level of NADH decreases slightly until this trend is gradually reversed owing to substrate limitation.

The rate of the production reaction first rises steeply until the correct NADH/NAD ratio has been reached. After this the reaction rate remains practically constant over a very wide conversion range (reaction of the zero order); a high degree of substrate saturation is present here. The reaction rate drops again only at very high conversions owing to substrate limitation.

The course of the NADH formation rate is somewhat more complex. NADH is first formed from NAD at a great velocity until the "correct" NADH/NAD ratio has been reached. This ratio is then changed in favor of

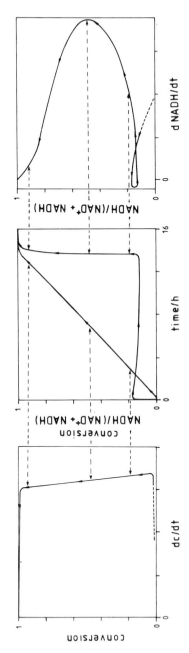

FIGURE 8.7 Concentration–time curves together with their derivation with respect to time (results in a membrane-loop reactor). 100 mmol/l α-ketoisocaproate, 400 mmol/l NH_4-formate, 0.5 mmol/l PEG-20,000–NAD(H), 2.8 U/ml FDH, 3.1 U/ml LeuDH, $pH_0 = 8.0$, $T = 25°C$.

NAD (negative NADH formation rate), since the substrate excess inhibition by α-keto acid is gradually reduced. The NADH formation rate becomes positive once again if the competition to the NADH-consuming reaction gradually recedes owing to substrate limitation in comparison to the NADH-forming reaction (regeneration reaction). Owing to this, the NAD concentration in the reaction mixture drops increasingly. By draining the NAD reservoir a "substrate limitation" gradually results, so the NADH formation rate is reduced again, whereas the NADH concentration continues to increase until all the NAD is reduced to NADH.

A course formally results for the production reaction that seems as though it could be described with Michaelis-Menten kinetics, although the ratios are considerably more complex here owing to coupling of the regeneration reaction. Since the reaction rate of the production reaction decreases only at very high conversions, the membrane reactor, which in principle displays the behavior of a continuous stirred tank reactor, proves to be a suitable reactor configuration. For given enzyme and coenzyme as well as substrate concentrations the achievable conversion can be numerically determined by plotting the corresponding convection straight lines (Wichmann et al., 1981). The kinetics of the coupled reaction measured in the membrane loop reactor can thus be easily used for designing the reactor.

The initial phase in the membrane loop reactor is shown once again in Fig. 8.8 with a greater temporal resolution. The course of the NAD concentration relative to the total concentration of the coenzyme is also plotted. It can be

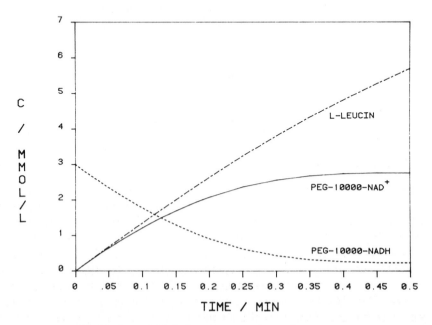

FIGURE 8.8 Initial phase in a membrane-loop reactor.

seen that, starting from pure PEG–NAD, the quasi-stationary state with respect to NAD and NADH is achieved fairly rapidly. Only when this is the case does the amino acid concentration rise practically linearly.

Both measurements of the initial reaction rates (cf. Fig. 8.3) as well as the curves from Fig. 8.7 can be used to determine the kinetic parameters. In both types of experimental procedure the reaction rates can be determined practically on-line (Wichmann and Wandrey, 1980). In order to evaluate the curves according to Fig. 8.7, the set of differential equations corresponding to Eqs. (8.4)–(8.8) must be solved simultaneously, for example, with a Runge Kutta procedure (Wichmann, 1981).

$$\frac{d\text{NAD}}{dt} = -\frac{d\text{NADH}}{dt} = R_1 \cdot E_1 - R_2 \cdot E_2 \tag{8.4}$$

$$\frac{dS_1}{dt} = \frac{(S_{10} - S_1)}{\tau} - R_1 \cdot E_1 \tag{8.5}$$

$$\frac{dS_2}{dt} = \frac{(S_{20} - S_2)}{\tau} - R_1 \cdot E_1 \tag{8.6}$$

$$\frac{dS_3}{dt} = \frac{(S_{30} - S_3)}{\tau} - R_2 \cdot E_2 \tag{8.7}$$

$$\frac{dP}{dt} = \frac{(P_0 - P)}{\tau} + R_1 \cdot E_1 \tag{8.8}$$

E_1	LeuDH (leucine dehydrogenase)
E_2	FDH (formate dehydrogenase)
R_1	Rate of LeuDH
R_2	Rate of FDH
S_1	α-ketoacid (α-ketoisocaproate)
S_2	ammonium
S_3	formate
P	amino acid (L-Leucine)

VI. REACTOR DESIGN

In designing the reactor the pH value and temperature are first defined on the basis of the activity and stability of the enzymes and coenzyme involved so that optimum product volume-specific enzyme costs and coenzyme costs result. Maximizing the activity can be dispensed with if stability can be increased in this way. In the case of the FDH/LeuDH–NADH/NAD system, pH 8 and T = 25°C were used. In principle, determination of the pH value and the temperature is an iterative process that must be aligned to the limiting components. The pH value is primarily determined by the pH optimum of the enzyme with low specific activity (FDH) and the temperature by the temperature stability of the coenzyme.

Optimum activity utilization of the enzymes depends on the correct ratio of the activity of the production enzyme to the activity of the regeneration enzyme. If this ratio is predetermined, then the space–time yield can be linearly increased in a first approximation by raising the enzyme concentration, as long as a substrate limitation does not begin at very high conversions (cf. Fig. 8.7). The conversion increase can, of course, also be achieved by extending the residence time, since the conversion is in the first approximation a function of the enzyme concentrations and residence time (cf. Eq. (8.9)):

$$U = f(E_1 \cdot \tau, \quad E_2 \cdot \tau) \tag{8.9}$$

The space–time yield can be increased in the range of coupled kinetics only by increasing the coenzyme concentration if there is not yet any substrate saturation at the enzymes with respect to the coenzyme (cosubstrate!). Owing to the small K_m values with respect to NADH and NAD, even slight coenzyme concentrations are sufficient for substrate saturation (cf. Fig. 8.3).

If the space–time yield is to be raised, for example, by doubling the enzyme concentrations, then if substrate saturation with respect to the coenzyme is already present, the coenzyme concentration practically need not be increased because, taking the molar ratios (enzymes and coenzyme) into consideration, the bulk phase concentration of the coenzyme hardly decreases when the enzyme concentrations are doubled.

These ratios are elucidated once again by Figs. 8.9 and 8.10. At high enzyme concentration and extremely small coenzyme concentration a queue of enzymes is formed waiting for reaction with the next free coenzyme molecule. However, in practice these conditions hardly ever occur because of the molar ratios. On the other hand, at a comparatively high coenzyme concentration a queue of coenzyme molecules is formed waiting for the next free enzyme molecule. In this case the space–time yield can be increased by raising the enzyme concentration (cf. above). Figure 8.10 describes this once again with a different representation. The achievable conversion for the predefined residence time is plotted against the activity ratio of the activity of the regeneration enzyme and the total of the activities of the regeneration and production enzymes. The parameter is the coenzyme concentration. Since the activities are defined under initial reaction rate conditions and since the activity of the two enzymes involved changes differently with increasing conversion, the optimum activity ratio is shifted farther and farther away from the value of 0.5 with increasing conversion. Furthermore, Fig. 8.10 shows schematically how the maximum achievable conversion can be clearly increased at first by raising the coenzyme concentration until a further increase results in only a slight conversion rise.

These ideas can be experimentally corroborated if at a given activity ratio and given residence time, even with saturation of the enzymes with coenzyme, the range of the reaction of zero order for the coupled reaction has not yet been left. Under these conditions an increase in conversion as a function of the coenzyme concentration corresponds to a coenzyme saturation curve of the

Surplus of enzymes

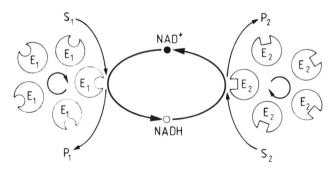

Surplus of coenzyme

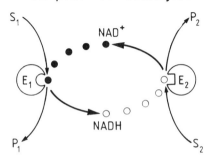

FIGURE 8.9 Influence of enzyme concentration and coenzyme concentration.

coupled reaction (cf. Fig. 8.11). The results shown in Fig. 8.11 were obtained with the FDH/HicDH system (see below). It becomes apparent from Fig. 8.11 that a total coenzyme concentration of approx. 0.3 mmol/l already effects an extensive saturation of the enzymes with coenzyme.

The optimum enzyme and coenzyme concentrations were calculated as a function of the residence time for the FDH/AlaDH system, taking into consideration the activity and stability of the catalytic components involved (Fiolitakis and Wandrey, 1982).

As can be seen from Fig. 8.12, if the residence time is reduced, the concentration of the enzymes involved must be increased approximately proportionately (cf. Eq. (8.9)) in order to maintain an economic conversion (substrate utilization!). On the other hand, the coenzyme concentration hardly needs to be raised if the residence time is reduced. The increase in coenzyme concentration resulting during the optimization computation at extremely short residence times can be traced back to the product volume specific enzyme costs, since at very short residence times and at correspondingly high enzyme

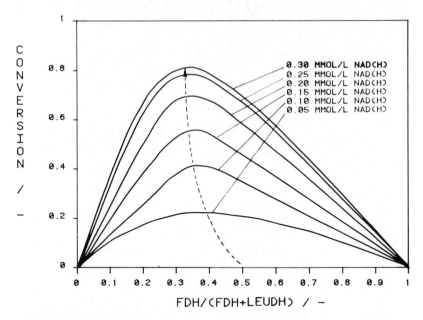

FIGURE 8.10   Influence of enzyme concentration and coenzyme concentration (schematic).

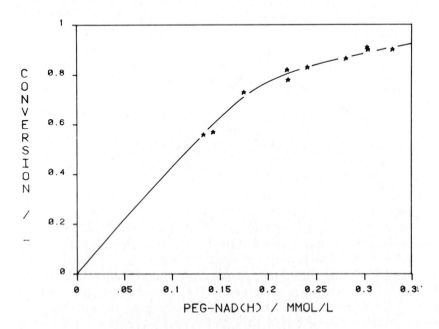

FIGURE 8.11   Coenzyme saturation.

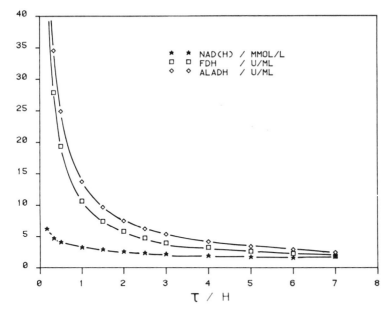

FIGURE 8.12 Optimization of enzyme and coenzyme concentration as function of residence time (system: FDH/AlaDH).

concentration the loss of enzyme rises approximately proportionately to the enzyme concentration (because retention is never 100%). Owing to the high number of cycles achievable with respect to the coenzyme (see below), economic optimization indicates that under such extreme conditions it is favorable to still slightly increase the activity exploitation of the enzymes by the most extensive possible coenzyme saturation.

An excess of coenzyme (queuing effect!) means that each coenzyme molecule is used less frequently than at a lower coenzyme concentration, and therefore the specification of turnover numbers with respect to the coenzyme alone provides only limited information. Conditions for the FDH/AlaDH system are illustrated in Fig. 8.13.

The turnover numbers referring to the production enzyme and the coenzyme are plotted against the conversion and the coenzyme/enzyme ratio in a three-dimensional representation. The highest turnover numbers with respect to the coenzyme are achieved with a coenzyme deficit and low conversion. Once again, the moderate substrate excess inhibition by α-keto acid can be seen from the representation.

For economic optimization, specification of a cycle number is more important than the specification of turnover numbers. What is meant by the cycle number is the number of product molecules formed on the average per coenzyme molecule until the coenzyme molecule is deactivated on a statistical average or becomes lost via the membrane.

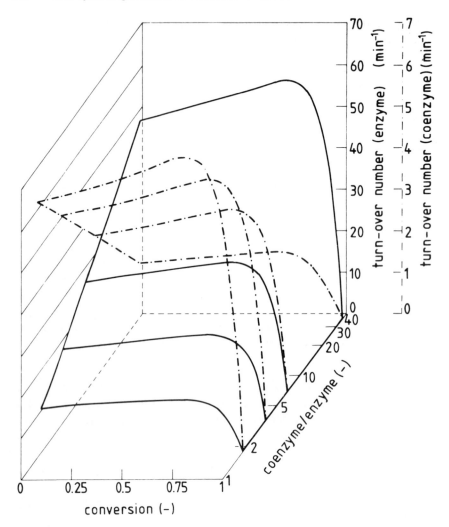

FIGURE 8.13   Turnover numbers as function of conversion and coenzyme/enzyme ratio (system: FDH/AlaDH).

With respect to the substrate concentrations, limitations usually result owing to the solubility of the substrates or products. Substrate excess inhibitions are not of any significance if the range of substrate excess inhibition can be "skipped over" by adjusting a high conversion in the membrane reactor, which behaves like a continuous stirred tank reactor with high back-mixing. Owing to the comparatively high K_m value for formate (cf. Fig. 8.3), the reaction rate of the regeneration reaction can be increased by high formate concentrations. At the same time this leads to a constantly sufficient excess of formate. The velocity of the production reaction can also be increased during reductive amination by raising the ammonium concentration. In the FDH/LeuDH system, for example, the ammonium concentration enters into the kinetics practically linearly. In other cases (FDH/AlaDH) a high excess of ammonia no longer increases activity. A saturation effect with respect to ammonium concentration thus formally results.

The membrane reactor with a sterile filter permits, in principle, sterile operation by sterilizing the reaction chamber before loading the reactor with enzymes and coenzymes and only adding the catalytic components afterwards via a sterile filter through which the enzymes and coenzyme can pass and which at the same time provides protection against microbial contamination. During partial deactivation of one of the catalytic components the required activity of the corresponding component can be supplemented in running operation at a constant residence time via a sterile filter so that operation can be maintained at a constant space–time yield for long periods of time (see below).

In adding enzymes or coenzyme in order to maintain a constant volume activity the total protein concentration or the total coenzyme concentration increases, in the course of which (at constant activity) the active fraction with respect to the total concentration is relatively reduced. In this way, technological limits result as a function of the initial concentration of the catalytic components if the effect of concentration polarization comes more and more into play by enzyme supplementation. Owing to these conditions, an optimum initial concentration of the catalytic components exists and thus also a certain reactor size for a given conversion. With low enzyme concentrations, continuous operation can indeed be maintained for a long time, but for a given conversion a large reactor volume then becomes necessary. Experience shows that low protein concentrations and large reactor surfaces cause low catalyst stability. On the other hand, too high initial enzyme concentration results in a rapid formation of secondary membranes during redosing (concentration polarization) (Flaschel et al., 1983). Concentration polarization can at a given volume flow also be limited by using large membrane surfaces. An optimum for the membrane surface to be used per unit volume flow also exists here because of the membrane costs and because of potential deactivation of part of the catalyst at the membrane surface. Finally, when hollow-fiber membranes or thin-channel systems are used, the channel length assumes considerable significance.

The channel length should theoretically be as short as possible in order to effect a small pressure drop along the channel and thus to achieve a uniform utilization of the membrane surface along the flow channel. In practice, channel lengths of 0.2–1.0 m have proved to be effective.

Scaling-up presents no basic problems, since the membrane reactor behaves in principle like a continuous stirred-tank reactor so that there is no primary scale influence. However, in scaling-up the ratio of surface to volume is favorably changed in such a way that potentially deactivating surfaces decrease in relation to volume.

For downstream processing, coenzyme technology in the membrane reactor makes it possible to provide an ultrafiltered product flow at a constant product concentration so that the processing steps can be optimally adjusted. Since, for example, in the reductive amination of α-keto acids the substrate can be practically quantitatively used, as opposed to the case of methods for racemate resolution of amino acids, problems with the derivatization of amino acid and reracemization of the nonconverted enantiomer do not occur.

VII. EXAMPLES OF EXPERIMENTAL APPLICATION

After defining the theoretically optimum reaction conditions for the continuous production of optically active amino or hydroxy acids in a membrane reactor with respect to the pH value, the temperature, the substrate and cosubstrate concentrations, the coenzyme concentration, the enzyme activities, and the residence time, the practical stability and capacity of the reaction system were experimentally determined in a small model membrane reactor. A flowchart of the model reactor is shown in Fig. 8.14, and a photograph in Fig. 8.15.

The substrate solution is fed into the reactor through a sterile filter. The product solution leaves the reactor through an ultrafiltration membrane (type YM 5, Amicon, Lexington, MA, U.S.A.) installed in the reactor outlet, which retains the enzymes and the polymer-bound coenzyme in the reactor.

A sterile addition of enzyme or coenzyme solution in order to maintain corresponding catalyst activities in the reactor can be undertaken between the substrate pump and the sterile filter. A measurement of pressure in front of the inlet to the reactor makes it possible to monitor flushing of the enzymes and coenzyme into the reactor. Furthermore, an increased coating of the membrane with polymers can be observed after enzyme supplementation.

Coating of the ultrafiltration membrane is limited in the reactor by magnetically coupled agitation. The reaction solution of the reactor is pumped through a photometer cuvette in an external loop. The solution in this external circuit of the reactor is of the same composition as in the reactor; in this way it becomes possible to determine the concentration of the reduced coenzyme in the reactor by measuring the extinction at 338 nm, in comparison to the ultrafiltered product solution. If the two three-way valves are now reversed, the cuvette becomes a small batch reactor in which the reaction solution can react out up to a state of equilibrium. A reaction course as already depicted in Fig. 8.7 thus results. The active coenzyme concentration and the enzyme activities, constituting the optimizing parameters, are obtained as a result of calculating a corresponding concentration–time course from the mass balance for the reaction system and adjustment to the measured data. A sample was also taken from the reactor for calibration in order to determine the effectiveness factor of the enzymes (Wichmann and Wandrey, 1982).

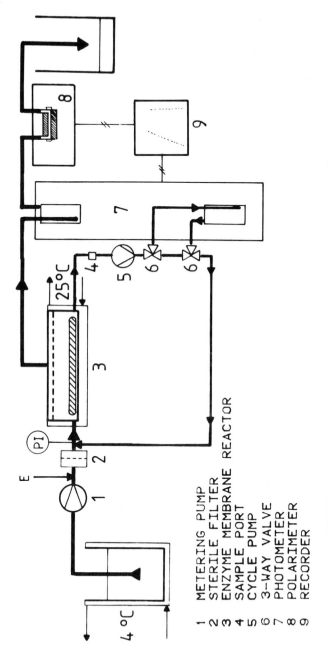

1 METERING PUMP
2 STERILE FILTER
3 ENZYME MEMBRANE REACTOR
4 SAMPLE PORT
5 CYCLE PUMP
6 3-WAY VALVE
7 PHOTOMETER
8 POLARIMETER
9 RECORDER

FIGURE 8.14 Flowchart of a laboratory-scale membrane reactor experiment.

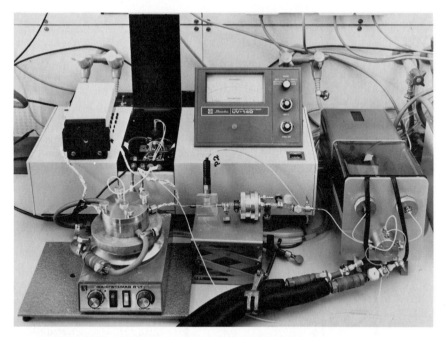

FIGURE 8.15 Membrane reactor (left) with substrate pump (right) and spectrophotometer for the continuous analysis of NADH.

The optical rotary value of the product solution is measured in a polarimeter in order to determine the conversion in the reactor. In the production of optically active hydroxy or amino acids from the corresponding keto acids this is quite simple, since the product is the only optically active compound in the product solution.

Before the experiment is started, the reactor is chemically sterilized with 70% aqueous ethanol or 0.1% peracetic acid, then flushed with water and substrate solution. The experiment is started by adding the enzymes and coenzyme, each in the form of an aqueous solution. The volume flow of the substrate solution, and thus also the residence time, is kept constant.

The results of continuous L-leucine production from α-keto isocaproate are shown in Fig. 8.16. Conversion as a function of time is shown for each of three experiments. Parameters and results are given for each test in tabular form.

In the first test (Fig. 8.16 top) a conversion of almost 100% was achieved over a period of 4 weeks. This was achieved by using a higher coenzyme concentration, higher enzyme activities, and a longer residence time in comparison with the other two tests. Since no enzyme or coenzyme was supplemented in the course of the experiment, the conversion dropped almost exponentially in the last few days of the experiment. In this case the drop in conversion was

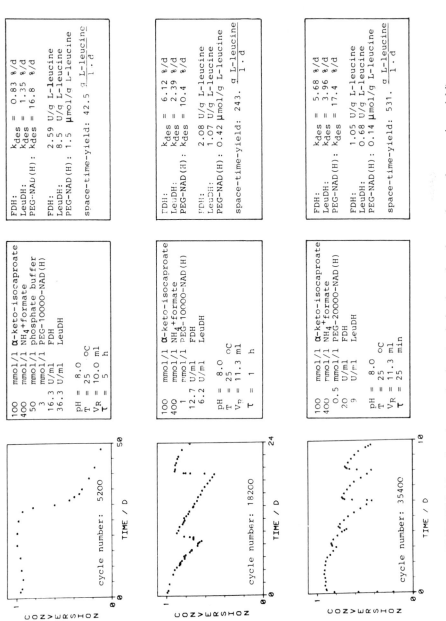

FIGURE 8.16 L-Leucine production (optimization of cycle number and space–time yield).

201

caused by a loss of active coenzyme in the reactor, which largely proceeds by deactivation depending on a reaction of the first order. In 48 days, 20.3 g of L-leucine was produced.

In contrast to this experiment a significantly shorter residence time and lower enzyme activities and coenzyme concentration were used in the second experiment (Fig. 8.16 center). The maximum conversion was initially also almost 100%, but it soon dropped in accordance with deactivation of the enzymes and the coenzyme. After the ninth and nineteenth days the deactivated enzymes and coenzyme were supplemented. The overall lower enzyme protein concentration did lead to a somewhat higher deactivation rate of the enzymes, but this can be contrasted with a higher space–time yield. In 24 days, 65.6 g of L-leucine were produced.

With a further shortened residence time, increased enzyme activities, but reduced coenzyme concentration, a further increase in the space–time yield was successfully achieved in another experiment (Fig. 8.16 bottom). Deactivation of the enzymes did not change significantly, whereas the effectiveness of supplementation of the coenzyme became less and less. This means that in spite of only a slight decrease in enzyme activities the effect of coenzyme supplementation became smaller and smaller. Since the largest fraction of deactivated coenzyme is not flushed out of the reactor, the deactivated coenzyme might function as a competitive inhibitor on the enzymes. The K_m values of FDH and LeuDH are apparently increased by this; however, the activity of the enzymes drops. In 9.5 days, 57.0 g of L-leucine was produced.

If ammonia containing the stable nonradioactive nitrogen isotope ^{15}N is used for L-leucine production, labeled ^{15}N L-leucine can be produced (Dolabdjian et al., 1978). In the first part of the experiment (Fig. 8.17) a substrate solution containing ^{15}N ammonium chloride was used. After this solution had been consumed, the experiment was switched over to a substrate solution without ammonium. The conversion thus dropped to 0%. A substrate solution with ^{14}N ammonium was then fed into the reactor. Consequently, the conversion increased up to almost 100%. In contrast to the tests depicted in Fig. 8.16, only a slight excess of ammonium salt was used in this experiment in order to use the valuable ^{15}N ammonium salt as completely as possible for synthesizing ^{15}N L-leucine. An excess of ammonium was nevertheless necessary, as discovered in preliminary experiments, since otherwise the activity of the LeuDH would be very poorly utilized and the residence time would have to be increased very considerably.

Using α-hydroxycarboxylic acid dehydrogenases instead of α-amino acid dehydrogenases, optically active α-hydroxycarboxylic acids can be fabricated from optically inactive α-keto acids using formate hydrogenase for coenzyme regeneration. As an example, the production of L-hydroxy isocaproic acid (L-2-hydroxy-4-methyl pentane acid) from α-keto isocaproic acid is shown in Fig. 8.18.

The same reactor was used as in the tests described above. The deactivation of the coenzyme and the enzymes is at the same level as in the experiments

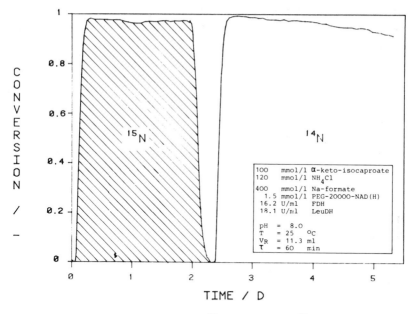

FIGURE 8.17   Production of ^{15}N-L-leucine and ^{14}N-L-leucine.

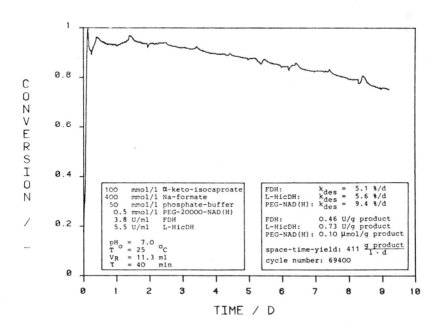

FIGURE 8.18   Production of α-L-hydroxy-isocaproate.

for synthesizing L-leucine. No coenzyme or enzymes were supplemented during the test.

Figure 8.19 presents a survey of the reaction systems already investigated; regeneration of the coenzyme was catalyzed in each case by formate dehydrogenase.

A particularly interesting form of NADH regeneration results during production of L-amino acids from α-hydroxy acids. The NADH consumed in the reductive amination of α-keto acids is regenerated by oxidizing hydroxy acid to α-keto acid (Wandrey et al., 1983).

If, for example, in addition to alanine dehydrogenase a L-lactate dehydrogenase and a D-lactate dehydrogenase are employed in the membrane reactor, then according to this principle the corresponding L-amino acid can be obtained from racemic hydroxy acids.

Analogously to NAD, ATP can also be covalently bound to polyethylene glycol. This results in the possiblity of continuous ATP regeneration in the membrane reactor (Berke et al., 1984). This technique can be used for phosphorylation reactions and also for nonribosomal peptide synthesis.

VIII. OUTLOOK

The cycle numbers achieved with the example of NADH regeneration in the membrane reactor prove that the coenzyme costs for continuous production are not limiting. Technical feasibility has also been successfully demonstrated up to a product scale of 0.5 kg/day (W. Leuchtenberger, Degussa, Hanau, West Germany, personal communication, 1982). However, a production process involving coenzyme regeneration will gain acceptance only if similarly favorable

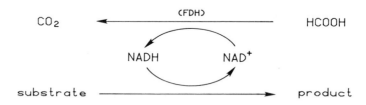

pyruvic acid	(AlaDH)	L-alanine
2-oxo-4-methyl pentanoic acid	(HicDH)	L-2-hydroxy-4-methyl pentanoic acid
phenyl pyruvic acid	(PheDH)	L-phenylalanine
phenyl pyruvic acid	(D-LDH)	D-phenyl lactic acid
2-oxo-4-methyl pentanoic acid	(LeuDH)	L-leucine (^{14}N, ^{15}N)
2-oxo-3-methyl pentanoic acid	(LeuDH)	L-isoleucine + L-allo isoleucine
2-oxo-4-(methylthio) butanoic acid	(LeuDH)	L-methionine
2-oxo-3,3-methyl butanoic acid	(LeuDH)	2-amino-3,3-dimethyl butanoic acid
dehydro proline	(LeuDH)	L-proline

FIGURE 8.19 Systems that have been operated with cofactor regeneration by PEG–NAD(H).

results are achieved at specific catalyst costs on an even larger scale. On the one hand, suitable membranes of an appropriate size must be available; developments in this direction are currently emerging. On the other hand, the enzymes required must be available at costs that can be met by the value added during the reaction step. The enzyme requirements of a production facility should influence the enzyme price per catalytic unit extremely favorably. This has also been experienced with previous processes using hydrolytic enzymes. Even today, enzyme consumptions of less than 1 U/g product are possible. Experience has indicated that these requirements drop further in scaling-up of membrane reactor technology owing to the ratio of surface to volume becoming more favorable.

Efforts at coenzyme regeneration have previously been concentrated on the NAD/NADH, ADP/ATP, and NADP/NADPH systems. It would be extremely desirable to extend this concept to further coenzymes; however, the covalent coupling to water-soluble polymers is apparently very much more difficult in systems such as Coenzyme A, so little is yet known about such approaches.

Since the carrier fixation of whole microorganisms has made great progress in the past few years, it represents ever-growing competition to the field of cofactor regeneration with isolated enzymes, so this process approach should also always be taken into consideration. Apart from large-scale chemical production, coenzyme regeneration is also of significance in analytical or medical applications.

Carrier fixation of principally hydrolytically active enzymes may in the past have narrowed one's vision for applying cofactor-dependent enzymes. The function of the coenzyme as a transport metabolite limits the possibility of applying carrier-fixed enzymes. Since the large majority of enzymatically catalyzed reactions are cofactor-dependent, important advances in enzyme technology should be expected in the future from the further development of carrier-fixed whole microorganisms as well as from the membrane reactor technique.

Symbols

C	mmol/l	concentration
E	mg/ml	enzyme concentration
K	d^{-1}	deactivation constant
P	mmol/l	product concentration
R	–	retention
R	U/ml	rate
S	mmol/l	substrate concentration
t	h, d	time
U	–	conversion
τ	h	residence time

Indices	
0	initial
Elu	elution
F	filtrate
R	retentate

Abbreviations	
AlaDH	alanine dehydrogenase
FDH	formate dehydrogenase
HicDH	hydroxyisocaproate dehydrogenase
LeuDH	leucine dehydrogenase

REFERENCES

Aizawa, M., Coughlin, R. W., and Charles, M. (1976) *Biotech. Bioeng. 18*, 209.

Asada, M., Nakanishi, K., Matsuno, R., Kariya, Y., Kimura, A., and Kamikubo, T. (1978) *Agr. Biol. Chem. 42*(8), 1533.

Berke, W., Wandrey, C., Morr, M., and Kula, M.-R. (1984) in *Enzyme Engineering*, Vol. VII (Laskin, A. I., Tsao, G. T., and Wingard, L. B., Jr., eds.) p. 257, New York Academy of Sciences.

Bhagat, A. K., and Wilke, C. R. (1966) Lawrence Radio Laboratory Report UCRL-16574.

Bueckmann, A. F. (1979) German Patent 28.41.414.

Bueckmann, A. F., Kula, M.-R., Wichmann, R., and Wandrey, C. (1981) *J. Appl. Biochem. 3*, 301.

Bueckmann, A. F., Morr, M., and Kula, M.-R. (1983) paper presented at the Fifteenth FEBS Meeting, Brussels.

Chambers, R. P., Ford, J. R., and Allender, J. H. (1974) in *Enzyme Engineering*, Vol. II (Pye, E. K., and Wingard, L. B., Jr., eds.), p. 195, Plenum Press, New York.

Chambers, R. P., Baricos, W. H., and Cohen, W. (1975) in *Enzyme Technology Grantees—Users Conference* (Pye, E. K., ed.), p. 26, John Wiley, New York.

Coughlin, R. W., Aizawa, M., Charles, M., and Alexander, J. H. (1975) *Biotech. Bioeng. 17*, 515.

DiCosimo, R., Wong, C.-H., Daniels, L., and Whitesides, G. M. (1981) *J. Org. Chem. 46*(22), 4622.

Dolabdjian, B., Grenner, G., Kirch, P., and Schmidt, H.-L. (1978) in *Enzyme Engineering*, Vol. IV (Broun, G. B., Manecke, G., and Wingard, L. B., Jr., eds.), p. 399, Plenum Press, New York.

Fink, D. J., and Rodwell, V. W. (1975) *Biotech. Bioeng. 17*, 1029.

Fiolitakis, E., and Wandrey, C. (1982) in *Proceedings of the Third Rothenburger Symposium* (Lafferty, R. M., ed.), Springer, New York.

Flaschel, E., Wandrey, C., and Kula, M.-R. (1983) *Adv. Biochem. Eng. 26*, 73.

Furukawa, S., Katayama, N., Iizuka, T., Urabe, I., and Okada, H. (1980) *FEBS Lett. 121*(2), 239.

Gardner, C. R. (1978) in *Enzyme Engineering*, Vol. IV (Broun, G. B., Manecke, G., and Wingard, L. B., Jr., eds.), p. 95, Plenum Press, New York.

Gardner, C. R., Colton, C. K., Langer, R. S., Hamilton, B. K., Archer, M. C., and Whitesides, G. M. (1974) in *Enzyme Engineering*, Vol. II (Pye, E. K., and Wingard, L. B., Jr., eds.), p. 209, Plenum Press, New York.

Gerhardt, P., and Gallup, D. M. (1963) *J. Bacteriol. 86*, 919.

Gerhardt, P., and Schultz, J. S. (1966) *J. Ferment. Technol. 44*, 349.

Grunwald, J., and Chang, T. M. S. (1979) *J. Appl. Biochem. 1*(2), 104.

Grunwald, J., and Chang, T. M. S. (1981) *J. Mol. Catal. 11*(1), 83.

Herbert, D. (1961) in *Society of Chemical Industry Monograph* 12, p. 21, Page Brothers, London.

Jones, J. B., Beck, J. F. (1976) in *Application of Biochemical Systems in Organic Chemistry* (Jones, J. B., Shih, C. J., Perlman, G., eds.), p. 107, Wiley-Interscience, New York.

Jones, J. B., Sneddon, D. W., Higgins, W., and Lewis, A. J. (1972) *J. Chem. Soc. Chem. Comm.*, No. 15, 856.

Julliard, M., and Le Petit, J. (1982) *Photochem. Photobiol. 36*, 283.

Kroner, K. H., Schuette, H., Stach, W., and Kula, M.-R. (1982) *J. Chem. Technol. Biotechnol. 32*(1), 130.

Le Goffic, F., Sicsic, S., and Vincent,C. (1980) *Biochimie 62*, 375.

Manson, M. O., Larsson, P.-O., and Mosbach, K. (1982) *Meth. Enzymol. 89*, 457.

Margaritis, A., and Wilke, C. R. (1972) in *Developments in Industrial Microbiology*, p. 159, American Institute of Biological Sciences, Washington, D.C.

Mathews, M. B., and Vennesland, B. (1950) *J. Biol. Chem. 186*, 667.

Membrana (1982), in *ACHEMA 82: Pollution Control* (Behrens, D., ed.), p. C.55. DECHEMA, Frankfurt, West Germany.

Miyawaki, O., Nakamura, K., and Yano, T. (1982a) *Agr. Biol. Chem. 45*(11), 2725.

Miyawaki, O., Nakamura, K., and Yano, T. (1982b) *J. Chem. Eng. Japan 15*(3), 224.

Mosbach, K., Larsson, P. O., and Lowe, C. (1976) in *Methods in Enzymology*, Vol. XLIV (Mosbach, K., ed.), p. 859, Academic Press, New York.

Nishida, Y., Sato, T., Tosa, T., and Chibata, I. (1979) *Enzyme Microbial Technol. 1*, 95.

Oda, G., Samejima, H., and Yamada, T. (1983) in *Biotech 83*, p. 597, Online, Northwood, England.

Pace, G. W., Yang, H. S., Tannenbaum, S. R., and Archer, M. C. (1976) *Biotech. Bioeng. 18*, 1413.

Samejima, H., Kimura, K., Ado, Y., Suzuki, Y., and Tadokoro, T. (1978) in *Enzyme Engineering*, Vol. IV (Broun, G. B., Manecke, G., and Wingard, L. B., Jr., eds.), p. 237, Plenum Press, New York.

Schmarkel, C. O., Santhanam, K. S. V., and Elving, P. T. (1974) *J. Electrochem. Soc. 21*, 345.

Schuette, H., Flossdorf, J., Sahm, H., and Kula, M.-R. (1976) *Eur. J. Biochem. 62*, 151.

Shaked, Z., Barber, J. J., and Whitesides, G. M. (1981) *J. Org. Chem. 46*(20), 4100.

Simon, H., Guenther, H., Bader, J., and Tischer, W. (1981) *Angew. Chem. Int. Ed. Engl. 20*, 861.

Stinson, R. A., and Holbrook, J. J. (1973) *Biochem. J. 131*, 719.

Vandecasteele, J.-P. (1980) *Appl. Environ. Microbiol. 39*(2), 327.

Wandrey, C., Wichmann, R., Leuchtenberger, W., Kula, M.-R., and Bueckmann, A. F. (1981) U.S. Patent 430,4858.

Wandrey, C., Wichmann, R., Leuchtenberger, W., Kula, M.-R., and Bueckmann, A. F. (1982) U.S. Patent 432,6031.

Wandrey, C., Fiolitakis, E., Wichmann, U., and Kula, M.-R. (1984) in *Enzyme Engineering*, Vol. VII (Laskin, A. I., Tsao, G. T., and Wingard, L. B., Jr., eds.) p. 91, New York Academy of Sciences.

Wang, S. S., and King, C. K. (1979) *Adv. Biochem. Eng. 12*, 119.

Weibel, M. K., Fuller, C. W., Stadel, J. M., Bueckmann, A. F., Doyle, T., and Bright, H. J. (1974) in *Enzyme Engineering*, Vol. II (Pye, E. K., and Wingard, L. B., Jr., eds.), p. 203, Plenum Press, New York.

Wichmann, R. (1981), *Spez. Ber. Kernforschungsanlage Juelich*, p. 119, Juelich, West Germany.

Wichmann, R., and Wandrey, C. (1980) in *Enzyme Engineering*, Vol. V (Weetall, H. H., and Royer, G. P., eds.), p. 259, Plenum Press, New York.

Wichmann, R., and Wandrey, C. (1982) in *Enzyme Engineering*, Vol. VI (Chibata, I., Fukui, S., and Wingard, L. B., Jr., eds.), p. 311, Plenum Press, New York.

Wichmann, R., Wandrey, C., Bueckmann, A. F., and Kula, M.-R. (1981) *Biotech. Bioeng. 23*, 2789.

Wratten, C. C., and Cleland, W. W. (1963) *Biochemistry 2*, 935.

Wykes, J. R., Dunill, P., and Lilly, M. D. (1972) *Biochim. Biophys. Acta 286*, 260.

Wykes, J. R., Dunill, P., and Lilly, M. D. (1975) *Biotech. Bioeng. 17*, 51.

Yamamoto, K., Sato, T., Tosa, T., and Chibata, I. (1980) *Biotech. Bioeng. 22*, 2045.

Yamazaki, Y., and Maeda, H. (1981) *Agr. Biol. Chem. 45*(9), 2091.

Yamazaki, Y., Maeda, H., and Suzuki, H. (1976) *Biotech. Bioeng. 18*, 1761.

Zapelli, P., Rossodivita, A., and Re, L. (1975) *Eur. J. Biochem. 54*, 475.

Zapelli, P., Rossodivita, A., and Re, L. (1976) *Eur. J. Biochem. 62*, 211.

Immobilized Enzymes for Clinical Analysis

Isao Karube
Shuichi Suzuki

I. INTRODUCTION

Most analyses of organic compounds are currently performed by spectrophotometric methods. Recently, many of these methods have utilized enzyme-catalyzed reactions because of the specificity of such reactions. However, these methods required complicated procedures and a long reaction time. Therefore electrochemical monitoring of these reactions may have definite advantages. For example, wide concentration ranges are measurable with dilution simply by changing flow rates, and the test sample does not need to be optically clear. Various enzyme electrodes have been described for the assay of organic compounds (Guilbault, 1976; Suzuki et al., 1977; Mosbach, 1976; Gulberg and Christian, 1981; Ianniello and Yacynych, 1981; Nikolelis et al., 1981; Hassan et al., 1981; Matsumoto et al., 1981, 1982; Macholan et al., 1981; Alexander and Joseph, 1981; Tipton et al., 1981; Johnson et al., 1982; Chen et al., 1982; Neujahr, 1982; Mizutani et al., 1983; Kovach and Meyerhoff, 1982; DiPaolantonio and Rechnitz, 1982; Ianniello et al., 1982; Winartasaputra et al., 1982; Vadgama et al., 1982; Nagy et al., 1982; Mizutani and Tsuda, 1982; Alexander and Maitra, 1982; Mascini et al., 1982; Watanabe et al., 1983). These electrodes are based on continuous amperometric or potentiometric detection of enzyme reaction products. Enzyme sensors for clinical analyses are described in this paper.

II. ENZYME SENSOR FOR CARBOHYDRATES

A. Glucose Sensor

Glucose oxidase (20 mg·g^{-1} of carrier) was immobilized on a porous polyvinylchloride (PVC) membrane. The stability of the glucose oxidase was improved with immobilization. The glucose oxidase–PVC membrane was applied to a glucose sensor (Suzuki et al., 1977; Mosbach, 1976; Gulberg and Christian, 1981; Ianniello and Yacynych, 1981; Nikolelis et al., 1981; Hassan et al., 1981; Matsumoto et al., 1981, 1982; Macholan et al., 1981; Alexander and Joseph, 1981; Tipton et al., 1981; Johnson et al., 1982; Chen et al., 1982; Neujahr, 1982; Mizutani et al., 1983; Kovach and Meyerhoff, 1982; DiPaolantonio and Rechnitz, 1982; Ianniello et al., 1982; Winartasaputra et al., 1982; Vadgama et al., 1982; Nagy et al., 1982; Mizutani and Tsuda, 1982; Alexander and Maitra, 1982; Mascini et al., 1982; Watanabe et al., 1983; Hirose et al., 1979).

A schematic diagram of the glucose sensor is given in Fig. 9.1. The sensor consisted of a double membrane, one layer of which was a glucose oxidase–PVC membrane and the other an oxygen-permeable Teflon membrane (40 μm), and alkaline electrolyte (KOH 30%), a platinum cathode, and a lead anode. The double membrane was attached directly to the platinum cathode and was tightly secured to the electrode with rubber rings. A nylon net was used as a support.

Six ml of 0.066 M phosphate buffer (pH 5.6) was placed in a suitable vessel and saturated with oxygen by bubbling with air. After the current became steady, 0.5 ml of various concentrations of β-D-glucose were injected into the vessel. The current changes were displayed on a recorder.

Figure 9.2 shows typical response curves obtained with various concentrations of β-D-glucose. The current at zero time is that obtained in a solution saturated with dissolved oxygen. Oxygen consumption by the enzymatic reaction, which resulted in a decrease in dissolved oxygen around the enzyme membrane, began when a sample solution containing glucose was injected into the system. As a result, the electrode current decreased markedly with time until a steady state was reached. The steady state indicated that the consumption of oxygen by the enzymatic reaction and the diffusion of oxygen from the solution to the membrane were in equilibrium. Assays were completed within 1 min by the steady state method and within 15 s by the kinetic method. The steady state method was employed in subsequent work.

The basic properties of the porous membrane, such as thickness and structure, affect the current output of the glucose sensor. An increase in thickness of the enzyme membrane resulted in a decrease of electrode current. Conse-

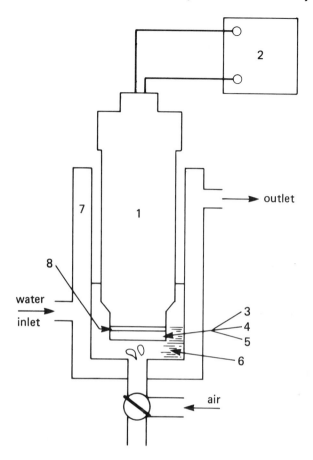

FIGURE 9.1 Schematic diagram of the enzyme electrode for the determination of glucose. 1: Oxygen electrode, 2: recorder, 3: Teflon membrane, 4: glucose oxidase–polyvinylchloride membrane (40 μm), 5: nylon net, 6: sample solution, 7: water jacket, 8: O-ring.

quently, a thin enzyme membrane is desirable for the glucose sensor. However, a thin membrane (e.g., 25-μm thickness) was too weak for practical purposes, and a 40-μm membrane was employed.

The glucose concentration was determined by rate and steady state assay. Figure 9.3 shows the relation between the current or the rate of current change and the glucose concentration below 400 mg·dl^{-1}. The results obtained by each method seem to be identical at glucose concentrations lower than 400 mg·dl^{-1}. Therefore a sample solution containing more than 400 mg·dl^{-1} of glucose must be diluted with buffer.

The current reproducibility was examined by using the same sample. The current was reproducible within ±7% of the relative error when a sample solu-

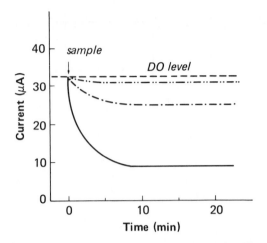

FIGURE 9.2 Response curves of the enzyme electrode. The measurements were carried out under standard conditions. Glucose concentrations were: 300 mg·dl^{-1} (solid curve); 100 mg·dl^{-1} (dashed curve); 20 mg·dl^{-1} (dot-dashed curve).

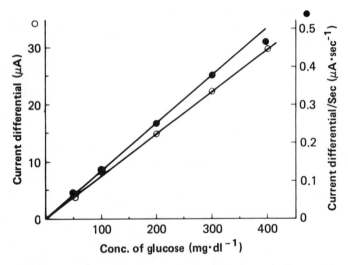

FIGURE 9.3 Calibration curves of the enzyme electrode. Preparation of the enzyme membrane and the determination were performed under standard conditions.

tion containing 400 mg·dl^{-1} of glucose was employed. The standard deviation for 77 experiments was 23 mg·dl^{-1}.

The reusability of the glucose sensor was examined by using a 400 mg·dl^{-1} glucose solution. The current output of the sensor remained almost constant for two months and 85 assays. Therefore the glucose oxidase–PVC membrane sensor can be used for a prolonged period for the assay of glucose.

Glucose concentrations of fresh sera were determined by the proposed method and by the conventional method. Satisfactory comparative results were obtained. The correlation coefficient was 0.93. This indicated that the immobilized enzyme systems proposed provide an economical and reliable method for the enzymatic assay of glucose in serum. Furthermore, many sensors have been developed for the determination of glucose (Hirose et al., 1979; Updike et al., 1979; Liu et al., 1979, 1981; Koyama et al., 1980; Sokol et al., 1980; Enfors, 1981; Tsuchida and Yoda, 1981; Gondo et al., 1981).

B. Sucrose Sensor

The measurement of sucrose is important not only in industrial analysis but also in clinical analysis. A sucrose sensor using a collagen membrane containing invertase, mutarotase, and glucose oxidase and an oxygen electrode was described by Satoh et al. (1976).

The hydrolysis of sucrose in the presence of invertase is well known (see Eq. (9.1)). The glucose component of sucrose is α-D-glucose. The specific substrate of glucose oxidase, however, is β-D-glucose. The mutarotation of α-D-glucose is attained via membranes containing mutarotase (see Eq. (9.2)). The oxidation of β-D-glucose with dissolved oxygen in the presence of glucose oxidase is shown in Eq. (9.3).

$$\text{sucrose} + H_2O \xrightarrow{\text{Invertase}} \alpha\text{-D-glucose} + \text{D-fructose} \tag{9.1}$$

$$\alpha\text{-D-glucose} \xrightarrow{\text{Mutarotase}} \beta\text{-D-glucose} \tag{9.2}$$

$$\beta\text{-D-glucose} + O_2 \xrightarrow[\text{oxidase}]{\text{Glucose}} \text{D-glucono-}\delta\text{-lactone} + H_2O_2 \tag{9.3}$$

The principle of the assay is based on monitoring the decrease in dissolved oxygen resulting from these three enzyme reactions in the presence of sucrose. Dissolved oxygen in the saturated solution is responsible for the current at time 0 in each sample solution. When the enzyme sensor is dipped in solutions of sucrose, the current that results from a reduced amount of dissolved oxygen working the electrode is observed to decrease very rapidly.

The steady state current is proportional to the sucrose concentration. With the enzyme, steady state currents are obtained because of an equilibrium between the dissolved oxygen consumption by the enzymatic reaction and the supply of oxygen from the bulk solution to the enzyme layer on the surface of the electrode. Three minutes were required to obtain a steady state.

A linear relationship is obtained up to 10 mM of sucrose concentration (Fig. 9.4). The reproducibility was determined using the same samples and was found to be 75 $\pm$ 0.2 A (7% of the relative error) in the case of 5 mM sucrose.

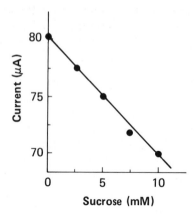

FIGURE 9.4   Calibration curve for sucrose; 0.1 M acetate buffer, pH 5.0, 25°C.

The effect of pH on the electrode response was studied from pH 4 to pH 8 for 5 mM of sucrose solution. Buffer solutions in which the ionic strength was fixed at 0.1 were used to adjust the pH of the sample. No differences in current output were observed from pH 4 to pH 8. pH profiles of invertase, glucose oxidase, and mutarotase are bell-shaped curves, and the optimum pH for these enzymes are 4.8, 6.0, and 5.7, respectively. The flat pH profile from 4 to 8 may be the result of a severe diffusion-limited reaction or a stabilization of enzymes with the collagen fibril matrix.

The stability was examined by testing the response of the sensor to 5 mM sucrose in 0.1 M acetate buffer (pH 5.0). No decrease in the output current was observed until the sensor had been used 85 times over a ten-day observation period.

C. Other Sugar Sensors

Glucoamylase and glucose oxidase were immobilized by adsorption on a porous acetylcellulose membrane and combined with a Pt electrode (for hydrogen peroxide) to provide a sensor for maltose. The response time is 6–7 min. Maltose (10^{-2}–10^3 mg$\cdot$l^{-1}) was determined by the sensor.

A sensor consisting of immobilized galactose oxidase and a Pt electrode (for hydrogen peroxide) has been also developed for the determination of galactose.

III. ENZYME SENSOR FOR LIPIDS

A. Free Cholesterol Sensor

The determination of cholesterol in serum is very important in clinical analysis. Quantitative determinations of free cholesterol can be made colorimetrically; however, pretreatment such as lipid extraction and separation from other lipid components is required before the determination. Therefore cholesterol measurement requires long times and many treatments.

The oxidation of free cholesterol with dissolved oxygen in the presence of cholesterol oxidase is shown in Eq. (9.4):

$$\text{Cholesterol} + O_2 \xrightarrow{\text{Cholesterol oxidase}} \text{Cholest-4-en-3-one} + H_2O_2 \qquad (9.4)$$

The principle of the assay is based on monitoring the decrease in dissolved oxygen resulting from the enzyme reaction in the presence of the free cholesterol. The sensor consisted of a cholesterol oxidase–collagen membrane and an oxygen electrode (Satoh et al., 1977a). The serum sample containing free cholesterol was injected into the system. The current decreased rapidly, and then steady state current was obtained. The steady state current is proportional to the free cholesterol concentration. These steady state currents result from an equilibrium between the consumption of dissolved oxygen by the enzymatic reaction and the supply of oxygen from the bulk solution to the enzyme layer. A few minutes were required to obtain a steady state current. A linear relationship is obtained up to 0.2 mM of free cholesterol concentration in serum.

B. Total Cholesterol Sensor

Total cholesterol in serum is an important indicator of abnormality in lipid metabolism, arteriosclerosis, and hypertension. An enzymatic method based on cholesterol esterase and cholesterol oxidase has been proposed:

$$(9.5)$$

$$(9.5)$$

However, these methods involve complicated and delicate procedures. Assay times are rather long because of the multistep reactions, and the cost is high because expensive enzymes must be used in each assay. An immobilized enzyme reactor containing cholesterol esterase and cholesterol oxidase was coupled with an amperometric detector system (Karube et al., 1982). The sensor system was applied to the determination of total cholesterol in human sera. The characteristics of the system were studied.

Cholesterol esterase and cholesterol oxidase were immobilized on octylagarose gel, activated with cyanogen bromide, and placed in a reactor. The sensor system for total cholesterol was assembled with the immobilized enzyme reactor, a hydrogen peroxide electrode, and a peristaltic pump.

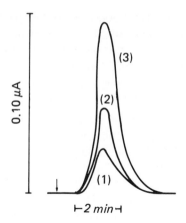

0.10 μA

(3)

(2)

(1)

⊢2 min⊣

FIGURE 9.5 Response curves. At the arrow, 10 μl of cholesterol palmitate solution containing 5% (w/v) Triton X-100 (pH 8.0) was injected. Flow rate was 1.0 ml·min^{-1}; 37°C. Cholesterol palmitate concentration. 1: 162, 2: 323, 3: 646 mg·dl^{-1}.

Typical response curves at various concentrations of cholesterol palmitate are shown in Fig. 9.5. The peak current increased with increasing cholesterol palmitate concentration below 1000 mg·dl^{-1}. As shown in Fig. 9.5, one sample could be assayed in 5 min.

The effect of the buffer flow rate on the current was examined (Fig. 9.6). At higher flow rates the current decreased with increasing flow rate because of incomplete hydrolysis of cholesterol palmitate. A decrease of the flow rate below 0.8 ml·min^{-1} also decreased the peak current because of dilution of the hydrogen peroxide produced by the enzymatic reaction. On the basis of these results the flow rate was adjusted to 1.0 ml·min^{-1} in all further experiments.

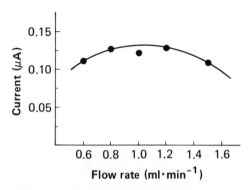

FIGURE 9.6 Effect of flow rate on the peak current; 10 μl of cholesterol palmitate (970 mg·dl^{-1}), 5% Triton X-100, pH 8.0, 37°C.

Human serum was diluted and examined in the system. The concentration of cholesterol in the samples was also determined with the cholesterol test kit. A linear relationship was obtained between the peak current and the cholesterol concentration in the range 100–400 mg · dl^{-1}; a 200 mg · dl^{-1} sample gave a current of 0.05 μA. The standard deviation for the determination of 300 mg · dl^{-1} cholesterol was 6 mg · dl^{-1} (2%, 50 experiments). Since human serum contains less than 400 mg · dl^{-1} of cholesterol, the flow system can be applied for such determinations.

The activity of the immobilized enzymes was retained for at least one month at 4°C. No appreciable decrease of their activity was observed after 300 successive assays. Thus the various components presented in serum had no interfering effects on the sensor.

Total cholesterol in human sera was assayed by the chemical method and the present system. The correlation coefficient was 0.94 for 27 samples. Therefore the present system is applicable to a rapid determination of total cholesterol in human sera. When the free cholesterol sensor described above and the system proposed here are used, the determination of cholesterol esters in sera is also possible.

C. Phosphatidyl Choline Sensor

Determination of phospholipids in serum is important in clinical analysis. Phosphatidyl choline is a main serum phospholipid; various procedures have been described for its determination. To improve the accuracy, selectivity, and rapidity of the assay in serum, enzymatic methods involving a combination of phospholipase D and choline oxidase have been proposed. The reactions involved are:

$$\text{Phosphatidyl choline} \xrightarrow{\text{Phospholipase D}} \text{phosphatidic acid} + \text{choline} \quad (9.6)$$

$$\text{Choline} + 2O_2 + H_2O \xrightarrow{\text{Choline oxidase}} \text{betaine} + 2H_2O_2 \quad (9.7)$$

In order to achieve a rapid and simple serum phospholipid assay, electrochemical monitoring of these reactions may have definite advantages. The use of immobilized enzymes with direct amperometric measurement of the hydrogen peroxide liberated appeared to be the best approach (Karube et al., 1979).

Phospholipase D and choline oxidase were immobilized together on cyanogen bromide–activated hydrophobic agarose gel. The measurement system for phosphatidyl choline consisted of an immobilized enzyme reactor containing phospholipase D and choline oxidase and the Pt electrodes positioned close to the reactor. Figure 9.7 shows a schematic diagram of the sensor for phospholipid determination. The hydrogen peroxide liberated enzymatically was monitored with a voltammetric system based on a platinum electrode (0.5 cm^2) at + 0.60 V vs. S.C.E., and a recorder.

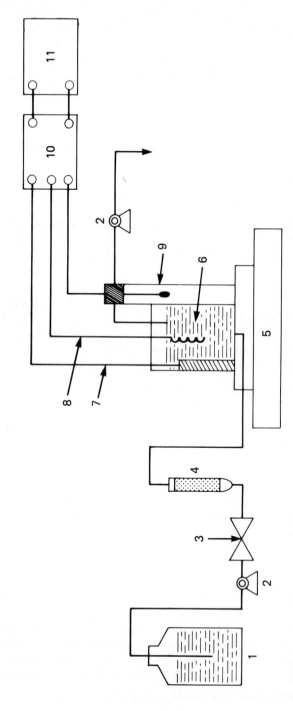

FIGURE 9.7 Schematic diagram of the sensor for phospholipid determination. 1: buffer reservoir, 2: peristaltic pump, 3: sample inlet, 4: immobilized enzyme reactor, 5: magnetic stirrer, 6: sensing chamber (0.4 ml), 7: anode (Pt plate, 0.5 cm^2), 8: cathode (Pt wire), 9: SCE, 10: voltammetric monitoring system, 11: recorder.

An assay could be completed within 4 min if the phosphatidyl choline concentration was 3 g·l^{-1}. If at least 0.3 I.U. of phospholipase was used, a reliable assay of phosphatidyl choline in serum was obtained. The proportion of the two immobilized enzymes is an improtant factor in obtaining maximum reaction rate and complete reaction. When the phospholipase/choline oxidase weight ratio after immobilization was 0.9, the reaction was completed within 4 min.

The calibration graph is linear up to 3 g·l^{-1} for phosphatidyl choline (Fig. 9.8). More concentrated sample solutions should be diluted with the buffer. The standard deviation for determination of 3 g·l^{-1} of phosphatidyl choline was 0.15 g·l^{-1} (50 experiments).

The phosphatidyl choline concentrations of fresh sera were determined by the proposed flow method and the conventional method. The results showed satisfactory agreement (correlation coefficient 0.90) for 16 assays of phosphatidyl choline in the range 1.5–3.5 g·l^{-1}. This indicated that the proposed immobilized enzyme system provides an economical and reliable method for enzymatic assay of phospholipids in serum. About 4% of the activity of the immobilized enzymes had been lost after 50 assays.

The activity of the reactor can be adjusted by varying the amount of immobilized enzymes placed inside the column. The rate of enzymatic catalysis can be regulated by varying the flow rate of the buffer. These, of course, control the overall reaction rate, which is very useful for enzymes having a low specific activity.

D. Neutral Lipid Sensor

Neutral lipids in serum are solubilized as lipoproteins and remain mainly as chyromycron and low-density lipoproteins. Lipoprotein lipase hydrolyzes

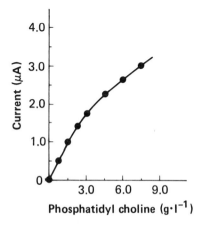

FIGURE 9.8 Calibration graph for flow analysis.

neutral lipids in serum to glycerol and fatty acids. The fatty acids can be determined with a pH electrode, so neutral lipids can be determined indirectly by
lipoprotein lipase and a pH electrode. A neutral lipid sensor with a reactor
containing a lipase–collagen membrane and a combined glass electrode for
determinations in serum has been described previously (Satoh et al., 1977b).
However, the time required for each determination was 30 min because of the
low activity of the immobilized lipase. Therefore lipoprotein lipase was covalently bound to polystyrene sheets coated with γ-aminopropyltriethoxysi-
lane, and a new flow-through pH electrode was employed for the neutral lipid
sensor (Satoh et al., 1979).

Various amounts of olein were injected into the sensor system. The potential of the glass electrode increased with time until a maximum was reached.
The time required to reach the maximum was 1 min; the potential returned to
its initial level within 3 min (Fig. 9.9).

The relationship between the logarithm of the concentration of the olein
and the potential difference is linear, changing by 8 mV over the range 5–50 M.

The effect of flow rate on the response of the sensor to two lipid concentrations was examined. A decrease in the potential difference with increase in flow
rate was observed. If the flow rate was $< 70 \ ml \cdot h^{-1}$, the immobilized lipase
had sufficient time to hydrolyze cholesterol esters measurably. Therefore a
flow rate of $72 \ ml \cdot h^{-1}$ is recommended for lipid determinations if cholesterol
esters are present.

The reusability of the sensor was examined with various concentrations of
the trioleate. Lipid determinations were done 20–25 times a day, and no decrease
in the response was observed over a 10-day period. The potential difference
was reproducible within 5%.

Neutral lipids were isolated from various human sera with isopropanol,
and determined by the conventional acetyl–acetone method and the elec-

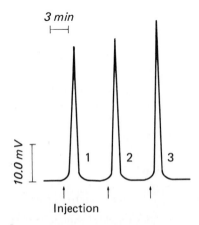

FIGURE 9.9   Response curves for glyceryl trioleate (olein). 1: 15.5 μM, 2:
30.0 μM, 3: 43.5 μM, at 20°C and pH 8.30; flow rate 72 $ml \cdot h^{-1}$.

trochemical method. The results show a good correlation. Therefore the sensor can be used for the determination of neutral lipids in human sera.

A rapid and continuous determination of lipid is possible with the lipid sensor.

IV. ENZYME SENSOR FOR CHOLINE

Choline is oxidized by choline oxidase to betaine with the simultaneous production of hydrogen peroxide:

$$\text{choline} + 2O_2 \xrightarrow{\text{Choline oxidase}} \text{betaine} + 2H_2O_2 \qquad (9.8)$$

We have studied the possibility of utilizing an enzyme sensor for the assay of choline. Choline oxidase was immobilized on a partially aminated polyacrylonitrile membrane that was applied to an enzyme electrode for choline (Matsumoto et al., 1980). Furthermore, the choline sensor can be applied to determine the phosphatidyl choline fraction in serum.

In the choline assay the average response time based on the endpoint method was 1–2 min, and that for the rate assay method was 7 sec. After measurement the sensor was washed with fresh buffer to recover the DO level of the electrode. Washing time for the sensor was 30–45 s. Accordingly, the time required for one assay of choline was about 1 min.

A calibration curve was prepared using 20 μl of choline chloride as a standard solution (1–10 mM). A linear relationship was observed at choline concentrations below 0.1 mM. Therefore a sample solution containing choline above 0.1 mM must be diluted with buffer. The reproducibility of the response was examined by using the same samples (0.2 mM). The response was reproducible within ± 2.3% of the relative error when a medium containing 0.2 mM of choline chloride solution was used. The standard deviation was 4.6 μM in 100 experiments.

The long-time stability of the immobilized enzyme and the choline sensor was examined. A standard choline solution (10 mM) was employed. The sensor was stored in 10 mM of Tris-HCl buffer (pH 8.0) at 4°C or room temperature (25°C). Even after 20 days the decrease of the current output of the sensor was below 5% at 4°C and about 40% at room temperature. However, the sensor was still useful for the choline assay. More than 1000 samples could be determined by using this enzyme sensor.

The choline sensor was applied to the determination of phosphatidyl choline. A calibration curve was prepared by using phospholipase D and 20 μl of phosphatidyl choline standard emulsion (1–10 mM). A linear relationship was observed at phosphatidyl choline concentrations below 0.1 mM. Therefore a sample solution containing phosphatidyl choline above 0.1 mM must be diluted with the buffer. The reproducibility of the response was examined by using the same samples (0.2 mM). The response was reproducible within

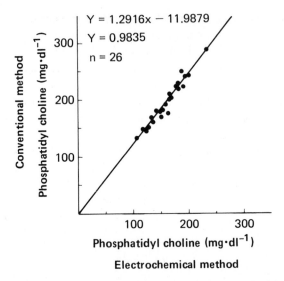

FIGURE 9.10   Comparison of the electrochemical method with the conventional method.

±2.8% of the relative error when a medium containing 0.2 mM of phosphatidyl choline emulsion was used. The standard deviation was 5.6 mM in 50 experiments.

Phosphatidyl choline in 26 serum samples was measured by the choline sensor and phospholipase D. The results of these comparisons are shown in Fig. 9.10. The calculated correlation coefficient was 0.9855, and the regression equation was $Y = 1.2916X - 11.9879$. The relative error of the determination by the choline sensor was ±3.5% when 20 μl of serum containing 178 mg·dl^{-1} phosphatidyl choline were used. The standard deviation was 6.23 mg·dl^{-1} in 50 experiments. The enzyme sensor method gave good agreement with values obtained by the enzymatic method. Therefore the choline sensor is a suitable method for the continuous analysis of phospholipid in serum.

V. ENZYME SENSOR FOR MONOAMINE

The method for the determination of monoamines requires complicated operations, such as extraction and centrifugation. Therefore simple, reproducible, and continuous methods for monoamine determination are still required. It is known that several kinds of amines are produced in the body, but tyramine and histamine are the main products. They are oxidized to the corresponding aldehydes by monoamine oxidase. The oxidation of monoamine with dissolved oxygen in the presence of monoamine oxidase is shown in Eq. (9.9):

$$R-CH_2NH_2 + O_2 + H_2O \xrightarrow[\text{oxidase}]{\text{Monoamine}} R-CHO + H_2O_2 + NH_3 \quad (9.9)$$

The principle of the assay is based on monitoring the decrease in dissolved oxygen resulting from the enzymatic reaction in the presence of monoamines, and the oxygen electrode can be used to determine dissolved oxygen. A monoamine sensor consisted of a monoamine oxidase–collagen membrane and an oxygen electrode (Karube et al., 1980).

When a sample solution containing amine was injected into the buffer solution, oxygen consumption by amine oxidase began in the collagen membrane. The consumption of oxygen for amine oxidation caused a decrease in dissolved oxygen around the membrane, resulting in a marked decrease in the sensor current with time until steady state was reached. In the case of histamine the response time of the sensor was about 6 min at 30°C. The response of the sensor to various amines is summarized in Table 9.1. Amine oxidase from *Aspergillus niger* was employed for the sensor. As shown in Table 9.1, the difference in current obtained from the sensor depended on the type of monoamine. As was reported previously, diffusion rates of low-molecular-weight substances were almost the same (Karube et al., 1978). This may therefore be caused by differences in the oxidation rate of various amines by the immobilized enzymes.

A linear relationship was obtained between the difference in current and concentrations of histamine, tyramine, and isobutylamine below 1 mM. The current was reproducible within 8% of the relative error when a solution containing 0.5 mM of histamine was used.

The reusability of the sensor was tested on benzylamine (1.7 mM) and the fungal amine oxidase sensor. Benzylamine was determined at least three times each day. No decrease in the output current was observed over an observation period of one week (S.D. = 0.07 mM).

The concentration of amines in pork could be determined by the enzyme sensor. Figure 9.11 shows the time course of amine content in meat paste as determined by the enzyme sensor compared with that determined by the conventional method (as basic nitrogen).

TABLE 9.1   Response of the Electrode to Various Amines

Amines (0.1 mM)	$\Delta I (\mu A)$
n-Amylamine	6.9
n-Butylamine	4.2
n-Hexylamine	10.9
Histamine	3.1
Isobutylamine*	1.0
Propylamine	2.2
Tyramine	5.0

* 0.25 mM.

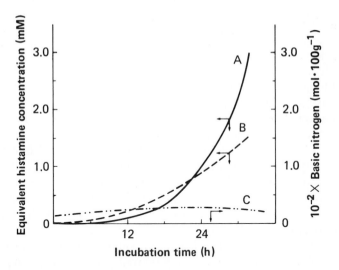

FIGURE 9.11 Determination of amines in meat. A: Fungal MAP electrode method, B: plasma MAO electrode method. The reactions were carried out at $30 \pm 0.1°C$ and pH 7.4 (0.1 M phosphate buffer). C: Titration method.

VI. OTHER ENZYME SENSORS

Many reports on enzyme sensors have been published for the determination of organic and inorganic compounds. Table 9.2 summarizes various kinds of enzyme sensors and their characteristics. The sensors for glucose, sucrose, lactose, lactic acid, L-amino acid, uric acid, urea, and α-amylose have been commercialized.

REFERENCES

Alexander, P. W., and Joseph, J. P. (1981) *Anal. Chim. Acta 131*, 103.

Alexander, P. W., and Maitra, C. (1982) *Anal. Chem. 54*, 68.

Chen, A. K., Starzmann, J. A., and Liu, C. C. (1982) *Biotech. Bioeng. 24*, 971.

DiPaolantonio, C. L., and Rechnitz, G. A. (1982) *Anal. Chim. Acta 141*, 1.

Enfors, S. O. (1981) *Enzyme Microb. Technol. 3*, 29.

Gondo, S., Osaki, T., and Morishita, M. (1981) *J. Mol. Catal. 12*, 365.

Guilbault, G. G. (1976) *Handbook of Enzymatic Methods of Analysis*, Marcel Dekker, New York.

Gulberg, E. L., and Christian, G. D. (1981) *Anal. Chim. Acta 123*, 125.

Hassan, S. S. M., and Rechnitz, G. A. (1981) *Anal. Chem. 53*, 512.

Hirose, S., Hayashi, M., Tamura, N., Suzuki, S., and Karube, I. (1979) *J. Mol. Catal. 6*, 251.

Ianniello, R. M., and Yacynych, A. M. (1981) *Anal. Chem. 53*, 2090.

TABLE 9.2 Properties of Enzyme Sensors

Sensor	Enzyme	Immobilization	Electrochemical Device	Stability (day)	Reaction Time (min)	Range $(mg \cdot l^{-1})$
Glucose	Glucose oxidase	Covalent	Oxygen elec.	100	1/6	$1\text{-}5 \times 10^2$
Ethanol	Alcohol oxidase	Cross-linked	Oxygen elec.	120	1/2	$5\text{-}10^3$
Pyruvate	Pyruvate oxidase	Adsorption	Oxygen elec.	10	2	$10\text{-}10^3$
Uric acid	Uricase	Cross-linked	Oxygen elec.	120	1/2	$10\text{-}10^3$
L-Amino acid	L-Amino acid oxidase	Covalent	Ammonia gas elec.	70	–	$5\text{-}10^2$
L-Glutamine	Glutaminase	Adsorption	Ammonium ion elec.	2	1	$10\text{-}10^4$
L-Glutamic acid	Glutamate dehydrogenase	Adsorption	Ammonium ion elec.	2	1	$10\text{-}10^4$
L-Asparagine	Asparaginase	Entrapment	Ammonium ion elec.	30	1	$5\text{-}10^3$
L-Tyrosine	L-Tyrosine decarboxylase	Adsorption	CO_2 gas elec.	20	1-2	$10\text{-}10^4$
L-Lysine	L-Lysine decarboxylase, amine oxidase	Cross-linked	Oxygen elec.	–	1-2	$10^3\text{-}10^4$
L-Alginine	Alginine decarboxylase, amine oxidase	Cross-linked	Oxygen elec.	–	1-2	$10^3\text{-}10^4$
L-Phenylalanine	L-Phenylalanine ammonia lyase	–	Ammonia gas elec.	–	10	$5\text{-}10^2$
L-Methionine	Methionine ammonia lyase	Cross-linked	Ammonia gas elec.	90	1-2	$1\text{-}10^3$

Ianniello, R. M., Lindsay, T. J., and Yacynych, A. (1982) *Anal. Chem. 54*, 1980.

Johnson, J. M., Halsall, H. B., and Heineman, W. R. (1982) *Anal. Chem. 54*, 1394.

Karube, I., Suzuki, S., Kusano, T., and Satoh, T. (1978) *J. Solid-Phase Biochem. 2*, 273.

Karube, I., Hara, K., Satoh, I., and Suzuki, S. (1979) *Anal. Chim. Acta 106*, 243.

Karube, I., Satoh, I., Araki, Y., and Suzuki, S. (1980) *Enzyme Microbiol. Technol. 2*, 117.

Karube, I., Hara, K., Matsuoka, H., and Suzuki, S. (1982) *Anal. Chim. Acta 139*, 127.

Kovach, P. M., and Meyerhoff, M. E. (1982) *Anal. Chem. 54*, 217.

Koyama, M., Sato, Y., Aizawa, M., and Suzuki, S. (1980) *Anal. Chim. Acta 116*, 307.

Lui, C. C., Wingard, L. B., Jr., Wolfson, S. K., and Yao, S. J. (1979) *Bioelectrochem. Bioeng. 6*, 19.

Liu, C. C., Weaver, J. P., and Chen, A. K. (1981) *Bioelectrochem. Bioeng. 8*, 379.

Macholan, L., Londyn, P., and Fischer, J. (1981) *Collect. Czech. Chem. Comm. 46*, 2871.

Mascini, M., Zolesi, F., and Palleschi, G. (1982) *Anal. Lett. 15*, 101.

Matsumoto, K., Seijo, H., Karube, I., and Suzuki, S. (1980) *Biotech. Bioeng. 22*, 1071.

Matsumoto, K., Yamada, K., and Osajima, Y. (1981) *Anal. Chem. 53*, 1974.

Matsumoto, K., Naotsuka, M., Shirasaka, Y., Nomura, T., and Osajima, Y. (1982) *Agr. Biol. Chem. 46*, 2749.

Mizutani, F., and Tsuda, K. (1982) *Anal. Chim. Acta 139*, 359.

Mizutani, F., Sasaki, K., and Shimura, Y. (1983) *Anal. Chem. 55*, 35.

Mosbach, K. (ed.) (1976) *Methods in Enzymology*, Vol. XLIV, Academic Press, New York.

Nagy, G., Rice, M. E., and Adams, R. N. (1982) *Life Sci. 31*, 2611.

Neujahr, H. Y. (1982) *Appl. Biochem. Biotechnol. 7*, 107.

Nikolelis, D. P., Papastathopoulos, D. S., and Hadjiioannou, T. P. (1981) *Anal. Chim. Acta 126*, 43.

Satoh, I., Karube, I., and Suzuki, S. (1976) *Biotech. Bioeng. 18*, 269.

Satoh, I., Karube, I., and Suzuki, S. (1977a) *Biotech. Bioeng. 19*, 1095.

Satoh, I., Karube, I., and Suzuki, S. (1977b) *J. Solid-Phase Biochem. 2*, 1.

Satoh, I., Karube, I., Suzuki, S., and Aikawa, K. (1979) *Anal. Chim. Acta 105*, 429.

Sokol, L., Garber, C., Shults, M. C., and Updike, S. J. (1980) *Clin. Chem. 26*, 89.

Suzuki, S., Karube, I., and Satoh, I. (1977) *Biomedical Application of Immobilized Enzymes and Proteins*, Vol. II, p. 177, Plenum Press, New York.

Tipton, K. F., McCrodden, J. M., and Bardsley, M. E. (1981) *Biochem. Soc. Trans. 9*, 324.

Tsuchida, T., and Yoda, K. (1981) *Enzyme Microbial Technol. 3*, 326.

Updike, S. J., Shults, M. C., and Busby, M. (1979) *J. Lab. Clin. Med. 93*, 518.

Vadgama, P. M., Alberti, K. G. M. M., and Covington, A. K. (1982) *Anal. Chim. Acta 136*, 403.

Watanabe, E., Ando, K., Karube, I., Matsuoka, H., and Suzuki, S. (1983) *J. Food Sci. 48*, 496.

Winartasaputra, H., Kuan, S. S., and Guilbault, G. G. (1982) *Anal. Chem. 54*, 1987.

Enzyme, Microbial, and Immunochemical Electrode Probes

George G. Guilbault
Graciliano de Olivera Neto

I. INTRODUCTION

The advent of over 50 new "ion-selective electrodes" into the analytical market has brought about a revolution in the area of electroanalytical chemistry. These electrodes generally work well for all types of cations and anions but show good selectivity for inorganic species only. The ion-selective electrodes for organic species are in most cases nonselective, irregular in response, and not very useful.

In clinical analysis today, all electrolytes (Na^+, K^+, Ca^{2+}, Cl^-, H^+, HCO_3^-) are measured with ion-selective electrodes in almost all instruments. What then could be more natural than to place into these instruments electrode probes for metabolites, such as glucose, urea, uric acid, cholesterol, or creatinine? Such "enzyme electrodes" could be fashioned by taking an ion-selective electrode for O_2, CO_2, or NH_3, and coating it with a layer of stabilized, immobilized enzyme. The substrate to be measured diffuses into the enzyme layer, where it is converted to a product with the uptake or release of

O_2, CO_2, or NH_3. The latter is then measured with an appropriate ion-selective electrode, with the potential or current produced linearly related to the product concentration.

II. ENZYME ELECTRODE PROBES

A. General Discussion

In addition to determinations of enzyme activity, electrochemical methods have been combined with enzymatic systems to provide highly selective and sensitive probes for the determination of the concentration of a given substrate. This is possible because under controlled conditions the rate of an enzyme-catalyzed reaction is proportional to the concentration of substrate. The concept of using an enzyme as a reagent in conjunction with an electrode was introduced by Clark and Lyon (1962), and the first working enzyme electrode was reported by Updike and Hicks (1971) using glucose oxidase immobilized in a gel over a polarographic oxygen electrode to measure the concentration of glucose in biological solutions and in tissues. By immobilizing the enzyme the amount of the material required to perform routine analysis is greatly reduced, and frequent assay of the enzyme preparation is not necessary. Furthermore, the stability of the enzyme is often improved when it is incorporated in a suitable gel matrix. An electrode for the determination of glucose prepared by covering a platinum electrode with glucose oxidase chemically bound has been used for over 300 days (Guilbault and Lubrano, 1973).

Of the two methods used to immobilize an enzyme—(a) the chemical modification of the molecules by the introduction of insolubilizing groups and (b) the physical entrapment of the enzyme in an inert matrix such as starch or polyacrylamide gels—the technique of chemical immobilization is the best to make electrode probes. The method of chemical bonding with the bifunctional reagent, glutaraldehyde, is very simple and quite useful: the enzyme is simply treated with the aldehyde and an inert support (albumin, glass beads, etc.); a rigid layer of bound enzyme results, which is quite stable for several months and thousands of assays (Guilbault, 1977). The immobilized enzyme is then placed over the sensor of an electrode that is sensitive to the product of the enzyme–substrate reaction. When the enzyme electrode is placed in a solution that contains the substrate for which the electrode is designed, the substrate diffuses into the enzyme layer where the enzyme-catalyzed reaction takes place, producing an ion that is detected by the electrode. Excellent chemical analysis can be performed with enzymes, which are biological catalysts; the real advantages of immobilized enzymes are many in analyses using electrochemical probes or other methods of analysis.

One advantage of the immobilized enzyme is a pH shift; that is, the pH optimum can be shifted to that region at which one wants to make a measurement, by choosing the right support for immobilization. Take an enzyme with a narrow pH range of, say, 6–8; this can be shifted on immobilization down to the acidic side or, conversely, up to the basic side. The enzymes are furthermore much more stable. In some work at Edgewood Arsenal, Maryland, we actually heated our enzyme probes to 150°F and brought them back down to room temperature, with very little loss of activity. No soluble enzyme could be treated in this fashion (Guilbault, 1977).

One advantage often overlooked is that better selectivity can be realized with the enzyme when immobilized; this insolubilized reagent becomes much more selective for an inhibitor, and only the most powerful inhibitor can actually attack the enzyme. We demonstrated this several years ago in an immobilized cholinesterase alarm for the assay of organophosphorus compounds in air and water. No other common interferants disturbed the alarm — it responded only to organophosphorus compounds. This factor is quite important for an electrode probe.

In 1961 at Edgewood Arsenal I first experimented with some soluble enzymes, such as glucose oxidase, and developed an electrochemical assay for glucose. This led to the use of immobilized enzymes with a commercially available ion-selective electrode sensor to form one self-contained sensor that could be used to measure either organic or inorganic compounds that are primary or secondary substrates for the immobilized enzyme. The base sensor can be glass; that is, the cation response can be measured (the ammonium ion, for example), or the pH change in a penicillin electrode can be measured, as was done by Nilsson et al. (1973) and Papariello et al. (1973). Or a gas membrane can be used as a base sensor, such as the ammonia or the CO_2 membrane. Next are the polarographic sensors that measure peroxide or oxygen, or any of the solid membrane electrodes, that is, the cyanide or iodide electrodes. For example, the enzyme can be placed on top of a flat glass electrode sensor; a membrane is then put over the outside of this sensor to hold the enzyme in and keep things like catalase and bacteria out. This protects the enzyme from bacterial spoilage, which is one of the primary reasons for loss of enzyme activity.

With potentiometric devices the response can be measured either by a steady state (i.e., equilibrium) method measuring millivolts or microamperes or by a rate method that senses the change in millivolts or microamperes per minute. Measurements of substrate can be performed by either a steady state or a rate method. But measurements of enzyme activity must be done by a rate method. This is a point often hazy in the literature; one can find many claims of measuring enzymes by steady state methods. This is impossible by the basic definition of enzyme activity. Enzymes are catalysts and have to be measured by a rate method, but this may be either an interrupted or a continuous measurement of rate. Table 10.1 presents a compilation of enzyme electrodes. It is an expansion of a table that was published in *Handbook of Enzymatic Analysis* (Guilbault, 1977). In this table are listed the enzymes that act on these various materials and some of the base sensors that might be useful. Take as a typical example glucose, which can be assayed with glucose oxidase:

$$O_2 + \text{Glucose} \xrightarrow{\text{Oxidase}} H_2O_2 + \text{Gluconic acid} \qquad (10.1)$$

TABLE 10.1 Various Electrodes and Their Characteristics

Type	Sensor	Immobilization[a]	Stability	Response Time	Amount of Enzyme (U)	Range
1. Urea	Cation	Physical	3 weeks	30 s–1 min	25	10^{-2}–5×10^{-5}
	Cation	Physical	2 weeks	1–2 min	75	10^{-2}–10^{-4}
	Cation	Chemical	>4 months	1–2 min	10	10^{-2}–10^{-4}
	pH	Physical	3 weeks	5–10 min	100	5×10^{-3}–5×10^{-5}
	Gas (NH_3)	Chemical	>4 months	2–4 min	10	5×10^{-2}–5×10^{-5}
	Gas (NH_3)	Chemical	20 days	1–4 min	0.5	10^{-2}–10^{-4}
	Gas (CO_2)	Physical	3 weeks	1–2 min	25	10^{-2}–10^{-4}
2. Glucose	pH	Soluble	1 week	5–10 min	100	10^{-1}–10^{-3}
	Pt (H_2O_2)	Physical	6 months	12 s kinetic[c]	10	2×10^{-2}–10^{-4}
	Pt (H_2O_2)	Chemical	>14 months	1 min, steady state	10	2×10^{-2}–10^{-4}
	Pt (H_2O_2)	Soluble	<1 week[d]	1–2 min	10	10^{-2}–10^{-4}
	Pt (Quinone)	Soluble	>4 months	3–10 min	10	2×10^{-2}–10^{-3}
	Pt (O_2)	Chemical	>1 month	1 min	10	10^{-1}–10^{-5}
	I^-	Chemical		2–8 min	10	10^{-3}–10^{-4}
	Gas (O_2)	Physical	3 weeks	2–5 min	20	10^{-2}–10^{-4}
	Gas (O_2)	Chemical	>3 weeks	2–5 min	10	2×10^{-2}–10^{-4}
3. L-amino acids General	Pt (H_2O_2)	Chemical	4–6 months	12 s kinetic[c]	10	10^{-3}–10^{-5}
	Gas (O_2)	Chemical		2 min	10	10^{-2}–10^{-4}
	Pt (O_2)	Chemical	>4 months	1 min	10	10^{-2}–10^{-4}
	Cation	Physical	2 weeks	1–2 min	10	10^{-2}–10^{-4}
	NH_4^+	Chemical	>1 month	1–3 min	10	10^{-3}–10^{-4}
	I^-	Chemical	>1 month	1–3 min	10	
L-Tyrosine	Gas (CO_2)	Physical	3 weeks	1–2 min	25	10^{-1}–10^{-4}

	Detection	Entrapment[a]	Lifetime	Response time		Concentration range[b]
L-Glutamine	Cation	Soluble	2 days[d]	1 min	50	$10^{-1} - 10^{-4}$
L-Glutamic acid	Cation	Soluble	2 days[d]	1 min	50	$10^{-1}-10^{-4}$
L-Asparagine	Cation	Physical	1 month	1 min	50	$10^{-2}-5 \times 10^{-5}$
4. D-Amino acids (general)	Cation	Physical	1 month	1 min	50	$10^{-2}-5 \times 10^{-5}$
5. Lactic acid	Pt [Fe(CN)$_6^{4-}$]	Soluble	<1 week	3–10 min	2	$2 \times 10^{-3}-10^{-4}$
6. Succinic acid	Pt (O$_2$)	Physical	<1 week	1 min	10	$10^{-2}-10^{-4}$
7. Acetic, formic acids	Pt (O$_2$)	Chemical	>4 months	30 s	10	$10^{-1}-10^{-4}$
8. Alcohols	Pt (H$_2$O$_2$)	Soluble	1 week	12 s kinetic[c]	10	0.5–100 mg %
	Pt (H$_2$O$_2$)	Soluble	1 day[d]	1 min	~1	0.5–50 mg %
	Pt (O$_2$)	Chemical	>4 months	30 s	10	0.5–100 mg %
9. Penicillin	pH	Physical	1–2 weeks	0.5–2 min	400	$10^{-2}-10^{-4}$
		Soluble	3 weeks	2 min	~1000	$10^{-2}-10^{-4}$
10. Uric acid	Pt (O$_2$)	Chemical	4 months	30 s	~10	$10^{-2}-10^{-4}$
11. Amygdalin	CN$^-$	Physical	3 days[e]	10–20 min	100	$10^{-2}-10^{-5}$
12. Cholesterol	Pt (H$_2$O$_2$)	Soluble		2 min		$10^{-2}-10^{-4}$
13. Phosphate	Pt (O$_2$)	Chemical	4 months	1 min	10 each	$10^{-2}-10^{-4}$
14. Nitrate	NH$_4^+$	Soluble		2–3 min	10	$10^{-2}-10^{-4}$
15. Nitrite	NH$_3$ (Gas)	Chemical	3–4 months	2–3 min	10	$5 \times 10^{-2}-10^{-4}$
16. Sulphate	Pt	Chemical	1 month	1 min	10	$10^{-1}-10^{-4}$

a "Physical" refers to polyacrylamide gel entrapment in all cases; "chemical" is attachment to glutaraldehyde with albumin, to polyacrylic acid, or to acrylamide chemically followed by physical entrapment.
b Analytical useful range either linear or with reasonable change if curvature is observed.
c "Kinetic," "rate of change in current measured after 12 s; "steady state," current reaches a maximum in 1 min.
d Preparation lacks stability as evidenced by constant decrease in signal each day.
Electrode responds to L-cysteine, L-leucine, L-tyrosine, L-tryptophan, L-phenylalanine, and L-methionine.
Electrode responds to D-phenylalanine, D-alanine, D-valine, D-methionine, D-leucine, D-norleucine, and D-isoleucine.
e Time required for signal to return to baseline before reuse.

One can measure the uptake of oxygen with a gas membrane electrode, a technique pioneered by Clark and Lyons (1962) and perfected by Updike and Hicks (1971) or record the peroxide or oxygen polarographically. There are other ways: One can measure the gluconic acid by a pH change, as Nilsson et al. (1973) showed very nicely at low buffer capacity or use an iodide membrane (the latter is much less recommended). The point I would like to make is that there are many ways to measure a particular substrate, and one should choose the one best for the application. For example, one would not choose to measure urea in biological fluid with an ammonium cation electrode because of the interference of potassium and sodium. One would choose, preferentially, an ammonia gas membrane electrode in which there is no interference from sodium and potassium.

Among the basic characteristics of enzyme electrodes is the five-step process of their operation. First, the substrate must be transported to the surface of the electrode. Second, the substrate must diffuse through the membrane to the active site. Third, reaction occurs at the active site. Fourth, product formed in the enzymatic reaction is transported through the membrane to the surface of the electrode, where, fifth, it is measured. The first step—transport of the substrate—is most critically dependent on the stirring rate of the solution, so rapid stirring will bring the substrate very rapidly to the electrode surface. If the membrane is kept very thin, using a highly active enzyme, then steps 2 and 4 are eliminated or minimized; since step 3 is very fast, the theoretical response of an enzyme electrode should approach the response time of the base sensor.

B. Construction of Enzyme Electrodes

There are four steps to follow in the construction of an enzyme electrode. Let us consider each of these factors in detail.

Step 1.   Select an enzyme that reacts with the substance to be determined. From information in standard reference books on enzymology, such as *Biochemists Handbook*, find an enzyme system that is suitable for your determination. In the ideal case this will involve the use of the primary function of the enzyme, that is, the main substrate–enzyme reaction. For example, for a glucose electrode, glucose oxidase would be used; for a urea electrode, urease; for a L-glutamic acid electrode, L-glutamate dehydrogenase. In other cases this may necessitate using an enzyme that acts on the compound of interest as a secondary substrate, for example, alcohol oxidase for acetic acid (Guilbault et al., 1969). Of course, this latter case will introduce more interferences and less selectivity into the assay.

Note that in some cases there are several enzymes that act on the substrate of interest, via different reactions. For example, L-tyrosine could be determined by using L-tyrosine decarboxylase and measuring the CO_2 liberated (Guilbault and Shu, 1972) or by using L-amino acid oxidase using a Pt electrode

(Guilbault and Lubrano, 1974) or an NH_4^+ electrode (Guilbault and Hrabankova, 1970a, 1971). The latter enzyme, although less selective, can be obtained commercially in high purity; the former is available in low purity and has to be purified before use. Hence the scientific capabilities of one's laboratory might dictate the choice of enzyme.

Step 2.   Obtain the enzyme. Having found the enzyme to be used for your application, check the catalogs of commercial suppliers, Boehringer, Calbiochem, Sigma, Worthington, etc. (see the complete list (Guilbault, 1985)) to see whether the enzyme can be purchased and its purity. The latter may or may not pose a problem. Many enzymes are stable in an impure state (e.g., jack bean urease and glucose oxidase from the food industry (General Mills)) and can be used satisfactorily in a pseudo-"immobilized" form, for example, as a liquid covered with a dialysis membrane, for up to a week. In other cases the impure enzyme has too low an activity to be useful without further purification (many of the decarboxylases available from Sigma), involving further work and possibly assistance from others.

In still other cases, one may find that the enzyme that one wants to use is not available commercially. In this case there are two possibilities:

1. Contact a large biochemical supply house and inquire whether it will isolate and purify the enzyme you want. Many will, for a suitable fee. New England Enzyme, for example, specializes in such a service.
2. Look up the enzyme in the literature or standard biochemistry–enzymology reference books, ascertain the isolation and purification methods used, and perform the purification yourself. This we have done ourselves, in many cases with excellent results, and in most cases the techniques are simple enough to be carried out by a person with reasonable scientific training. The results are frequently well worth the effort.

Step 3.   Immobilize the enzyme. A simple rule of thumb to follow is that the better the enzyme is immobilized, the more stable it is and hence the longer it can be useful and the more assays are possible from one batch. Let us now consider the various possibilities and the characteristics of the product.

C. Performance Characteristics of Electrodes

Now that the electrode has been made, let us consider some of the factors that affect the response and stability of the electrode.

1. Stability.   Some of the factors that affect the stability of enzyme electrodes are listed in Table 10.2. The stability of an enzyme electrode is a difficult term to define, since an enzyme can lose some activity, resulting in a shift of the

TABLE 10.2   Factors That Affect the Stability of Enzyme
Electrodes
_____

1. Type of entrapment
   a. Soluble + dialysis membrane−1 week or 25–50 assays
   b. Physical−3–4 weeks, 50–100 assays
   c. Chemical−4–14 months, 200–10,000 assays
2. Content of enzyme in gel and purity
3. Optimum conditions of enzyme
4. Stability of base sensor
_____

calibration curve downwards. Yet if the slope remains constant, as is frequently the case, the electrode is still useful, needing only calibration daily. This is seldom a problem, since all who use electrodes of any type (e.g., glass electrode) reset the pH or potential of their electrode at least once a day. This should be done with the enzyme electrode also using serum (e.g., Monitrol, Dade, Miami). Another problem in the definition of stability is that many workers measure the potential of their electrode occasionally over a long period of time and report these data as the stability. This may mean that the electrode was used one time a day or week, ten times a day, or a hundred times a day. Naturally, the more the use, the shorter will be the overall lifetime. The first factor that affects the stability is the type of entrapment used. As a general rule, a "soluble" electrode is useful for about a week or 25–50 assays, provided that the electrode is kept refrigerated between uses. The physically entrapped "polyacrylamide" electrodes are good for about three weeks or 50–100 assays, depending crucially on the degree of care exercised in the preparation of the polymer. The chemically attached enzyme can be kept if not used very much even at room temperature (see Table 10.2 and Fig. 10.1) as long as 14 months for glucose oxidase, greater than 4–6 months for L-amino acid oxidase or uricase. One can expect to get about 200–1,000 assays per each electrode, again depending on how a synthesis is effected. Although the electrode can be stored at room temperature, it is recommended that all electrodes be kept in a refrigerator and covered with a dialysis membrane to prevent the action of bacteria, which tend to feed on the enzyme, destroying its activity. The dialysis membrane (mol. wt. exclusion about 1500, pore size about 35 μm) prevents the enzyme from getting out and bacteria from getting in.

The stability of the physically entrapped enzyme varies greatly with experimental conditions, and a thorough study of these factors has been made by Guilbault and Montalvo (1970).

To determine the effect of physical immobilization parameters on the stability of the urea electrode, a series of enzyme electrodes was prepared in which one immobilization parameter was varied and all of the parameters were maintained constant. To determine the stability of the immobilized urease coating on the surface of the cation electrode, the steady state potential was obtained for a given urea substrate concentration at periodic time intervals. If

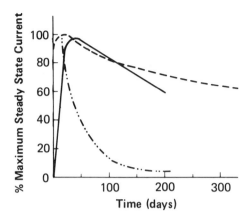

FIGURE 10.1 Long-term stability of glucose electrodes by the steady state method. Type 1: chemically bound — glutaraldehyde (dashed curve); type 2: chemically bound — polyamide (solid curve); type 3: physically bound (dotted curve). (From Guilbault and Lubrano (1973); reprinted with permission of the author.)

the steady state potential is constant within a certain period of time, no loss of activity of the immobilized enzyme has occurred. All stability data reported were obtained with the electrode stored at 25°C in tris buffer between measurements.

The maximum stability that could be achieved with the physically entrapped enzyme electrode was obtained with the following immobilization parameters: photopolymerizing for 1 hour at 28°C with a No. 1 150-W photoflood lamp and a gel-layer thickness of 350 μm; and an enzyme concentration in the gel of 175 mg/cm^3 gel (10 units). The slope of the stability curve, $\Delta mV/\Delta t$, shows that the measured stability depends on the substrate concentration used in the stability measurements. When the urea concentration used in the stability measurements is high enough that the steady state response is independent of the substrate concentration, $\Delta mV/\Delta t$ was 0.2 mV/ day over a 14-day period. At lower substrate concentrations, as for example, 1 $\times$ 10^{-3} M urea, the steady state response is first order in urea concentration, and a much smaller loss in activity was obtained, 0.05 mV/day over a 14-day period. At 8.33 $\times$ 10^{-2}M urea, the upper limit of substrate concentration that can be measured with the enzyme electrode, the steady state response falls by only 0.7 mV during 14 days of operation at 25°C. After 14 days the loss in activity was much greater for both substrate concentrations.

To study the effect of the activity of immobilized urease on enzyme gel stability, physically entrapped enzyme electrodes were prepared with activity of enzyme from 375 to 3500 Sumner units per gram of enzyme. No appreciable change in stability occurred with this relatively large change in enzyme activity.

In contrast, highly purified urease is known to be very unstable in solution. A similar trend in stability would be expected with immobilized urease.

Greater stability with the enzyme electrode was always obtained when the gel solution was less than 2 days old. Gel solutions were stored without added polymerization catalysts when the storage period was greater than 2 days. The solutions were always stored in the dark at room temperature. The stability of the urease electrode was studied as a function of enzyme gel-layer thickness in the range 30–350 μm. The stability increased with thickness of the enzyme gel layer, but response time increased.

Several experiments were run to determine quantitatively the effect of photopolymerization light intensity and time on enzyme electrode stability. When a high-intensity photoflood lamp is substituted with a 60-W domestic lamp, the loss in activity rises from 0.2 to 4.2 mV/day for 8.33×10^{-2} M urea. A similar loss in activity for the electrode was obtained when only the photopolymerization time was reduced from 1 hour to 15 min.

To study the effect of photopolymerization temperature and water content of the gel layer during photopolymerization on the electrode stability, a series of enzyme electrodes were prepared with temperatures ranging from 4°C to 43°C. The water content of the gel layer over the electrode surface was also varied when the photopolymerization temperature was changed; this is because the rate of evaporation of water from the thin enzyme gel layer varies directly with temperature. When the photopolymerization temperature and water content of the gel were varied to study electrode stability, the other immobilization parameters were adjusted to give maximum stability. The stability, measured with 8.33×10^{-2} M urea, showed a loss of only 0.2 mV/day at a 28°C photopolymerization temperature; upon lowering the immobilization or photopolymerization temperature to 6°C, the loss in electode activity is much higher, 3.7 mV/day. At 6°C the rate of evaporation of water from the enzyme gel layer during photopolymerization is so slow that the gel layer is still damp to the touch after immobilization is complete. This large loss in activity is due to leaching of enzyme from the gel layer. Enzyme that had leached out of the gel layer could easily be detected in the buffer solution used to store the electrode. At 28°C the rate of evaporation of water from the gel layer is sufficiently rapid that when the polymerization is complete, the electrode is dry to the touch. The enzyme electrode is now more stable because a less porous polymer is formed. At a higher polymerization temperature, such as 43°C, the resulting electrode is again less stable than when the polymerization temperature is 28°C. Therefore, maximum stability is obtained with this enzyme electrode when the photopolymerization temperature is 25–28°C.

To determine the effect of a film of cellophane (dialysis membrane) on enzyme electrode stability, an electrode was made by placing a thin film of cellophane over the enzyme gel layer. The cellophane was permeable to the urea substrate but not to the high-molecular-weight enzyme. Polymerization parameters were the same as those used to obtain the maximum stability for the cellophane electrode. Enzyme electrode stability, measured with either 8.33

$\times$ 10^{-2} or 1×10^{-3} M urea, showed no measurable loss in activity for 21 days (electrode stored between measurements in tris buffer at 25°C). After 21 days the electrode began to lose activity. The increased stability of this electrode over the membraneless electrode is apparently due to the cellophane, which prevents any enzyme from leaching out of the enzyme gel layer. The membrane also keeps bacteria from getting into the enzyme gel.

Another factor that affects the apparent stability of all electrodes, especially the "soluble" and physically entrapped electrodes, is the content of enzyme in the reaction layer. As will be shown later, a certain amount of enzyme is required to yield a Nernstian calibration curve. Many times it is advantageous to add more enzyme, for example, twice as much; in this case, more enzymatic activity can be lost, yet a linear Nernstian plot is still obtained.

Still another factor that affects the stability of an electrode is the choice of operating conditions. An example of this is the comparison of the results obtained by Rechnitz and Llenado (1971a, b) and those of Mascini and Liberti (1974) for the amygdalin electrode. Amygdalin is cleaved by β-glucosidase to give a CN^- ions, which are sensed by a CN^- ion-selective electrode. Since this electrode responds best to free CN^- ions, obtainable only at pH $>$ 10, Rechnitz and Llenado used this high pH for operation of their electrode. Even though the enzyme was physically bound, it lost activity continually and showed a lifetime of only a few days. It is known that almost all enzymes will lose activity at pH $<$ 3 and $>$ 9, and undoubtedly this was one contributor to the poor stability. Mascini and Liberti used only a soluble enzyme at a pH of 7 and found not only better stability (1 week, which is still all that can be expected from a soluble enzyme) but also faster response times. Another reason for the poor stability of Rechnitz and Llenado (1971b) is the "sausage" polymerization these authors tried, in which large pieces of physically entrapped enzyme are prepared and slices are cut for each assay. From our own experience and that of others with such a technique, the sausage obtained is like a roast beef placed in a very hot oven for 30 min — it is well done on the outside and raw on the inside. The reader is advised not to attempt such a large-scale entrapment but should prepare individual small batches of polyacrylamide enzyme gels.

A thorough comparison of the stability of the three types of "immobilized" electrodes — physically entrapped (type 3), and chemically bound (types 1 and 2) — was shown by Guilbault and Lubrano (1973) for glucose oxidase (Fig. 10.1). The long-term stabilities of the types 1, 2, and 3 electrode were studied by testing the response of each type of electrode to 5×10^{-3} M glucose in phosphate buffer, pH 6.6, at least once a week for several months. When not in use, the electrodes were stored in phosphate buffer at 25°C.

The results, shown in Fig. 10.1, show that the long-term stability decreased in the order chemically bound $>$ physically bound. Not only did the type 3 electrode response decrease with time, it also decreased with each determination of glucose. This is a serious problem, and as a result, the type 3 electrode is of limited use analytically, except with frequent calibration and use of a large excess of enzyme. This problem is not encountered with the type 1 and type 2 electrode consisting of immobilized enzyme. The activity of these two electrodes actually increases for the first 20–40 days before beginning to decrease; this is probably due to the establishment of diffusion channels in the matrix over a period of time, with concomitant increase in apparent activity until the channel formation ceases and only denaturation is observed. Likewise, it could be due to changes in the conformation of the fraction of enzyme immobilized in a nonactive conformation to the more stable and preferred conformation. Immobilization in an unfavorable conformation can be due to pH, temperature, or stirring effects during the immobilization process. The decrease in response is due to a decrease in activity of the enzyme layer because of slow denaturation and possibly also slow irreversible inhibition. The type 1 and type 2 electrodes eventually reach a stability change of -0.25 and -0.08% of maximum response per day, respectively. The physically bound enzyme lost half its activity in 7 months, but the chemically bound enzyme had lost only 30% of its activity in 400 days (13 months). Of course, this stability would have been much less if the electrodes had been subjected to considerable use each day; actually, about 200 assays for the type 3 electrode and almost 1000 for the type 1 or 2 enzyme electrode are possible.

Still another factor affecting the stability of some enzyme electrodes is the leaching out of a loosely bound cofactor from the active site, a cofactor that is needed for the enzymatic activity. Such was found by Guilbault and Hrabankova (1971) in the case of D-amino acid oxidase in a polyacrylamide membrane. The bond between protein and coenzyme (flavine adenine dinucleotide, FAD) is very weak in D-amino acid oxidase, and FAD is easily removed by dialysis against buffer without FAD. Without FAD in the solution used to store the electrode, all activity is lost in 1 day; using a 4×10^{-4} M solution of FAD in tris buffer, pH 8.0, to store the electrode between use resulted in a 3-week stability, with little loss in activity.

Finally, the stability of the enzyme electrode will depend on the stability of the base sensor. This, in most cases, is not the limiting factor in the stability, the sensor having a longer stability than the immobilized enzyme. This factor should be considered, however, in use of some of the shorter lifetime electrodes, such as the liquid membrane electrodes.

2. Response Time. There are many factors that affect the speed of response of an enzyme electrode; these are listed in Table 10.3. To obtain a response, (1) the substrate must diffuse through solution to the membrane surface, (2) the substrate must diffuse through the membrane and react with the enzyme at the active site, and (3) the products formed must then diffuse to the electrode surface where they are measured. Let us consider each of these factors in detail and see how the response time can be optimized. Mathematic models describing this effect can be derived, as was done by Blaedel et al. (1972) and by Mell and Maloy (1975).

TABLE 10.3   Factors Affecting the Response Time of an Enzyme Electrode

1. Stirring rate of solution
2. Concentration of substrate $10^{-1} > 10^{-3} > 10^{-5}$
3. Concentration of enzyme
4. pH optimum
5. Temperature (most effect on rate)
6. Dialysis membrane

*A fast response is defined as a low response time.

a. Rate of Diffusion of the Substrate.   In simplest, practical terms the rate of substrate diffusion depends on the stirring rate of the solution, as has been shown experimentally by Mascini and Liberti (1974) for the amygdalin electrode and as is described in Fig. 10.2. In an unstirred solution the substrate gets to the membrane surface, albeit slowly, so long response times are observed. At high stirring rates the substrate quickly diffuses to the membrane surface, where it can react. The difference can be a decrease of response time as much as from 10 min to 1–2 min, or less. With rapid stirring for the urea electrode (Guilbault and Montalvo, 1969a, 1970) a response time less than 30 s was achieved. Of importance also is the relationship of stirring rate to the

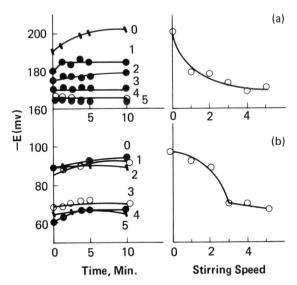

FIGURE 10.2   Stirring effect. Response time and equilibrium value with different stirring speed. The number 0 corresponds to unstirred solutions; numbers 1–5 represent increasing stirring speed and are arbitrary numbers. [Amygdalin] = (a) 10^{-2} M, (b) 10^{-4} M; enzyme amount = 1 mg. (Reproduced from Mascini and Liberti (1974) by permission of the authors.)

equilibrium potential observed. As shown in Fig. 10.2, the potential shifts as a function of stirring rate due to the changes in the amount of substrate brought to the electrode surface and the degree of its reactivity. Hence for fast response time and steady reproducible values it is recommended that a fast stirring of the solution be effected, yet a constant stirring rate be used (i.e., set the speed on your stirrer and use this same setting for all readings).

b. Reaction with Enzyme in Membrane. The rate of reaction will depend, according to the Michaelis-Menten equation:

$$V = \frac{k_3 [E][S]}{K_m + [S]}$$

on the activity of enzyme and factors that affect it, that is, pH, temperature, inhibitors, and the concentration of substrate. The equilibrium potential obtained, however, should be dependent only on the substrate concentration and the temperature (since this term appears in the Nernst equation). The response rate also depends on the thickness of the membrane layer in which reaction occurs (see Mell and Maloy, 1975) and on the size of the dialysis membrane used to cover the enzyme layer, if one is used. Let us consider each of these factors separately.

3. Factors that Affect Response Time

a. Effect of Substrate. Two typical examples of the effect of substrate concentration on response rate are shown in Figs. 10.3 and 10.4. Figure 10.3 shows the response of a β-glucosidase membrane electrode to amygdalin at various concentrations (Mascini and Liberti, 1974), and Fig. 10.4 shows the response of a glucose oxidase membrane electrode to glucose (Guilbault and Lubrano, 1973). In both cases the rate of reaction increases (as indicated by the increased inflection of the E versus time or i versus time curve) as the substrate concentration increases, and a faster response time is observed, that is, 1 min for 10^{-1} M amygdalin and 5 min for 10^{-4} amygdalin. As an alternative to waiting until an equilibrium potential or current is reached, the rate of change in the current or potential (Δi or $\Delta E/\Delta t$) can be measured and equated to the concentration of substrate. This was done by Guilbault and Lubrano (1973) in the case of the glucose electrode (Fig. 10.4).

b. Effect of Enzyme Concentration. The activity of enzyme in the gel will have two effects on an enzyme electrode: (1) it will ensure that a Nernstian calibration plot is obtained, and (2) it will affect the speed of response of the electrode. However, this effect is a tricky one, inasmuch as an increase in the amount of enzyme also affects the thickness of the membrane. This is demonstrated in Fig. 10.5, taken from the results of Mascini and Liberti (1974), for the amygdalin electrode. As the amount of enzyme is increased

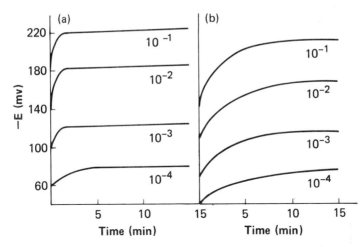

FIGURE 10.3 Amygdalin response-time curves for an electrode containing 1 mg of β-glucosidase immobilized by a dialysis paper (a) at pH 7 and (b) at pH 10. (From Mascini and Liberti (1974); reprinted with permission of authors.)

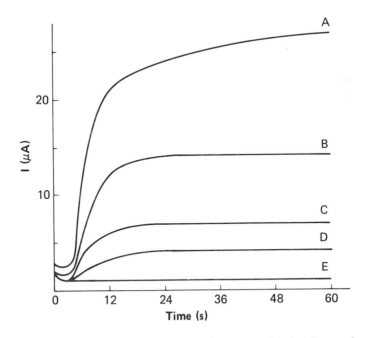

FIGURE 10.4 Family of current versus time curves for the glucose electrode poised at 0.6 V. Glucose solutions are in phosphate buffer, pH 6.0; ionic strength 0.1. A: 2.0×10^{-2} M; B: 1.0×10^{-2} M; C: 5.0×10^{-3} M; D: 2.5×10^{-3} M; E: 5.0×10^{-4} M. (Reproduced from Guilbault and Lubrano (1972) by permission of the authors.)

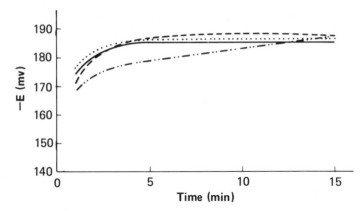

FIGURE 10.5   Response-time curves with different amounts of enzyme. pH 7; [Amygdalin] $= 10^{-2}$ M. (---) 0.1 mg enzyme; ($\cdots$) 1 mg enzyme; ($-$) 2.5 mg enzyme; ($-\cdots-$) 5 mg enzyme.

from 0.1 to 2.5 mg of β-glucosidase, a shorter response time is observed; yet when 5 mg of enzyme was used, the response time became much longer. This latter effect is due to a further thickening of the membrane layer by the use of more weight of enzyme, resulting in an increase in the time required for the substrate to diffuse through the membrane. (See Mell and Maloy (1975) for a mathematical discussion of this effect.) If one weight of enzyme had been chosen and the activity of enzyme increased at constant mass, a steady increase in the rate of response would have been observed, followed by a gradual leveling off in response time. Hence for best results it is recommended that as active an enzyme be used as possible, to ensure rapid kinetics, in as thin a membrane as is obtainable.

c. Effect of pH.   Every enzyme has a maximum pH at which it is most active and a certain range of pH in which it demonstrates any reactivity. The immobilized enzyme has a different pH range from the range of the soluble enzyme because of its environment, as has been discussed. The pH range for immobilized glucose oxidase is about 5.8–8.0 (Guilbault and Lubrano, 1973) (solution enzyme 5–7); that for β-glucosidase is about 5–8 (Mascini and Liberti, 1974). Hence for fastest responses, one should work at the pH optimum. This is not always possible, however, since the sensor electrode may not respond optimally at the pH of the enzyme reaction. Thus a compromise is generally necessary between these two factors. However, one should be careful not to be trapped into forcing the enzyme system to conform with the requirements of the sensor, as was done by Llenado and Rechnitz in the case of the amygdalin electrode (Rechnitz and Llenado, 1971a, b). These authors tried a pH of 10, which has been shown to be optimum for the electrode sensor, the CN$^-$ electrode. Longer response times were obtained at pH 10 (Fig. 10.3) compared to pH 7, since the enzyme has very little activity at this pH. Furthermore, the enzyme rapidly loses activity at high pH (>9–10), and Rechnitz and Llenado found their immobilized enzyme electrode to be very unstable.

Yet by working at pH 7, Mascini and Liberti (1974) found their soluble enzyme electrode to be still useful after 1 week. Similar effects have been noted in other studies; Guilbault and Shu (1971), for example, found the response time of a glutamine electrode decreased from pH 6 to pH 5, the latter being optimum for the enzyme reaction.

Further examples of making the electrode conform to the enzyme system, instead of vice versa, are the results of Anfalt et al. (1973) and Guilbault and Tarp (1974) on the NH_3 sensor used to monitor the urea–urease reaction. Although at the optimum pH for this enzyme reaction, 7–8.5, there is very little free NH_3 present to be sensed by a gas-type electrode, which would predict poor results for urea assay, both groups found that the sensitivity of the sensor was more than sufficient for each measurement at these low pH. This is partly ascribed to the fact that there is a buildup of larger amounts of product in the reaction layer than in solution, and hence an increase in sensitivity is obtained for the sensor, which sits close to the enzyme layer.

D. Assay of Organic Species

1. Urea.   An enzyme electrode was prepared for the substrate urea by immobilizing urease in a polyacrylamide matrix on 100-μm dacron and nylon nets. These nets were placed over the Beckman 39137 cation-selective electrode (which responds to NH_4^+ ion) (Guilbault and Montalvo, 1969a). The resulting "enzyme" electrode responds only to urea. The urea diffuses to the urease membrane, where it is hydrolyzed to NH_4^+ ion. This NH_4^+ ion is monitored by the ammonium ion–selective electrode, the potential observed being proportional to the urea content of the sample in the range 1.0–30 mg of urea/100 ml of solution. This enzyme electrode appears to possess stability (the same electrode has been used for weeks with little change in potential readings or drift), sensitivity (as little as 10^{-4} mol/l urea is determinable) and specificity. Results are available to the analyst in less than 100 s after initiation of the test, and the electrode can be used for individual samples or in continuous operation.

In a later publication, Montalvo and Guilbault (1969) described an improved urea-specific enzyme electrode that was prepared by placing a thin film of cellophane around the enzyme gel layer to prevent leaching of urease into the surrounding solution. The electrode could be used continuously for 21 days with no loss of activity. A full discussion of the parameters that affect the polymerization of urease as well as of the stability of four types of urease electrode was published by Guilbault and Montalvo (1970).

Guilbault and Montalvo (1969c) described the preparation of a sensitized cation-selective electrode. By placing a film of urease over the outside of an ordinary cation-selective glass electrode these workers obtained an electrode with increased sensitivity.

Because sodium and potassium ions interfered in the measurement, Guilbault and Hrabankova (1970b) used an uncoated NH_4^+ ion electrode as reference electrode to the urease-coated NH_4^+ electrode and added ion-exchange resin in attempts to develop an urea electrode useful for assay of blood and urine. Good precision and accuracy were obtained.

In attempts to improve the selectivity of the urea determination, Guilbault and Nagy (1973a) used a silicone rubber-based nonactin ammonium ion–selective electrode as the sensor for the NH_4^+ ions liberated in the urease reaction. The selectivity coefficients of this electrode were 6.5 for NH_4^+/K^+, 7.50×10^2 for NH_4^+/Na^+, and much higher for other cations. The reaction layer of the electrode was made of urease enzyme chemically immobilized on polyacrylic gel. A still further improvement was described by Guilbault et al. (1973) using a three-electrode system, which allowed dilution to a constant interference level. Analysis of blood serum showed good agreement with spectrophotometric methods, and the enzyme electrode was stable for 4 months at 4°. Schindler and Gülich (1981) described an urea electrode based on the use of an NH_4^+-selective electrode. By using catheter electrodes, continuous measurements could be made at precisely localized positions of the living organism. Problems of toxicology, sterility, and safety were discussed.

Still further improvement in the selectivity of this type of electrode was obtained by Anfalt et al. (1973), who polymerized urease directly onto the surface of an Orion ammonia gas electrode probe by means of glutaraldehyde. Sufficient NH_3 was produced in the enzyme reaction layer even at pH as low as 7–8 to allow direct assay of urea in the presence of large amounts of Na^+ and K^+. A response time of 2–4 min was observed.

Guilbault and Tarp (1974) described a still better, total interference-free, direct-reading electrode for urea, using the air-gap electrode of Ruzicka and Hansen (1974). A thin layer of urease chemically bound to polyacrylic acid was used, at a solution pH of 8.5, where good enzyme activity was still obtained, yet where sufficient NH_3 is liberated to yield a sensitive measurement with the air-gap NH_3 electrode. The urea diffuses into the gel; the NH_3 produced diffuses out of solution to the surface of the air-gap electrode, where it is measured. A linear range of 3×10^{-2} to 5×10^{-5} M was obtained, with a slope of 0.75 pH unit per decade. The electrode could be used continuously for blood serum analysis for up to a month (at least 500 samples) with a precision and accuracy of less than 2%. The response time is 2–4 min at pH 8.5, and the electrode is washed under a water tap for 5–10 s after each measurement. Absolutely no interference from any levels of substances commonly presented in blood was observed (Na^+, K^+, NH_4^+, ascorbic acid, etc.).

A urea electrode using physically entrapped urease and a glass electrode to measure the pH change in solution was described by Nilsson et al. (1973). The response time of the electrode to urea was about 7–10 min and had a linear range from 5×10^{-5} to 10^{-2} M with a change of about 0.8 pH unit per decade. The electrode could be kept at room temperature for about 2–3 weeks. The ionic strength and pH were controlled using a weak (10^{-3} M) tris buffer and 0.1 M NaCl.

Still another possibility for a urea electrode is the use of a CO_2 sensor to measure the second product of the urea–urease reaction, HCO_3^-. Guilbault and Shu (1972) evaluated the use of the CO_2 sensor and found that a urea electrode, prepared by coupling a layer of urease covered with a dialysis net with a CO_2 electrode, had a linear range of 10^{-4} to 10^{-1} M, a response time of about 1–3 min, and a slight response only to acetic acid. Na^+ and K^+ ions had no interference.

A potentiometric enzyme electrode was reported by Alexander and Joseph (1981), in which an enzyme immobilized in polyvinyl chloride is used to coat an antimony metal electrode to detect changes in pH when the electrode is immersed in a solution of the enzyme substrate. As an example, urea is determined in solution by using immobilized urease on an antimony electrode, giving a linear concentration range of 5.0×10^{-4} to 1.0×10^{-2} M urea with a slope of 44 mV per decade change in urea concentration. The response slope is stable for about 1 week, with response times in the range 1–2 min, but with absolute potential changes occurring from day to day.

A highly specific, reproducible enzyme electrode for urea was developed by Mascini and Guilbault (1977) that is vastly superior to other electrodes. The enzyme urease is chemically bound and attached to a new, improved Teflon membrane, which is an integral part of the NH_3 gas membrane electrode. From 200 to 1000 assays can be performed on one electrode with a C.V. of 2.5% over the range 5×10^{-5} to 10^{-2} M. At least 20 assays/h can be made with excellent correlation with the results obtained by the spectrophotometric diacetyl procedure.

Enzyme electrodes for urea, using glutaraldehyde bound urease and either CO_2 or glass electrodes, have been described by Tran-Minh and Broun (1975). By using an NH_4^+ glass electrode a response of 10^{-1}–10^{-5} M was obtained, similar to the results of Guilbault and co-workers (Guilbault and Montalvo, 1969a, 1970; Guilbault and Hrabankova, 1970b; Guilbault and Nagy, 1973). By using a CO_2 electrode, linearity was observed from 5×10^{-4} to 10^{-2} M.

2. Glucose and Sugar Electrodes.   The first report of an enzyme electrode was given by Clark and Lyons (1962), who proposed that glucose could be determined amperometrically by using soluble glucose oxidase held between Cuprophane membranes. The oxygen uptake was measured with an O_2 electrode:

$$\text{Glucose} + O_2 + H_2O \xrightarrow{\text{Glucose oxidase}} H_2O + \text{Gluconic acid} \qquad (10.2)$$

The term "enzyme electrode" was introduced by Updike and Hicks (1971), who coated an oxygen electrode with a layer of physically entrapped glucose oxidase in a polyacrylamide gel. The decrease in oxygen pressure was equivalent to the glucose content in blood and plasma. A response time of less than a minute was observed. Clark (1970) proposed measuring the hydrogen peroxide produced in the enzymic reaction with a Pt electrode. An instrument based on this concept, using glucose oxidase held on a filter pad, is now sold by Yellow

Springs Instrument Co. (Yellow Springs, Ohio). Two platinum electrodes are used, one to compensate for any electro-oxidizable compounds in the sample, such as ascorbic acid, the second to monitor the enzyme reaction.

Guilbault and Lubrano (1972, 1973) described a simple, stable, rapid-reading electrode for glucose. The electrode consists of a metallic sensing layer (Pt-glass), covered by a thin film of immobilized glucose oxidase covered in place by means of cellophane. When poised at the correct potential, the current produced is proportional to the glucose concentration. The time of measurement required with this amperometric approach is less than 12 s when a kinetic method is used. The electrode is stable for over 1 year when stored at room temperature with only a 0.1% change from maximum response per day. The enzyme electrode determination of blood glucose compares favorably with commonly used methods with respect to accuracy, precision, and stability, and the only reagent needed for assay is a buffer solution.

Guilbault and Lubrano (1978) constructed enzyme electrodes using membranes containing immobilized glucose oxidase constructed in a simple, easily reproducible procedure. The enzyme is co-cross-linked with bovine serum albumin using the bifunctional agent, glutaraldehyde. The best response was obtained by using membranes of high activity and thin membrane construction. The electrodes could be used continuously for several months.

Nagy et al. (1973) described a self-contained electrode for glucose based on an iodide membrane sensor:

$$\text{Glucose} + O_2 \xrightarrow{\text{Glucose oxidase}} \text{Gluconic acid} + H_2O_2 \qquad (10.3)$$

$$H_2O_2 + 2I^- + 2H^+ \xrightarrow{\text{Peroxidase}} 2H_2O + I_2 \qquad (10.4)$$

The highly selective iodide sensor monitors the local decrease in the iodide activity at the electrode surface. The assay of glucose was performed in a stream and at stationary electrode. Pretreatment of the blood sample was required to remove interfering reducing agents, such as ascorbic acid, tyrosine, and uric acid.

Nilsson et al. (1973) described the use of conventional hydrogen ion glass electrodes for the preparation of enzyme–pH electrodes by entrapping the enzymes either within polyacrylamide gels around the glass electrode or as a liquid layer trapped within a cellophane membrane. In an assay of glucose, based on a measurement of the gluconic acid produced, the pH response was almost linear from 10^{-4} to 10^{-3} M with a ΔpH of about 0.85 per decade. Electrodes of this type were also constructed for urea and penicillin (see below). The ionic strength and pH were controlled by using a weak (10^{-3} M) phosphate buffer, pH 6.9, and 0.1 M Na_2SO_4.

Dr. Pierre Coulet and co-workers (Jacques Julliard, Danielle Gautheron, and others) at the Laboratory of Biology and Membrane Technology of University Claude Bernard in Lyon have been especially active in the development of enzyme electrodes that are stable, self-contained, and easily fabricated in large scale (Coulet et al., 1974, 1980; Thevenot et al., 1978a, b, 1979; Coulet and Gautheron, 1981). A generally useful mild coupling method for placement of the enzyme onto the collagen membranes is used, with acyl azide activation and coupling. For example, stable, very sensitive (10^{-8} M detectable) glucose sensors have been described that are now sold commercially by Tacussel, Lyon, and are highly reliable.

The glucose sensor was prepared by using glucose oxidase immobilized on collagen membranes by this technique (Thevenot et al., 1979) and consists of a modified gas electrode in which the pH detector is replaced by a platinum anode and the porous selective membrane by the enzyme membrane. The latter is tightly pressed against the anode by a screw cap and is thus easily removable.

The electrode is immersed in a small vessel with the selected buffer in which glucose containing samples are injected. Enzymatically generated peroxide is detected by anodic oxidation at $+650$ mV versus a Ag/AgCl reference. After an injection the current output increases and reaches a steady state within a few minutes; differentiation gives a signal peak in 1 min. Linearity of signal and glucose concentration is obtained over the range 10^{-7} to 2×10^{-3} M. A compensating electrode mounted with a noncollagen membrane permits the detection of, and correction for, electrochemical interferences when testing samples with high levels of electroactive species. The glucose probe was used for rapid assays in blood (Thevenot et al., 1978b) and food (Sternberg et al., 1980).

A two-enzyme electrode for maltose determination using the same electrochemical detector has been designed with membranes prepared by asymmetric coupling (Coulet and Bertrand, 1979). The two enzymes involved were glucoamylase and glucose oxidase. Glucoamylase above was immobilized on the membrane face exposed to the bulk phase into which the maltose containing samples were injected. The hydrolysis of maltose occurred according to the reactions:

$$\text{Maltose} + H_2O \xrightarrow{\text{Glucoamylase}} 2 \text{ D-Glucose} \qquad (10.5)$$

The glucose produced migrated through the membrane and was then oxidized on the inner face with immobilized glucose oxidase that was in close contact with the platinum disc. As with the monoenzyme system for glucose, the sensitivity and linearity were excellent.

The same base electrochemical sensor has also been used for the determination of various other species with collagen membranes bearing monoenzyme or multienzyme systems. A single multipurpose electrode (Bertrand et al., 1981), where selected membranes bearing different oxidases can be easily replaced, has been described for assay of galactose (galactose oxidase), starch (glucoamylase), and sucrose (invertase). In all electrodes it was found that asymmetric coupling improved the electrode performance, and all sensors could be used for hundreds of assays. Linearity was approximately 10^{-7} to 10^{-2} M for maltose or galactose, 10^{-4} to 2×10^{-3} M for sucrose, and 6×10^{-5} to 10^{-3} M for lactose.

3. Amino Acid Electrodes. Enzyme electrodes for the determination of L-amino acids were developed by Guilbault and Hrabankova (1970a), who placed an immobilized layer of L-amino acid oxidase over a monovalent cation electrode to detect the ammonium ion formed in the enzyme catalyzed oxidation of the amino acid. Two different kinds of enzyme electrodes were prepared by Guilbault and Nagy (1973b) for the determination of L-phenylalanine. One of the electrodes used a dual-enzyme reaction layer — L-amino acid oxidase with horseradish peroxidase — in a polyacrylamide gel over an iodide-selective electrode. The electrode responds to a decrease in the activity of the iodide at the electrode surface owing to the enzymatic reaction and subsequent oxidation of iodide.

$$\text{L-Phenylalanine} \xrightarrow{\text{L-Amino acid oxidase}} H_2O_2 \qquad (10.6)$$

$$H_2O_2 + 2H^+ + I^- \xrightarrow[\text{peroxidase}]{\text{Horseradish}} I_2 + H_2O \qquad (10.7)$$

The other electrode was prepared by using a silicone rubber–base nonactin-type ammonium ion–selective electrode covered with L-amino oxidase in a polyacrylic gel. The same principle of diffusion of substrate into the gel layer, enzymatic reaction, and detection of the released ammonium applies to the system. The CO_2 sensor was evaluated by Guilbault and Shu (1972) for response to tyrosine when coupled with tyrosine decarboxylase held in an immobilized form by a dialysis membrane. A linear range of 2.5×10^{-4} to 10^{-2} M was observed with a slightly faster response time than that observed with the urea electrode mentioned above. A slope of 55 mV per decade was obtained, compared to 59 mV per decade for the urea electrode.

Guilbault and Shu (1971) described an enzyme electrode for glutamine, prepared by entrapping glutaminase on a nylon net between a layer of cellophane and a cation electrode. The electrode responds to glutamine over the concentration range 10^{-1} to 10^{-4} M with a response time of only 1–2 min.

Guilbault and Lubrano (1974) prepared an electrode for L-amino acids by coupling chemically bound L-amino acid oxidase to a Pt electrode that senses the peroxide produced in the enzyme reaction:

$$\text{L-Amino acid} + O_2 + H_2O \longrightarrow \text{COCOOH} + NH_3 + H_2O_2 \quad (10.8)$$

The time of measurement is less than 12 s, using a kinetic measurement of the rate of increase in current per unit time, and the only reagent required is a phosphate buffer. The L-amino acids cysteine, leucine, tyrosine, phenylalanine, tryptophane, and methionine were assayed.

Guilbault and Nanjo (1974) proposed enzyme electrodes for L-amino acids based on the use of immobilized enzymes and a Pt-based O_2 electrode. The change in the dissolved O_2 level is monitored, and the electrode responds to L-methionine, L-leucine, L-phenylalanine, L-tyrosine, L-cysteine, L-lysine, and L-isoleucine. Excellent stability (> 4 months), fast response times (< 1 minute), and greater selectivity over the peroxide-based sensors were observed.

A totally specific enzyme electrode useful for the assay of ʟ-lysine in grains and foodstuffs was described by White and Guilbault (1978). No response is noted with any ᴅ-amino acid or any other ʟ-amino acid. The electrode can be used for assay of this amino acid in mixtures, without the necessity for extensive separations and expensive instrumentation (i.e., the amino acid analyzer). The electrodes are quite stable, with a linear range of ʟ-lysine concentration of 5×10^{-5} to 10^{-1} M. The only limitation is the long response time (5–10 min). After preparation, buffer is the only reagent needed.

Fung et al. (1979) have proposed a totally selective electrode probe for ʟ-methionine. The electrode is prepared by immobilizing purified ʟ-methionine γ-lyase (E.C.4.4.1.11) onto an ammonia-specific electrode. The α,γ elimination of ʟ-methionine proceeds with formation of α-ketobutyrate, methanethiol, and ammonia. Only ʟ-methionine reacts with the purified enzyme; a linear range of 10^{-2}–10^{-5} M is observed.

A potentiometric ʟ-tyrosine selective probe for the direct determination of ʟ-tyrosine in biological fluids and foods was described by Havas and Guilbault (1982). The sensor element of the probe, based on a CO_2 sensitive gas membrane, is a layer containing immobilized *apo*-ʟ-tyrosine decarboxylase. A linear range 2.6×10^{-3} to 4×10^{-5} M is observed, with no interference from any ʟ-amino acids or ᴅ-tyrosine. Kumar and Christian (1975) have used a gas O_2 electrode and the enzyme tyrosinase to measure ʟ-tyrosine. The decrease in the O_2 level serves as an indication of the concentration of ʟ-tyrosine.

4. Alcohol and Acid Electrodes.   Alcohol oxidase catalyzes the oxidation of lower primary aliphatic alcohols.

$$RCH_2OH + O_2 \xrightarrow[\text{oxidase}]{\text{Alcohol}} RCHO + H_2O_2 \qquad (10.9)$$

The hydrogen peroxide produced in these reactions may be determined amperometrically with a platinum electrode as in the determination of glucose above. Guilbault and Lubrano (1974) used the alcohol oxidase obtained from a basidiomycete to determine the ethanol concentration of 1-ml samples over the range 0–10 mg/100 ml, with an average relative error of 3.2% in the 0.5–7.5 mg/100 ml range. This procedure should be adequate for clinical determinations of blood ethanol, since normal blood from individuals who have not ingested ethanol ranges from 40 to 50 mg/100 ml. Methanol is a serious interference in the procedure, since the alcohol oxidase is more active for methanol than ethanol when H_2O_2 is measured (see below). However, the concentration of methanol in blood is negligible in comparison with that of ethanol.

A vastly improved alcohol electrode, selective for ethanol, was described by Guilbault and Nanjo (1975). The decrease in current as O_2 was depleted from solution in the enzymic reaction (see Eq. (10.9)) was measured at an applied potential of -0.6 V versus SCE.

From 0.4 to 50 mg % can be assayed with little interference. The electrode was also found to exhibit a totally different selectivity pattern from the H_2O_2 probe used by Guilbault and Lubrano (1974), ethanol being the preferred substrate. This is believed to be due to the total specificity of the O_2 probe, which provides a measurement independent of the chemical reaction that can occur between the aldehyde formed with peroxide to give the organic acid (which is also a substrate for the enzyme reaction). Thus all previous selectivity values given in the literature for this enzyme are believed to be in error; the correct values are given below:

Alcohol	Relative Reactivity		
	Janssen[a]	Enzyme O_2[b]	Electrode H_2O_2[c]
MeOH	100	3	0.98
EtOH	28	12,000	0.46
Allyl alcohol	17	190	1.14
n-Propanol	5.3	1,300	0.36
n-Butanol	2.1	2,500	0.21

[a] Colorimetric assay of H_2O_2 formed (Janssen and Ruelius, 1968; Janssen et al., 1965).
[b] Method of Guilbault and Nanjo (1975).
[c] Method of Guilbault and Lubrano (1974).

In addition the enzyme reacts quite well with aldehydes and acids, as shown:

Substrate	Relative Activity (O_2)
CH_3OH	1.0
HCOH	17,600
HCOOH	5,640
C_2H_5OH	1.00
CH_3COH	0.89
CH_3COOH	0.88
C_2H_5OH	1.00
Acetic acid	0.88
Formic acid	0.51
Lactic acid	0.37
Butyric acid	0.21
Pyruvic acid	0.03

Thus acids like acetic or formic and aldehydes like formaldehyde and acetaldehyde can be easily determined.

5. Cholesterol.   An enzyme electrode for free cholesterol was described by Satoh et al. (1977). The electrode comprised double membrane layers—of which one was a cholesterol oxidase–collagen membrane and the other an oxygen permeable membrane—a platinum, and lead anode. The assay involved monitoring the decrease in dissolved oxygen (see Eq. (10.11)).

$$\text{Cholesterol Esters} \xrightarrow{\text{Esterase}} \text{Cholesterol} \qquad (10.10)$$

$$\text{Cholesterol} + O_2 \xrightarrow{\text{Oxidase}} \Delta^4\text{-cholestenone-3} + H_2O_2 \qquad (10.11)$$

An assay of total cholesterol esters was proposed by Huang et al. (1975). Chemically, immobilization of the enzymes cholesterol esterase and cholesterol oxidase onto alkylamine glass beads provided a stable enzyme stirrer to convert completely all the total cholesterol esters into first cholesterol (see Eq. (10.10)) and then hydrogen peroxide (see Eq. (10.11)), which is measured by the current flow at a Pt electrode.

Clark et al. (1978) described a cholesterol electrode by use of a polarographic anode with multiple enzymes. Cholesterol ester hydrolase and cholesterol oxidase are used to produce peroxide, which is sensed by a Pt anode. By combining these two enzymes it is possible to obtain the benefits of enzyme specificity and devise a system that requires only small plasma samples. Since the enzyme electrode response is measured by the rate of reaction, results are achieved within 4 min. The Yellowsprings Model 23 glucose analyzer was used. The cholesterol oxidase was glutaraldehyde bound to a collagen membrane, but the esterase was soluble. Only a gradual small reduction in sensitivity was observed with time of the electrode. The recovery was 97%, and the C.V. was 4% with a range of 25–300 mg % cholesterol.

Coulet and co-workers (Coulet and Bertrand, 1979; Coulet et al., 1981) have described an enzyme electrode using collagen-immobilized cholesterol oxidase for the microdetermination of free cholesterol. The electrode is poised at a potential of $+650$ mV versus Ag/AgCl and detects the H_2O_2 produced in the enzymatic reaction. Very high sensitivity and a wide range of linearity (10^{-7} to 0.8×10^{-4} M) results. The use of a nonenzymatic electrode associated with the enzymatic one allowed the detection of, and correction for, electrochemical interferences when applied to human sera for free cholesterol determinations.

6. Penicillin Electrode.   The first attempt at the design of a penicillin electrode was made by Papariello et al. (1973). The electrode was prepared by immobilizing penicillin β-lactamase (penicillinase) in a thin membrane of polyacrylamide gel molded around, and in intimate contact with, a glass (H^+) electrode. The increase in hydrogen ion from the penicilloic acid liberated from penicillin is measured:

$$\text{Penicillin} \xrightarrow{\text{Penicillinase}} \text{Penicilloic acid} \qquad (10.12)$$

The response time of the electrode was very fast (< 30 s) and had a slope of 52 mV/decade over the range 5×10^{-2} to 10^{-4} M for sodium ampicillin. The reproducibility of the electrode was very poor, probably because no attempts was made to control the ionic strength and pH.

Nilsson et al. (1973) prepared a penicillin electrode by entrapping penicillinase as a liquid layer trapped within a cellophane membrane around a glass (H^+) electrode, yet controlled the ionic strength and pH by using a weak 0.005 M phosphate buffer, pH 6.8, and 0.1 M NaCl. Good results were obtained in comparison to the results of Papariello et al.; the calibration plot was linear from 10^{-2} to 10^{-3} M with a ΔpH of 1.4 and as little as 5×10^{-4} M sodium penicillin could be determined. The electrode could be stored for 3 weeks, and the average deviation was $\pm 2\%$ with a response time of about 2–4 min.

Cullen and co-workers (1974) reported a revised model of their original penicillin electrode. The authors claimed that it was critical that a membrane be placed between the enzyme layer and the glass electrode to achieve satisfactory results.

Nilsson, Mosbach, and others (1978) used a penicillin-sensitive enzyme electrode to analyze the concentration of benzylpenicillin in fermentation broth. The electrode response time was in the region of 2 min, and the response to penicillin concentration was linear within the range of 1–10 mM. At low buffer capacity the sensitivity of the enzyme electrode to penicillin was very high. Constant calibration curves were obtained with the electrode when used for 2 hours daily in a fermentation medium over a 6-day period.

Similarly, Enfors and Nilsson (1979) and Hewetson et al. (1979) have described the use of an immobilized penicillinase electrode in the monitoring of penicillin in fermentation broths. Enfors and Nilsson (1979) described the purpose of their research as the construction of an autoclavable enzyme electrode and to describe some characteristics of a penicillin electrode built according to this principle — the response time of the electrode was 1 min. Hewetson et al. (1979) described the development of an on-line electrode suitable for use in a fermentation environment. Both groups used chemically bound (glutaraldehyde) enzyme on a flat surface glass pH electrode.

Olliff et al. (1978) described electrodes sensitive to penicillin with a response time of < 2 min based on covalent linkage of pencillinase to the glass of a pH electrode.

7. Creatinine.   A new kinetic method for potentiometric determination of creatinine in serum, based on the creatinine–picrate reaction in alkaline solution (Jaffe reaction), was described by Diamandis and Hadjiioannou (1981). The reaction is monitored with a picrate electrode, and the increase in electrode potential during 270 s is measured and related directly to the creatinine concentration. Small cation exchange columns were used to separate creatinine from interfering substances.

Thompson and Rechnitz (1974) have described the use of unpurified creatininase and an NH_4^+ probe for an electrode for creatinine and a creatinine.

$$\text{Creatinine} \xrightarrow{\text{Creatininase}} \textit{N}\text{-Methyl hydantoin} + NH_3 \qquad (10.13)$$

An enzyme electrode was described by Chen et al. (1980).

The enzyme creatininase was immobilized onto alkylamine glass beads, then placed into a stirrer. The free NH_3 produced was measured with a gas membrane electrode. Low levels of creatinine could not be measured because of the NH_4^+ present in the sample—these could be removed by several techniques described. An improved direct-reading specific electrode for creatinine was developed by Guilbault and Coulet (1983) using a new enzyme from Carla-Erba. Linearity from 1 to 100 mg % was obtained.

An enzyme electrode system for the determination of creatinine and creatine was developed by utilizing three enzymes: creatinine amidohydrolase (CA), creatine amidinohydrolase (CI), and sarcosine oxidase (SO) (Tsuchida and Yoda, 1983). These enzymes were coimmobilized onto the porous side of a cellulose acetate membrane with asymmetric structure, which has selective permeability to hydrogen peroxide. Two kinds of multienzyme electrodes were constructed by combining a polarographic electrode for sensing hydrogen peroxide and an immobilized CA/CI/SO membrane or CI/SO membrane for creatinine plus creatine or creatine, respectively. The multienzyme electrodes responded linearly up to 100 mg of creatinine and creatine per liter in human serum. Response time was 20 s in the rate method, and the detection limit was 1 mg/L. Only 25 μL of serum sample is required. Analytical recoveries, precisions, and correlations with the Jaffe method were excellent, and the multienzyme electrodes were sufficiently stable to perform more than 500 assays. No loss of activity of immobilized enzymes was observed after nine months of storage at 4°C in air.

III. ELECTRODE PROBES UTILIZING WHOLE CELLS— MICROBIAL OR TISSUE ENZYMES

One of the newest areas in biological electrode probes has been the application of whole-cell microorganisms or tissue cells to the surface of an I.S.E. to form a bioselective sensor. Divies (1975) used bacteria with an electrochemical assay for ethanol, but the first potentiometric microbial probes was reported by Divies (1976).

Several excellent reviews of this area have been written by Suzuki et al. (1982) (Biosensors in Japan), by Kobos (1980), and by Rechnitz (1981). A rather complete coverage of research through 1982 is presented herein.

The use of microbes for electrode probes offers three main advantages:

1. Purified enzymes are not necessary; the tissue slice or whole cell could be used directly without extensive purification and separation steps.
2. The electrode can be regenerated by immersion in nutrient broth. The microbe is essentially "living" and can be fed and kept alive for long periods.
3. The whole cell can contain many enzymes and several cofactors that can catalyze extensive transformations that could be difficult, if not impossible, to effect with single immobilized enzymes. In addition, the cofactors necessary for enzymic reaction are held in a natural immobilized state.

Disadvantages include:

1. Poor selectivity can result because the bacteria or microbe contains several enzymes that can convert many different substrates in addition to the one desired.
2. Poor response times are often observed because the enzymes in the microbe are present at low concentrations and the electrode membrane is very thick, subject to slow diffusion processes.

As an example of these advantages and disadvantages, consider first the aspartate electrode of Kobos and Rechnitz (1977). This bacterial sensor was stable for 10 days, using the ability of the bacterial colonies to regenerate themselves in appropriate growth media; the purified aspartate ammonia lyase was stable for only three days. Thus the growth and replenishment of the inactive cells must be the explanation for the increased lifetime of the probe. However, typically 10–20 mg (10^9 cells) are placed on each electrode for good response; if only 1–2 mg of bacteria are used, very long response times of 20 min are encountered.

The arginine electrode produced (Divies, 1976) illustrates the lack of selectivity of the bacterial probe—not only arginine, but glutamine and asparagine also respond. This is typical of a bacterial probe, since frequently many enzymes are present in the cell. *Escherichia coli* has several decarboxylases present in the cell that could catalyze the decarboxylation of numerous amino acids. Purification can lead to great specificity for only one amino acid (White and Guilbault, 1978). Some bacterial electrodes, on the other hand, are highly selective. Consider the glutamine electrode, which is based on the activity of glutamine deaminase present in the bacteria. No interference was observed from alanine, arginine, asparagine, aspartic acid, histidine, or several other amino acids or amines tested.

An excellent example of the complex reaction sequences that can be catalyzed by the many enzymes and cofactors present in the microbe is the research of Kobos and Pyon (1981) who described an electrode for nitrilotriacetic acid (NTA). The bacterial cells carry out a four-step reaction sequence converting the NTA to ammonia. The greatest activity was obtained with bacterial cells harvested in the exponential growth phase; good sensitivity was obtained, but again there were considerable interferences with the probe that could severely limit its usefulness.

A list of bacterial and tissue based electrodes is given in Table 10.4. Over 30 different electrode probes using whole cells have been described. The characteristics of some selected microbial sensors are presented in Table 10.4.

Essentially, the sensors fall into two categories:

1. Those that are based on an uptake of O_2 in the respiratory process (I of Fig. 10.6). In this case the microbial sensor is constructed by placing the immobilized microorganisms directly onto an O_2 electrode.
2. The amperometric or potentiometric determination of electroactive products liberated in the enzymatic reaction (II of Fig. 10.6). Such base electrodes could be pH, NH_3, H_2S, CO_2, or lactate, onto which is placed a layer of the tissue or bacterial cells.

IV. IMMUNO ELECTRODE PROBES

A. General Discussion
Another possible application of biological probes would be the construction of sensor probes utilizing bound antibodies or antigens. The linkage of an enzyme directly to an antigen or antibody or the direct binding of the antigen or antibody to a carrier like glass or collagen can be effected as easily as the binding of an enzyme. The enzyme–antigen or enzyme–antibody linkage, called enzyme immunoassay (EIA), has many advantages over radioimmunoassay (RIA): elimination of expensive counting equipment, no radioactive waste to deal with, cheap and stable EIA reagents. The technique of EIA, together with spectrophotometric or fluorogenic assays, is widely used in many clinical analyzers (like the Dade Stratum, Abbott 100) for analysis of drugs and metabolites. In this section the application of antigens and antibodies linked to electrode probes is discussed.

B. Linked Antibodies
One application of bound antibodies is the assay of antigens that react specifically with the antibody.

Yuan et al. (1981a) have proposed a novel creatine kinase isoenzyme MB (CK-MB) electrode, based on the principle that after immunoinhibition with goat-anti-human CK-M antibodies the residual activity of CK-B in serum is detected with a platinum electrode by coupling the NADH (generated from the hexokinase-glucose-phosphate dehydrogenase reactions) to the ferricyanide-diaphorase indicator reaction. The electrochemical oxidation of ferrocyanide

TABLE 10.4 Characteristics of Microbial Sensors

Sensor	Species	Electrode	Response Time (min)	Stability (days)	Range (ng/L)
Acetic acid	T. brassicae	O_2	15	30	10–100
Ammonia	N. europea	O_2	8	15	0.2–1.2
Arginine	S. faecium	NH_3	10	21	2–200
Aspartate	B. cadaveris	NH_3	20	10	0–20
BOD	T. cutaneum	O_2	20	17	0–60
Cephalosporin	C. freundii	pH	10	7	60–500
Cholesterol	N. erythropolis	O_2	1	30	3–40
Ethanol	T. brassicae	O_2	10 steady state	30	3–22.5
			6 pulse method		
Formic acid	C. butyricum	Fuel cell	20	20	10–1000
Glucose	P. fluorescens	O_2	10	14	2–20
Glutamic acid	E. coli	CO_2	5	15	8–800
Glutamine	Porcine Kidney	NH_3	5–7	30	3–300
Nicotinic acid	L. arabinosus	pH	60	30	0.05–5
Nitrate	A. vinelandii	NH_3	10	20	0.6–50
Nystatin	S. cervisiae	O_2	60	20	0.5–80
Phenol	T. cutaneum	O_2	0.25	5	0–15
Phenylalanine	L. mesenteroides	Lactate	90	30	1–500
Sugars	B. lactofermentum	O_2	10	20	20–200
Vitamin B_1	L. fermenti	Fuel cell	360	60	1–500
	Yeast	O_2	3	5	0.01–0.5

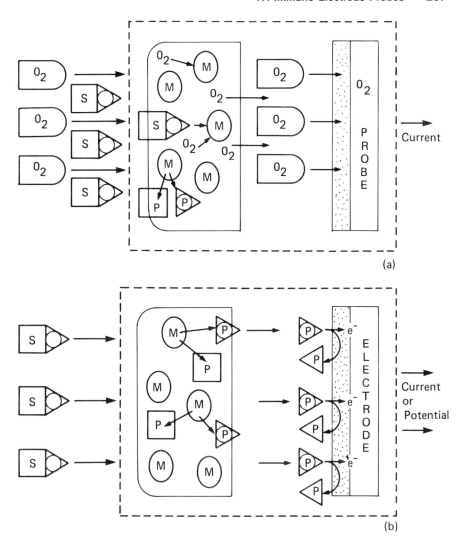

FIGURE 10.6 Principle of the microbial sensor. (a) Amperometric determination of respiration activity. (b) Amperometric or potentiometric determination of metabolites (electroactive substances). ⓢ▷, substrate; ⓟ▷ products (electrochemically inactive); ▷ product (electrochemically active); Ⓜ, immobilized microorganism.

ion is monitored at +0.36V versus S.C.E. The whole assay took only 10 min, and the linearity of a calibration plot of serum CK-MB enzyme activity extends up to 875 U/L. The C.V. and recovery value of this method were 3.0% and 97.8% respectively. The results obtained were in excellent agreement with the Helena electrophoresis method for the isoenzyme.

$$CK - M + B$$

$$\downarrow \text{Antibody CK-M}$$

$$CK\text{-}B$$

$$\text{CrP} \underset{\text{Cr}}{\overset{\text{ADP}}{\Huge)}} \overset{\text{ATP}}{\underset{}{}} \underset{\text{Glucose}}{\overset{\text{G-6-P}}{\Huge(}} \underset{\text{6-P-Gluconate}}{} \left(\begin{array}{c} \text{NADP} \\ +H_2O \\ \text{NADPH} \\ +H^+ \end{array} \right) \begin{array}{c} \text{Diaphorase} \\ \overbrace{\quad\quad} \text{Fe(CN)}_6^{4-} \\ \left(E = +0.36 \text{ V} \right) \\ \underbrace{\quad\quad} \text{Fe(CN)}_6^{3-} \end{array} \quad (10.14)$$

Yuan et al. (1981b) developed an immobilized immunostirrer for the determination of CK-MB isoenzyme in blood serum. The IgG antibodies are immobilized on alkylamine glass beads using glutaraldehyde as cross-linking agent, and the beads are packed into a rotating porous cell. After incubation with stirring, the CK-M isoenzymes in the blood serum sample are inhibited and are bound to the antibodies inside the stirrer. The residual CK-B isoenzyme activity is then determined electrochemically as described above in Eq. (10.14). The binding capacity of the immunostirrer to CK-M isoenzyme was estimated to be 800 U/liter, with an efficiency of 97.8%. The within-day and day-to-day coefficients of variation were 5% and 4%, respectively, over a period of 52 days. An immunostirrer loaded with antibodies attached to cyanogen bromide–activated cellulose beads was also characterized, but the antibodies were not as stable as on glass beads.

C. Bound Antigens

An enzyme immunoassay method using adenosine deaminase as the enzyme label was described by Gebauer and Rechnitz (1981). Potentiometric rate measurements were made with an ammonia gas–sensing electrode to determine the activity of enzyme label bound to agarose bead-immobilized second antibody. The activity is related to the concentration of either a model haptenic dinitrophenyl, DNP, or anti-DNP antibody. The detection limit is 50 ng antibody.

Aizawa and Suzuki (1977) and Aizawa et al. (1977) have described an immunosensor for determining specific protein. A liquid antigen containing cardiolipin, phosphatidyl choline, and cholesterol was immobilized to an acetyl cellulose membrane. The membrane-bound antigen retained immunological reactivity to Wasserman antibody. The asymmetric potential developed was dependent on the concentration of the antibody.

Suzuki (1979) bound this antigen and developed a method for assay of syphilis antibody in blood serum. The contact potential between bound antigen and antibody was measured, with very low ΔmV changes (eq. 1–3 mV) observed.

V. COMMERCIAL AVAILABILITY

Enzyme electrode probes for glucose are available from Tacussel (Lyon, France), and probes for glucose, urea, creatinine, uric acid, alcohols and amino acids are available from Universal Sensors (P.O. Box 736, New Orleans, Louisiana 70148).

Instruments utilizing immobilized enzymes and electrode sensors are sold by Leeds and Northrup (North Wales, Penn.), Yellow Springs Instrument Company (Ohio), and Technicon (Ardsley, N.Y.).

REFERENCES

Aizawa, M., and Suzuki, S. (1977) *Chem. Lett.* 7, 779–782.

Aizawa, M., Kato, S., and Suzuki, S. (1977) *J. Membrane Sci.* 2, 125.

Alexander, P. W., and Joseph, J. P. (1981) *Anal. Chim. Acta 131*, 103.

Anfalt, T., Granelli, A., and Jagner, D. (1973) *Anal. Lett.* 6, 969.

Bertrand, C., Coulet, P. R., and Gautheron, D. C. (1981) *Anal. Chim. Acta 126*, 23.

Blaedel, W. J., Kissel, T. R., and Bogaslaski, R. C. (1972) *Anal. Chem. 44*, 2030.

Chen, B., Kuan, S., and Guilbault, G. G. (1980) *Anal. Lett. 13*, 1607.

Chen, F., and Christian, G. (1979) *Clin. Chim. Acta 91*, 295.

Clark, L. C. (1970) U.S. Patent 3,529,455.

Clark, L., and Lyons, C. (1962) *Ann. N.Y. Acad. Sci. 102*, 29.

Clark, L. C., Emory, C., Glueck, C., and Campbell, M. (1978) in *Enzyme Engineering*, Vol. III (Pye, E. K., and Weetall, H. H., eds.), pp. 409–425, Plenum Press, New York.

Coulet, P. R., and Bertrand, C. (1979) *Anal. Lett. 12*, 581.

Coulet, P. R., and Gautheron, D. (1981) *J. Chromatogr. 215*, 65.

Coulet, P. R., Gautheron, D. C., and Bertrand, C. (1981) *Anal. Chim. Acta 126*, 23.

Coulet, P. R., Julliard, J. H., and Gautheron, D. C. (1974) *Biotech. Bioeng. 16*, 1055.

Coulet, P. R., Sternberg, R., and Gautheron, D. C. (1980) *Biochim. Biophys. Acta 612*, 317.

Cullen, L. F., Rusling, J. F., Schleifer, and Papariello, G. J. (1974) *Anal. Chem. 46*, 1955.

Diamandis, E., and Hadjiioannou, T. (1981) *Clin. Chem. 27*, 455.

Divies, C. (1975) *Ann. Microbiol.* (Paris) *126A*, 175.

Divies, C. (1976) *Chem. Eng. News 54*(44), 23.

Enfors, S., and Nilsson, H. (1979) *Enzyme Microbial Technol. 1*, 260.

Fung, K. W., Kuan, S. S., Sung, H. Y., and Guilbault G. G. (1979) *Anal. Chem. 51*, 2319.

Gebauer, C., and Rechnitz, G. (1981) *Anal. Lett. 14*, 97.

Guilbault, G. G. (1977) *Handbook of Enzymatic Analysis*, Marcel Dekker, New York.

Guilbault, G. G. (1985) *Handbook of Immobilized Enzymes*, Marcel Dekker, New York.

Guilbault, G. G., and Coulet, P. R. (1983) *Anal. Chim. Acta 152*, 223.

Guilbault, G. G., and Hrabankova, E. (1970a) *Anal. Lett. 3*, 53.

Guilbault, G. G., and Hrabankova, E. (1970b) *Anal. Chim. Acta 52*, 287.

Guilbault, G. G., and Hrabankova, E. (1971) *Anal. Chim. Acta 56*, 285.

Guilbault, G. G., and Lubrano, G. J. (1972) *Anal. Chim. Acta 60*, 254.

Guilbault, G. G., and Lubrano, G. J. (1973) *Anal. Chim. Acta 64*, 439.

Guilbault, G. G., and Lubrano, G. J. (1974) *Anal. Chim. Acta 69*, 183–189.

Guilbault, G. G., and Lubrano, G. J. (1978) *Anal. Chim. Acta 97*, 229.

Guilbault, G. G., and Montalvo, J. G. (1969a) *J. Amer. Chem. Soc. 91*, 2164.

Guilbault, G. G., and Montalvo, J. G. (1969b) *Anal. Lett. 2*, 283.

Guilbault, G. G., and Montalvo, J. G. (1970) *J. Amer. Chem. Soc. 92*, 2533.

Guilbault, G. G., and Nagy, G. (1973a) *Anal. Chem. 45*, 417.

Guilbault, G. G., and Nagy, G. (1973b) *Anal. Lett. 6*, 301.

Guilbault, G. G., and Nanjo, M. (1974) *Anal. Chim. Acta 73*, 367.

Guilbault, G. G., and Nanjo, M. (1975) *Anal. Chim. Acta 75*, 169.

Guilbault, G. G., and Shu, F. (1971) *Anal. Chim. Acta 56*, 333.

Guilbault, G. G., and Shu, F. (1972) *Anal. Chem. 44*, 2161.

Guilbault, G. G., and Tarp, M. (1974) *Anal. Chim. Acta 73*, 355.

Guilbault, G. G., McQueen, R., and Sadar, S. (1969) *Anal. Chim. Acta 45*, 1.

Guilbault, G. G., Nagy, G., and Kuan, S. S. (1973) *Anal. Chim. Acta 67*, 195.

Havas, J., and Guilbault, G. G. (1982) *Anal. Chem. 54*, 1991.

Hewetson, J. W., Jong, T. H., and Gray, P. P. (1979) *Biotech. Bioeng. Sympos. 9*, 125.

Huang, H., Kuan, S. S., and Guilbault, G. G. (1975) *Clin. Chem. 21*, 1605.

Janssen, W., and Ruelius, H. W. (1968) *Biochem. Biophys. Acta 151*, 330.

Janssen, W., Kerwin, R. M., and Ruelius, H. W. (1965) *Biochim. Biophys. Res. Comm. 20*, 630.

Kobos, R. (1980) in *Ion-Selective Electrodes in Analytical Chemistry*, Vol. II (Freiser, H., ed.), pp. 69–84, Plenum Press, New York.

Kobos, R. K., and Pyon, H. Y. (1981) *Biotech. Bioeng. 23*, 627.

Kobos, R. K., and Rechnitz, G. A. (1977) *Anal. Lett. 10*, 751.

Kumar, A., and Christian, G. (1975) *Clin. Chem. 21*, 325.

Mascini, M., and Guilbault, G. G. (1977) *Anal. Chem. 49*, 795.

Mascini, M., and Liberti, A. (1974) *Anal. Chim. Acta 68*, 177.

Mell, L., and Maloy, J. (1975) *Anal. Chem. 47*, 299.

Montalvo, J. G., and Guilbault, G. G. (1969) *Anal. Chem. 41*, 1897.

Nagy, G., von Storp, L. H., and Guilbault, G. G. (1973) *Anal. Chim. Acta 66*, 443.

Nilsson, H., Akerlund, Å., and Mosbach, K. (1973) *Biochim. Biophys. Acta 320*, 529.

Nilsson, H., Mosbach, K., Enfors, S., and Molin, N. (1978) *Biotech. Bioeng. 20*, 527.

Olliff, C. J., Williams, R. T., and Wright, J. M. (1978) *J. Pharm. Pharmacol. 30*, 45.

Papariello, G. J., Mukerji, A. K., and Shearer, C. M. (1973) *Anal. Chem. 45*, 790.

Rechnitz, G. (1981) *Science 214*, 287.

Rechnitz, G. A., and Llenado, R. (1971a) *Anal. Chem. 43*, 283.

Rechnitz, G. A., and Llenado, R. (1971b) *Anal. Chem. 43*, 1457.

Ruzicka, J., and Hansen, E. H. (1974) *Anal. Chim. Acta 69*, 129.

Satoh, I., Karube, I., and Suzuki, S. (1977) *Biotech. Bioeng. 19*, 1095.

Schindler, J. G., and Gülich, M. (1981) *Biomed. Technik 26*, 43.

Sternberg, R., Apoteker, A., and Thevenot, D. R. (1980) in *Electroanalysis in Hygiene, Environmental, Clinical and Pharmaceutical Chemistry* (Smyth, W. F., ed.), p. 461, Elsevier, New York.

Suzuki, S. (1979) *J. Solid-Phase Biochem. 4*, 25.

Suzuki, S., Satoh, I., and Karube, I. (1982) *Appl. Biochem. Biotechnol. 7*, 147.

Thevenot, D. R., Coulet, P. R., Sternberg, R., and Gautheron, D. C. (1978a) in *Enzyme Engineering*, Vol. IV (Braun, G. B., Manecke, G., and Wingard, L. B., eds.) p. 221, Plenum Press, New York.

Thevenot, P. R., Coulet, P. R., Sternberg, and Gautheron, D. C. (1978b) *Bioelectrochem. Biotechnol. 5*, 541, 548.

Thevenot, D. R., Coulet, P. R., Sternberg, R., Laurent, J., and Gaucheron, D. C. (1979) *Anal. Chem. 51*, 96.

Thompson, H., and Rechnitz, G. (1974) *Anal. Chem. 46*, 246.

Tran-Minh, C., and Broun, G. (1975) *Anal. Chem. 47*, 1359.

Tsuchida, T., and Yoda, K. (1983) *Clin. Chem. 29*(1), 51.

Updike, S. J., and Hicks, G. P. (1971) *Nature* (London) *214*, 986.

White, W., and Guilbault, G. (1978) *Anal. Chem. 50*, 1481.

Yuan, C., Kuan, S., and Guilbault, G. (1981a) *Anal. Chem. 53*, 190.

Yuan, C., Kuan, S., and Guilbault, G. (1981b) *Anal. Chim. Acta 124*, 169.

11

Biomedical Applications of Artificial Cells Containing Immobilized Enzymes, Proteins, Cells and Other Biologically Active Materials

Thomas M. S. Chang

I. INTRODUCTION

Artificial cells containing immobilized enzymes, proteins, or biological cells have exciting potential for biotechnological and medical applications (Chang, 1972b, 1976e, 1977b, d, 1979a, 1984; Chang et al. 1982c). The initial concerns about safety in patients have been resolved largely as a result of the successful development for medical applications of artificial cells containing other materials. Thus artificial cells containing adsorbents developed in our laboratories have been used for some time to replace standard hemodialysis treatment for patients with accidental or suicidal poisoning. The use of such cells as an artificial liver support for acute liver failure looks very promising in clinical trials. Their use as an artificial kidney in patients is also progressing well. With these demonstrated clinical applications of artificial cells, increasing interest is now also being channeled toward the possible clinical applications of artificial cells containing more specific materials like enzymes, immunosorbents, other proteins, and cell cultures. Problems of availability of special enzymes, monoclonal antibodies, cell cultures, and other biologically active materials required for medical applications of artificial cells are being solved by recent advances in biotechnology. It is expected that progress in the medical applications of artificial cells that are based on enzymes, proteins, and cells will in the next few years have a direct relationship to progress in other areas of biotechnology. Also, as in other areas of biotechnology, many potential medical applications of artificial cells are still in the experimental stages. Some of these areas will be presented as examples.

Artificial cells are prepared in the laboratory by using synthetic or biological materials (Chang, 1957, 1964, 1965, 1970, 1971a, b, c, 1973b, 1976f, 1977b, 1984; Chang et al., 1971b). A typical artificial cell is about the same size as many biological cells. Each consists of a spherical ultrathin membrane enveloping biologically active material. The membrane of each artificial cell separates the contents from the external environment. External permeative molecules can equilibrate rapidly into the enclosed biologically active enzymes, proteins, cell culture, or other materials (Chang, 1964, 1965, 1972a, b; Chang and Poznansky, 1968b). The membrane composition, permeability, and other characteristics can be varied over a wide range. It is possible to enclose almost any combination of enzymes, multienzyme systems, cofactor-regenerating enzyme systems, cell extracts, whole cells, proteins, adsorbents, magnetic materials, antigens, antibodies, vaccines, hormones, multicompartmental systems, and other materials. A number of potential applications suggested earlier have now reached the stage of clinical trial or application, while other areas are in the experimental stage. Detailed reviews of artificial cells are available (Chang, 1972b, 1975a, 1976c, e, 1977c, d, 1978, 1979a, 1980b, c, 1981a, 1984).

II. RED BLOOD CELL SUBSTITUTES FOR TRANSFUSION

At present we have to depend on donor blood for use in transfusion. There are a number of problems. The amount of donor blood can barely supply the ordinary demands of hospitals. There are problems related to possible transmission of hepatitis and AIDS. Donor blood can be stored by standard procedure only for a limited time. The blood group antigens on red blood cell membranes require cross-matching of donor blood. In some cases it may be difficult to find donors with the required rare blood groups. These problems are minor in comparison with the sudden increased demands for transfusion blood in the case of war or major disasters.

Hemoglobin is the oxygen-carrying protein of red blood cells. In free solution it is broken down to dimers and excreted rapidly, and its affinity for oxygen is also too high. To solve these problems, the contents of red blood cells have been microencapsulated within artificial cells for use as red blood cell substitutes (Fig. 11.1) (Chang, 1957, 1964, 1965, 1972a, 1980c; Sekiguchi and

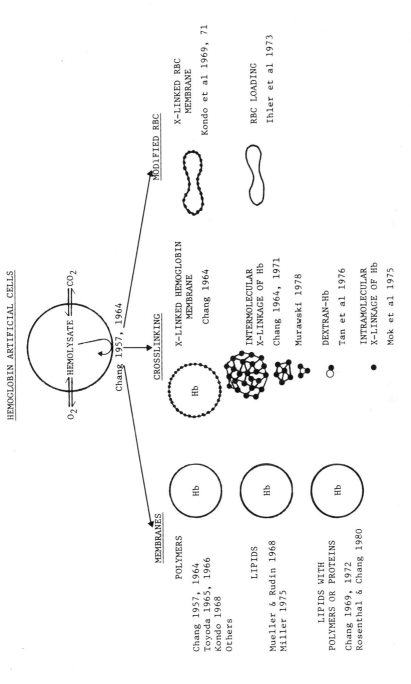

FIGURE 11.1 Development of artificial red blood cells based on hemoglobin. References refer to first publication. (From Chang, 1980c, by permission of the publisher.)

265

Kondo, 1966). The synthetic membrane, unlike the red blood cell membrane, does not carry blood group antigens and is much stronger, especially for storage. The hemoglobin inside the artificial cells is prevented from being broken down to dimers. However, artificial cells smaller in size than red blood cells are still filtered out by the liver and spleen as foreign particles (Chang, 1965, 1972a). A second approach is the use of cross-linking (Fig. 11.1) (Chang, 1964, 1965, 1971a, 1972a, b, 1980c). The hemoglobin can be cross-linked on the surface of the microdroplet into a polyhemoglobin membrane. All the hemoglobin of small microdroplets, down to 1 micron in diameter, can also be cross-linked into polyhemoglobin microspheres (Chang, 1964, 1965, 1972a, 1977a). Even with this, the artificial cells still do not have a sufficient retention time in the circulation. This, with the problem that cross-linking further increased the affinity of the hemoglobin for oxygen, was solved by further research. Thus Benesch et al. (1975) demonstrated that pyridoxylated hemoglobin has a lower affinity to oxygen, allowing the oxygen it carries to be released more easily when needed. Extension of the cross-linking approach using pyridoxylated hemoglobin resulted in soluble macromolecular polyhemoglobin, which has a lower oxygen affinity (Keipert et al., 1982; Keipert and Chang, 1983, 1984). Animal studies have demonstrated the effectiveness of this type of polyhemoglobin as a transfusion blood substitute for hemorrhagic shocks (Keipert and Chang, 1984). Further animal studies are required before this can be used in human patients. The results obtained so far indicated that it is much better than the fluorocarbon emulsion (Jamieson and Greenwalt, 1978) now being tested as blood substitute. Unlike fluorocarbon, which is retained by the liver for a long time, polyhemoglobin, after removal from the blood, can be metabolized in the body in the same way as hemoglobin from red blood cells. This is an exciting area that may one day help to solve the problem of donor blood in transfusion.

III. ARTIFICIAL CELLS IN ARTIFICIAL ORGANS

Artificial cells have a very large surface-to-volume ratio. Thus the total surface area in 33 ml of 100-micron-diameter artificial cells is two square meters, double the total surface area of a standard hemodialysis machine. Furthermore, the membrane thickness of artificial cells is only 200 Å, which is 1000 times thinner than the hemodialysis membrane being used in standard dialysis machines. This means that from a theoretical point of view the transport rate across the membranes of 33 ml of 100-micron-diameter artificial cells could be up to 2000 times higher than in the standard hemodialyzer (Chang, 1964, 1966, 1972a, 1974a). Enzymes, adsorbents, or other materials placed inside these artificial cells can efficiently remove or convert metabolites diffusing into the artificial cells. For example, 10 ml of these artifical cells containing urease retained in a small extracorporeal blood perfusion device converted 50% of the urea in a dog within 45 min (Chang, 1966). Tyrosinase artificial cells used this way could also lower systemic tyrosine to less than 50% of its original level within 1 hour (Shu and Chang, 1981). Similar studies were carried out by using artificial cells containing activated charcoal or ion exchange resin to remove other toxic materials (Chang, 1966, 1969a, b, 1972a).

The implications of artificial cells containing enzymes and cells in artificial organs can best be illustrated by the present clinical applications of artificial cells containing adsorbents.

Blood-compatible artificial cells containing activated charcoal do not adversely affect platelets or other formed elements of blood (Chang, 1966, 1969c, 1972b, 1974b, 1976b). An albumin coating could be applied to make the surface even more blood compatible (Chang, 1969c, 1972b, 1973b, 1974b, 1976b). An extracorporeal blood perfusion device with 100–300 grams of artificial cells containing activated charcoal has been studied in detail. Dogs that have received lethal doses of medications like barbiturate, salicylate, or Doridan, when treated with this artificial cell detoxifier device, recovered quickly as the drugs were rapidly removed from the circulation (Chang, 1969c, 1972b). When treated this way, patients who have taken suicidal doses of sleeping pills recovered quickly, as the drugs were rapidly removed (Chang et al., 1973a, b). The artificial cells are much more efficient than the standard hemodialyzer in removing these medications. This is now a routine treatment for poisoning in medical centers around the world (Agishi et al., 1980; Bonomini and Chang, 1982; Chang, 1975b, 1976a, 1978, 1980a; Gelfand et al., 1977; Piskin and Chang, 1982; Sideman and Chang, 1980; Vale et al., 1975), including treatment of infants (Chang et al., 1980; Chavers et al., 1980; Papadopoulous and Novello, 1982). It effectively removes aluminum (Chang and Barre, 1983). Artificial cells have also been used to microencapsulate ion exchange resin for blood-compatible resin hemoperfusion systems (Chang, 1966, 1972b). The albumin-coating technique developed here (Chang, 1969c, 1972b) has also been applied to resins (Sideman et al., 1981; Ton et al., 1979) to improve blood compatibility for application in patients with poisoning.

The first partial success with an artificial liver was realized here in 1972, using the artificial cell detoxification device (Chang, 1972c; Chang and Migchelsen, 1973). Treatments resulted in temporary recovery of consciousness in unresponsive grade IV hepatic coma patients. These results were supported by centers around the world (Agishi et al., 1980; Bonomini and Chang, 1982; Chang, 1975b, 1976a, 1978; Gelfand et al., 1978; Odaka et al., 1978; Sideman and Chang, 1980; Williams and Murray-Lyon, 1975). However, despite the temporary recovery of consciousness, improvement in the long-term survival rate of grade IV coma patients has not been conclusive. Our studies on a rat model showed that this treatment significantly increased the long-term survival rates (from 30% in control to 70% in treated) only if the animals were treated in the earlier stage of coma (Chang et al., 1978; Chirito et al., 1977; Tabata and Chang, 1980). This is supported by recent clinical trials (Gimson et al., 1982). With this first successful artificial liver available for

acute liver failure, many centers are now stimulated to develop more complicated artificial livers for chronic terminal liver failure (Chang, 1981b, 1982b,
1983). For this development the more actively metabolic aspects of liver function
will require the use of enzyme, multienzyme systems, and other biologically active systems (Chang, 1983). Tyrosinase artificial cells retained in extracorporeal blood perfusion devices effectively lowered elevated tyrosine levels in the
blood of rats with liver failure (Shi and Chang, 1982; Shu and Chang, 1980,
1981). Artificial cells containing multienzyme systems with cofactor recycling
are being studied for the sequential conversion of ammonia into amino acids
(Chang et al., 1979b). When further basic knowledge of the toxins responsible
for hepatic coma (Chang and Lister, 1981) is available, artificial cells with
more specific adsorbent or enzymes can be constructed.

Artificial cells are being developed as the basis of an artificial kidney to
replace the bulky and expensive hemodialysis machine (Chang, 1966, 1972b,
1974a, 1975b, 1976b, 1978, 1979b, 1981a, 1984; Chang and Malave, 1970).
The blood detoxifier, comprising artificial cells containing activated charcoal,
can maintain terminal renal failure patients alive and eliminate their uremic
symptoms of nausea, vomiting, fatigue, bleeding, and other problems (Chang,
1972b, 1976b, 1979b, 1981a; Chang et al., 1971a, 1972, 1974). To remove
water, electrolytes, and urea, the detoxifier was used in series with a hemodialysis machine (Chang, 1976a, 1978, 1979b; Chang et al., 1975, 1981, 1982a,
1982b). Because of the much more efficient removal of uremic toxins by artificial cells, much less treatment time is required than when using standard hemodialysis alone (Chang and Migchelsen, 1973; Chang et al., 1974). These results
are supported by further studies here and elsewhere (Bonomini and Chang,
1982; Bonomini et al., 1982; Martin et al., 1979; Odaka et al., 1978, 1980;
Stefoni et al., 1980; Winchester et al., 1976). To eliminate the need for a hemodialysis machine, the artificial cell blood detoxifier was combined in series with
a small ultrafiltrator (Chang, 1976a, 1978; Chang et al., 1975, 1977, 1979a).
This way, the artificial cells can remove the uremic waste metabolites and toxins while the small ultrafiltrator removes sodium chloride and water (Chang et
al., 1975, 1977, 1979a). In this approach, potassium and phosphate could be
removed by the oral administration of potassium adsorbent. However,
removal of urea is still a problem and may require the use of enzyme systems.
Artificial cells containing urease have been used in hemoperfusion to lower the
systemic urea level of dogs to 50% within 45 min (Chang, 1966). However,
urea is converted into ammonia, which can be removed by artificial cells containing ammonia adsorbent (Chang, 1966), although ammonia adsorbents
now available do not have the required capacity for ammonia. This principle
has now been developed for use in dialysate regeneration (Gordon et al.,
1969). Another approach we tested was the oral administration of urease artificial cells and ammonia adsorbent (Chang, 1972b, 1976a; Chang and Loa,
1970; Chang and Poznansky, 1968a). This way, urea from blood diffusing into
the intestine was converted into ammonia, which is removed by the ammonia
adsorbent. This study was extended (Gardner et al., 1971) and is now being

tested clinically in patients (Kjellstrand et al., 1981). The ammonia adsorbent still does not have sufficient capacity, and a large volume is required. Urease artificial cells have also been tested using a liquid membrane that is permeable to urea but impermeable to ammonium ions (Asher et al., 1975; May and Li, 1972). We are studying another approach using artificial cells containing urease and a complex multienzyme system for cofactor recycling and for the conversion of urea into ammonia which is then sequentially converted into different amino acids (Chang and Malouf, 1978; Chang et al., 1979b; Cousineau and Chang, 1977; Yu and Chang, 1981b, 1982).

IV. ARTIFICIAL CELLS CONTAINING SIMPLE ENZYME SYSTEMS

Most enzymes in the body carry out their functions inside cells. If enzymes, especially enzymes of nonhuman sources, are injected in free solution into the body, there may be a number of problems. These include hypersensitivity reactions, production of antibodies, rapid removal, and inactivation. Furthermore, enzyme in the free form cannot be directed to the specific site required for the enzymatic function. Enzymes in free solution are not stable, especially at a body temperature of 37°C. Furthermore, most of the important enzyme systems in the body function as multienzyme system with sequential enzymatic reaction requiring cofactors. Artificial cells have been investigated for solving some of the above problems.

With proper modifications the methods available (Chang, 1972b, 1976c, 1977a) can be used successfully to enclose most enzymes. Artificial cells containing a high concentration of protein, for example, a 10 g/100 ml hemoglobin solution, offer an intracellular environment that stabilizes the enclosed enzymes (Chang, 1965, 1971a). Further stabilization can be obtained by cross-linking with glutaraldehyde (Chang, 1971a).

Microencapsulated urease has been used as a model immobilized enzyme system for experimental therapy (Chang, 1964, 1965, 1966, 1972a; Chang et al., 1966). The basic results obtained pave the way for other types of enzyme replacement therapy. The first demonstration of the use of immobilized enzymes for replacement in hereditary enzyme-deficiency conditions involved microencapsulated catalase, effectively to replace an hereditary catalase deficiency in acatalasemic mice (Chang, 1972b; Chang and Poznansky, 1968a; Poznansky and Chang, 1974). Repeated injections did not result in the production of immunological reactions to the heterogenous enzyme in the artificial cells (Poznansky and Chang, 1974). Subsequently, liposome microencapsulated enzymes have been used for replacement in hereditary enzyme-deficiency conditions related to storage diseases (Gregoriadis, 1979; Gregoriadis et al., 1971). More recently, red blood cell–entrapped enzymes (Ihler et al., 1973) have also been tested for possible use in storage diseases. Artificial cells containing tyrosinase have been used in extracorporeal hemoperfusion to lower tyrosine levels in rats (Shu and Chang, 1980, 1981). Research into therapeutic applications of artificial cells containing enzymes also include the use of asparaginase for tumor suppression. The effectiveness of artificial cells containing asparaginase for experimental tumor suppression having been demonstrated (Chang, 1971b, 1972b), more detailed studies were carried out on the various aspects of artificial cells containing asparaginase (Chang, 1973a; Mori et al., 1973; Siu Chong and Chang, 1974). Artificial cells with immobilized phenylalanine ammonia lyase lowered phenylalanine levels in rats with phenylketonuria (Bourget and Chang, 1985).

As was described in the earlier section, artificial cells containing tyrosinase, urease, and other enzymes have also been studied for the construction of artificial kidneys, artificial livers, and detoxifiers (Chang, 1966, 1969b, 1972a; Chang and Malave, 1970; Chang et al., 1971a).

V. ARTIFICIAL CELLS CONTAINING MULTIENZYME SYSTEMS WITH COFACTOR RECYCLING

Most metabolic functions are carried out in biological cells by complex multienzyme systems with cofactor requirements. Artificial cells containing multienzyme systems with cofactor regeneration will have greater potential than those containing single-enzyme systems (Chang et al., 1982d).

Artificial cells containing hexokinase and pyruvate kinase recycled ATP for the continuous conversion of glucose into G-6-P and phosphoenol pyruvate into pyruvate (Campbell and Chang, 1976, 1977). Artificial cells containing alcohol dehydrogenase and malic dehydrogenase can recycle NADH, making use of NAD^+ (Campbell and Chang, 1976, 1977). Urease, glutamate dehydrogenase, and glucose-6-phosphate dehydrogenase, all within each artificial cell, can convert urea sequentially to an amino acid, glutamate, with glucose-6-phosphate dehydrogenase to recycle the cofactor (Cousineau and Chang, 1977). Artificial cells containing urease, glutamate dehydrogenase, and glucose dehydrogenase can carry out the same sequential conversions, but here glucose instead of glucose-6-phosphate is used to recycle the cofactor (Chang and Malouf, 1978). This has been optimized so that it can make use of glucose at concentrations normally present in the blood (Chang and Malouf, 1979). Artificial cells, each containing a multienzyme system of urease, glutamate dehydrogenase, alcohol dehydrogenase, and transaminase, convert the glutamic acid formed from urea into other amino acids (Chang et al., 1979b). Artificial cells containing multienzyme systems have also been studied for galactose removal (Chang and Kuntarian, 1978).

Thus it is feasible to prepare artificial cells containing multienzyme systems for the sequential conversion of substrates, with recycling of the required cofactors. Furthermore, waste metabolites like urea can be converted into useful products such as amino acids. However, for medical applications it would be more practical if cofactors could be retained inside the artificial cells for continuous recycling. Cofactors linked to soluble macromolecules like dextran can be retained within the artificial cells to be recycled in multienzyme reactions (Grunwald and Chang, 1978, 1979, 1981). These artificial cells have been used in small experimental continuous-flow reactors without the need to supply external cofactors (Grunwald and Chang, 1979). Another approach for retaining cofactors inside artificial cells for continuous recycling is being studied for lipid-soluble substrates. A lipid–polymer complex membrane artificial cell system (Chang, 1972b) has special permeability characteristics (Rosenthal and Chang, 1980). This type of artificial cell has very low permeability, even to small hydrophilic molecules, but very high permeability to lipid-soluble substrates. Ultrathin lipid–polymer membrane artificial cells containing multienzymes, cofactors, and substrates for multistep enzyme reactions (Yu and Chang, 1981a, b, 1982) can act as self-sufficient systems with external lipid-soluble substrates (Fig. 11.2). For example, artificial cells containing NAD^+,

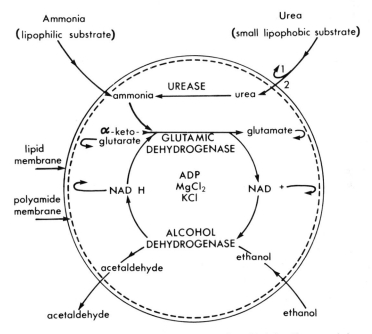

FIGURE 11.2 Schematic representation of artificial cells containing a multienzyme system with cofactor recycling. The ultrathin polymer membrane (200 Å) can be complexed with lipid. This way the cofactor NAD^+, α-ketoglutarate, and the enzymes are retained inside the artificial cells. External permeant lipophilic substrates like ammonia and ethanol or a very small hydrophilic substrate like urea can enter to take part in the reaction. If the external solutes are larger, hydrophilic substrates lipid should not be applied to the membrane. In these cases the cofactors have to be linked to soluble macromolecules like dextran to be retained inside. (From Yu and Chang, 1982, by permission of the publisher.)

alcohol dehydrogenase, glutamate dehydrogenase, α-ketoglutarate, $MgCl_2$, KCl, and ADP can act on a substrate solution containing only ammonia and alcohol. Ammonia is converted into an amino acid in the artificial cell, while alcohol is used to recycle the required cofactor in the artificial cells. These results bring the enzymatic functions of artificial cells a step closer to biological cells; furthermore, one is now no longer limited to using simple single-enzyme systems for possible medical applications.

VI. ARTIFICIAL CELLS CONTAINING BIOLOGICAL CELLS AND OTHER BIOLOGICAL MATERIALS

Artificial cells were prepared here to contain biological cells (Chang, 1965, 1972b, 1977a; Chang et al., 1966). This way, biological cells are separated from the external environment. It was proposed that in this form, cells, especially endocrine cells, could be implanted and still retain function, avoiding rejection by the immunological system of the body (Chang, 1965, 1972b). This principle has already been shown in the implantation of artificial cells containing enzymes (e.g., catalase and asparaginase) for successful in vivo action without causing immunological reactions (Poznansky and Chang, 1974; Siu Chong and Chang, 1974). More recently rat islet cells have been microencapsulated and then implanted intraperitoneally into diabetic rats (Lim and Sun, 1980). This way the microencapsulated islet cells can avoid rejection and can function to maintain normal glucose levels in the diabetic animals. This is a very promising approach that is being developed for possible clinical trials. Artificial cells containing fibroblasts or plasma cells have been used in in vitro cultures for the production of interferon and monoclonal antibodies, respectively (Damon Corporation, 1981). Other biologically active materials like hormones, antigens, antibodies, and vaccines can be enclosed within artificial cells (Chang, 1972b, 1975c, 1975d, 1976d, 1977a). Artificial cells containing antibodies to hormones have also been studied in clinical laboratories to measure plasma hormone levels (Ashkar et al., 1980). We have recently used artificial cells containing microorganisms and 3a-hydroxysteroid dehydrogenase for NAD recycling and stereospecific steroid oxidation (Ergan et al., 1984).

VII. IMMUNOSORBENT

Albumin has been applied to the collodion membranes of artificial cells containing activated charcoal (ACAC) to make the surface blood compatible, for hemoperfusion (Chang, 1969c, 1973b, 1974b, 1976b, 1976f, 1978). This albumin coating also acts to facilitate the transport of loosely albumin-bound molecules in the blood into the artificial cells (Chang et al., 1973a, b). Furthermore, this albumin on the surface of the ACAC can remove antibodies to albumin in the circulating blood of animals (Terman, 1980; Terman et al.,

1977). This latter observation has led to the incorporation of different antibodies or antigens into the collodion membrane, for use in the removal of specific antigen or antibody from the circulating blood (Terman, 1980). For example, albumin has been substituted with protein A on collodion-activated charcoal (Terman et al., 1981). When plasma was perfused over this column and then returned to patients, the result was a decrease in breast cancer in patients (Terman et al., 1981). These results are preliminary (Chang, 1982a).

Synthetic immunosorbent can be used to remove antibodies to blood groups. By applying an albumin collodion coating to the surface of the synthetic immunosorbent, such adverse effects as platelet removal and particulate embolism are eliminated (Chang, 1980d). Albumin collodion coating on this immunosorbent has now been applied, in clinical trials, to remove blood group antibodies before bone marrow transplantation (Bensinger et al., 1981a, b).

VIII. FUTURE DEVELOPMENTS OF ARTIFICIAL CELLS IN BIOTECHNOLOGY AND MEDICINE

Basic studies in artificial cells were started long before the present interest in biotechnology (Chang, 1964, 1972b). As a result, many of the basic ideas and methods have been waiting for development and application. With the present availability of many of the needed enzyme systems, special cell cultures, monoclonal antibodies, and other materials, artificial cells are now being rapidly developed for application. There is already substantial technology in artificial cells that can be developed for applications in biotechnology. Some of these applications will be discussed.

Numerous synthetic polymer membrane systems have been used to form the membranes of artificial cells (Fig. 11.3). Spherical ultrathin polymer membranes of numerous types can be formed using emulsification followed by interfacial polymerization or by interfacial organic phase separation (Aisina et al., 1976; Chang, 1957, 1964, 1965, 1972a, 1976c, 1977a; Chang et al., 1966, 1971b; Mori et al., 1973; Shiba et al., 1970). Multiple-compartment membrane systems consisting of small artificial cells enveloped within larger artificial cells have also been formed (Chang, 1965, 1972b). Another route, secondary emulsion, uses silastics, cellulose acetate, polylactic acids, and other polymers (Chang, 1965, 1966, 1972b, 1976c, 1977a). Liquid membranes can also be used in the secondary emulsion approach (May and Li, 1972). Biological and biodegradable membranes can also be used to microencapsulate enzymes and other biologically active materials. These include spherical ultrathin cross-linked protein membranes (Chang, 1964, 1965, 1972b, 1976c, 1977a; Chang et al., 1966); heparin-complexed polymer membranes (Chang, 1970, 1971c; Chang et al., 1967); spherical ultrathin lipid membranes (Mueller and Rudin, 1968); lipid-complexed membranes (Chang, 1969a, 1972b; Rosenthal and Chang, 1980); liposomes (Gregoriadis et al., 1971; Sessa and Weissman, 1970); erythrocyte-encapsulated enzymes (Ihler et al., 1973); and biodegradable polymer

ENCAPSULATION OF ENZYMES AND PROTEINS

Membrane Materials and Configurations

(REFERENCE - FIRST REPORT OF ENZYME ENCAPSULATION)

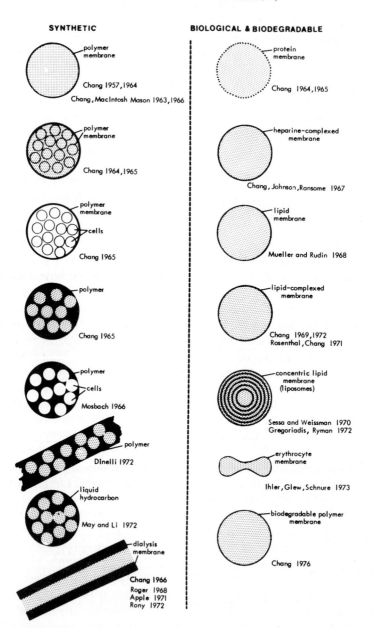

SYNTHETIC

polymer membrane

Chang 1957, 1964
Chang, MacIntosh Mason 1963, 1966

polymer membrane

Chang 1964, 1965

polymer membrane
cells

Chang 1965

polymer

Chang 1965

polymer
cells

Mosbach 1966

polymer

Dinelli 1972

liquid hydrocarbon

May and Li 1972

dialysis membrane

Chang 1966
Roger 1968
Apple 1971
Rony 1972

BIOLOGICAL & BIODEGRADABLE

protein membrane

Chang 1964, 1965

heparine-complexed membrane

Chang, Johnson, Ransome 1967

lipid membrane

Mueller and Rudin 1968

lipid-complexed membrane

Chang 1969, 1972
Rosenthal, Chang 1971

concentric lipid membrane (liposomes)

Sessa and Weissman 1970
Gregoriadis, Ryman 1972

erythrocyte membrane

Ihler, Glew, Schnure 1973

biodegradable polymer membrane

Chang 1976

FIGURE 11.3 Artificial cells: example of possible variations in membrane compositions and configurations. References refer to first publication. (From Chang, 1977d.)

274

membranes (Chang, 1976d). A cross-linked protein system can also be formed as cross-linked membranes, as cross-linked protein microspheres or soluble cross-linked macromolecules (Chang, 1964, 1965, 1971a, 1972b, 1977a).

Most biologically active material can be immobilized by using artificial cells (Fig. 11.4). This includes enzymes, multienzyme systems, cell extracts, and other proteins (Chang, 1957, 1964, 1972b, 1976c, 1977a; Chang et al., 1971b; Kitajima and Kondo, 1971; Østergaard and Martiny, 1973); granules of enzyme systems and proteins (Chang, 1957, 1972b); enzyme together with adsorbent (Chang, 1966, 1972b, 1977a; Gardner et al., 1971; Sparks et al., 1972); and biological cells and microbial cells (Chang, 1965, 1972b, 1977a; Chang et al., 1966; Ergan and Chang, 1984; Damon Corporation, 1981; Lim and Sun, 1980). Magnetic materials have been immobilized in artificial cells to allow external magnetic fields to direct the movements of the artificial cells (Chang, 1966, 1977a). Other materials include radioisotope-labeled enzymes and proteins (Chang, 1965, 1972b); insolubilized enzymes (Chang, 1969b, 1972b); cofactor recycling systems (Campbell and Chang, 1975, 1976; Chang, 1977b; Chang and Malouf, 1979; Chang et al., 1979b; Cousineau and Chang, 1977; Grunwald and Chang, 1978; Yu and Chang, 1981a, b, 1982); and antigens, antibodies, vaccines, and hormones (Chang, 1976c, d, 1977a, 1980d; Damon Corporation, 1981; Terman, 1980).

As was emphasized earlier (Chang, 1972b), the basic methods used for the preparation of artificial cells are, in fact, physical examples for demonstrating the principle of artificial cells. With the experimental and clinical demonstrations of this basic principle, many new physical systems can be developed based on the same principle. With biotechnology solving the problems of the availability of enzymes, cell cultures, and other biologically active materials, the technologies of artificial cells can now be more easily applied and extended.

REFERENCES

Agishi, T., Yamashita, N., and Ota, K. (1980) in *Hemoperfusion*, Part 1, *Kidney and Liver Support and Detoxification* (Sideman, S., and Chang, T. M. S., eds.), pp. 255–263, Hemisphere, Washington, D.C.

Aisina, R. B., Kazanskata, N. F., Lukasheva, E. V., and Berezin, V. (1976) *Biokhimiya 41*, 1656–1661.

Asher, W. J., Bovee, K. C., Frankenfeld, J. W., Hamilton, R. W., Henderson, L. W., Holzapple, P. G., and Li, N. N. (1975) *Kidney Int. 7*, S409–412.

Ashkar, F. S., Buehler, R. J., Chan, T., and Hourani, M. (1980) *J. Nucl. Med. 20*, 956–960.

Benesch, R. E., Benesch, R., Renthal, R. D., and Maeda, N. (1972) *Biochem. 11*, 3576–3582.

Bensinger, W. I., Baker, D. A., Buckner, C. D., Clift, R. A., and Thomas, E. D. (1981a) *New Engl. J. Med. 304*, 160–162.

Bensinger, W. I., Baker, D. A., Buckner, C. D., Clift, R. A., and Thomas, E. D. (1981b) *Transfusion 21*, 335–342.

MICROENCAPSULATION OF ENZYMES AND PROTEINS

Variations in Contents

(REFERENCE - FIRST REPORT OF ENZYME ENCAPSULATION)

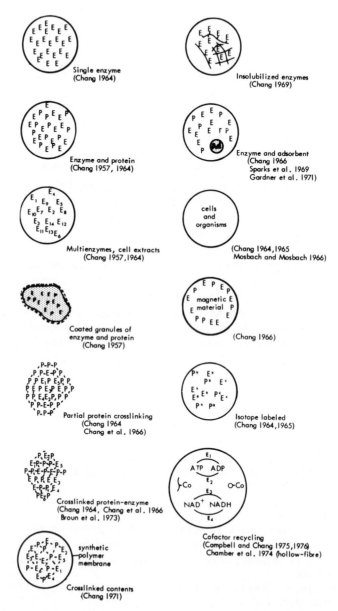

FIGURE 11.4   Artificial cells: examples of variations in compositions and contents of artificial cells. References refer to first publication. (From Chang, 1977d.)

Bonomini, V., and Chang, T. M. S. (1982) *Hemoperfusion, Contributions to Nephrology Series*. Karger, Basel, Switzerland.

Bonomini, V., Stefoni, S., Casciani, C. U., Taccone, M., Albertazzi, A., Cappelli, P., Mioli, V., Boggi, R., Mastrangelo, F., and Rizzelii, S. (1982) in *Hemoperfusion* (Bonomini, V., and Chang, T. M. S., eds.), pp. 133-142, Karger, Basel, Switzerland.

Bourget, L., and Chang, T. M. S. (1985) *FEBS Lett. 180*, 5-8.

Campbell, J., and Chang, T. M. S. (1975) *Biochim. Biophys. Acta 397*, 101-109.

Campbell, J., and Chang, T. M. S. (1976) *Biochem. Biophys. Res. Comm. 69*, 562-569.

Campbell, J., and Chang, T. M. S. (1977) in *Biomedical Applications of Immobilized Enzymes and Proteins*, Vol. II (Chang, T. M. S., ed.), pp. 281-302, Plenum Press, New York.

Chang, T. M. S. (1957) "Hemoglobin Corpuscles," Report of research project for B.Sc. Honours, McGill University, Montreal.

Chang, T. M. S. (1964) *Science 146*, 524-525.

Chang, T. M. S. (1965) "Semipermeable Aqueous Microcapsules," Ph.D. thesis, McGill University, Montreal.

Chang, T. M. S. (1966) *Trans. Amer. Soc. Artif. Intern. Organs 12*, 13-19.

Chang, T. M. S. (1969a) *Fed. Proc. 28*, 461.

Chang, T. M. S. (1969b) *Science Tools 16*, 33-39.

Chang, T. M. S. (1969c) *Can. J. Physiol. Pharmacol. 47*, 1043-1045.

Chang, T. M. S. (1970) U.S. Patent 3,522,346.

Chang, T. M. S. (1971a) *Biochem. Biophys. Res. Comm. 44*, 1531-1536.

Chang, T. M. S. (1971b) *Nature 229*(528), 117-118.

Chang, T. M. S. (1971c) Canadian Patent 876,100.

Chang, T. M. S. (1972a) in *Recent Development in Separation Science*, Vol. I (Li, N. N., ed.), pp. 203-216, CRC Press, Boca Raton, Fla.

Chang, T. M. S. (1972b) *Artificial Cells*, Thomas, Springfield, Ill.

Chang, T. M. S. (1972c) *Lancet 2*, 1371-1372.

Chang, T. M. S. (1973a) *Enzyme 14*(2), 95-104.

Chang, T. M. S. (1973b) U.S. Patent 3,725,113.

Chang, T. M. S. (1974a) *Separ. Purif. Methods 3*, 245-262.

Chang, T. M. S. (1974b) *Can. J. Physiol. Pharmacol. 52(2)*, 275-285.

Chang, T. M. S. (1975a) *Chem. Techn. 5*, 80-85.

Chang, T. M. S. (1975b) *Kidney Int. 7*, S387-392.

Chang, T. M. S. (1975c) in *Socio-Economic and Ethical Implications of Enzyme Engineering* (Heden, C-G., ed.), pp. 17-18, International Federation of Institutes for Advanced Studies, Stockholm, Sweden.

Chang, T. M. S. (1975d) in *Clinical Pharmacology, Proceedings of the Sixth International Congress of Pharmacology, International Society of Pharmacology, New York, NY, USA*, Vol. V (Matilla, M. J., ed.), pp. 81-90.

Chang, T. M. S. (1976a) *Kidney Int. 10*, S305-311.

Chang, T. M. S. (1976b) *Kidney Int. 10*, S218-224.

Chang, T. M. S. (1976c) in *Methods in Enzymology: Immobilized Enzymes*, Vol. XLIV (Mosbach, K., ed.), pp. 201-217, Academic Press, New York.

Chang, T. M. S. (1976d) *J. Bioeng. 1*, 25-32.

Chang, T. M. S. (1976e) in *Methods in Enzymology: Immobilized Enzymes*, Vol. XLIV (Mosbach, K., ed.), pp. 676-698, Academic Press, New York.

Chang, T. M. S. (1976f) Canadian Patent 982941.

Chang, T. M. S. (1977a) in *Biomedical Applications of Immobilized Enzymes and Proteins*, Vol. I (Chang, T. M. S. ed.), pp. 69–90, Plenum Press, New York.

Chang, T. M. S. (1977b) in *Biomedical Applications of Immobilized Enzymes and Proteins*, Vol. I (Chang, T. M. S., ed.), pp. 93–104, Plenum Press, New York.

Chang, T. M. S. (1977c) in *Biomedical Applications of Immobilized Enzymes and Proteins*, Vol. I (Chang, T. M. S., ed.), pp. 147–162, Plenum Press, New York.

Chang, T. M. S. (1977d) *Biomedical Applications of Immobilized Enzymes and Proteins*, Vols. I and II, Plenum Press, New York.

Chang, T. M. S. (1978) *Artificial Kidney, Artificial Liver, and Artificial Cells*, Plenum Press, New York.

Chang, T. M. S. (1979a) in *Drug Carriers in Biology and Medicine* (Gregoriadis, G., ed.), pp. 271–285, Academic Press, New York.

Chang, T. M. S. (1979b) *Clin. Nephrol. 11*, 111–119.

Chang, T. M. S. (1980a) *Clin. Toxicol. 17*, 429–542.

Chang, T. M. S. (1980b) *Enzyme Eng. 5*, 225–229.

Chang, T. M. S. (1980c) *Trans. Amer. Soc. Artif. Intern. Organs 26*, 354–357.

Chang, T. M. S. (1980d) *Trans. Amer. Soc. Artif. Intern. Organs 26*, 546–549.

Chang, T. M. S. (1981a) in *Advances in Basic and Clinical Nephrology, Proceedings of the Eighth International Congress of Nephrology*, pp. 400–406, Karger, Basel, Switzerland.

Chang, T. M. S. (1981b) in *Artificial Liver Support* (Brunner, G., and Schmidt, F. W., eds.), pp. 126–133, Springer, New York.

Chang, T. M. S. (1982a) *New Engl. J. Med. 306*, 936.

Chang, T. M. S. (1982b) *Lancet 2*, 1039.

Chang, T. M. S. (1983) *Int. J. Artif. Organs 24*, 95–103.

Chang, T. M. S. (1984) *Microencapsulation Including Artificial Cells*, Humana Press, New York.

Chang, T. M. S., and Barre, P. (1983) *Lancet 2*, 1051–1053.

Chang, T. M. S., and Kuntarian, N. (1978) in *Enzyme Engineering*, Vol. IV (Broun, G. B., Manecke, G., and Wingard, L. B., Jr., eds.), pp. 193–197, Plenum Press, New York.

Chang, T. M. S., and Lister, C. (1981) *Artif. Organs 4*, S169–172.

Chang, T. M. S., and Loa, S. K. (1970) *Physiologist 13*, 70.

Chang, T. M. S., and Malave, N. (1970) *Trans. Amer. Soc. Artif. Intern. Organs 16*, 141–148.

Chang, T. M. S., and Malouf, C. (1978) *Trans. Amer. Soc. Artif. Intern. Organs 24*, 18–20.

Chang, T. M. S., and Malouf, C. (1979) *Artif. Organs 3*(1), 38–41.

Chang, T. M. S., and Migchelsen, M. (1973) *Trans. Amer. Soc. Artif. Intern. Organs 19*, 314–319.

Chang, T. M. S., and Poznansky, M. J. (1968a) *Nature 218*(5138), 243–245.

Chang, T. M. S., and Poznansky, M. J. (1968b) *J. Biomed. Mater. Res. 2*, 187–199.

Chang, T. M. S., MacIntosh, F. C., and Mason, S. G. (1966) *Can. J. Physiol. Pharmacol. 44*, 115–128.

Chang, T. M. S., Johnson, L. J., and Ransome, O. (1967) *Can. J. Physiol. Pharmacol. 45*, 705–715.

Chang, T. M. S., Gonda, A., Dirks, J. H., and Malave, N. (1971a) *Trans. Amer. Soc. Artif. Intern. Organs 17*, 246–252.

Chang, T. M. S., MacIntosh, F. C., and Mason, S. G. (1971b) Canadian Patent 873815.

Chang, T. M. S., Gonda, A., Dirks, J. H., Coffey, J. F., and Burns, T. (1972) *Trans. Amer. Soc. Artif. Intern. Organs 18*, 465–472.

Chang, T. M. S., Coffey, J. F., Barre, P., Gonda, A., Dirks, J. H., Levy, M., and Lister, C. (1973a) *Can. Med. Assoc. J. 108*, 429–433.

Chang, T. M. S., Coffey, J. F., Lister, C., Taroy, E., and Stark, A. (1973b) *Trans. Amer. Soc. Artif. Intern. Organs 19*, 87–91.

Chang, T. M. S., Migchelsen, M., Coffey, J. F., and Stark, A. (1974) *Trans. Amer. Soc. Artif. Intern. Organs 20*, 364–371.

Chang, T. M. S., Chirito, E., Barre, P., Cole, C., and Hewish, M. (1975) *Trans. Amer. Soc. Artif. Intern. Organs, 21*, 502–508.

Chang, T. M. S., Chirito, E., Barre, P., Cole, C., Lister, C., and Resurreccion, E. (1977) *J. Dialysis 1*(3), 239–259.

Chang, T. M. S., Lister, C., Chirito, E., O'Keefe, P., and Resurreccion, E. (1978) *Trans. Amer. Soc. Artif. Intern. Organs 24*, 243–245.

Chang, T. M. S., Chirito, E., Barre, P., Cole, C., Lister, C., and Resurreccion, E. (1979a) *Artif. Organs 3*, 127–131.

Chang, T. M. S., Malouf, C., and Resurreccion, E. (1979b) *Artif. Organs 3*, S284–287.

Chang, T. M. S., Espinosa-Melendez, E., Francoeur, T. E., and Eade, N. R. (1980) *Pediatric 65*, 811–814.

Chang, T. M. S., Lacaille, Y., Picart, X., Resurreccion, E., Loebel, A., Messier, D., and Man, N. K. (1981) *Artif. Organs 5*, S200–203.

Chang, T. M. S., Barre, P., Kuruvilla, S., Man, N. K., Lacaille, Y., Messier, D., Messier, M., and Resurreccion, E. (1982a) in *Artificial Support Systems* (Belinger, J., ed.), pp. 63–67, W. B. Saunders Co. Ltd., London, England.

Chang, T. M. S., Barre, P., Kuruvilla, S., Messier, D., Man, N. K., and Resurreccion, E. (1982b) *Trans. Amer. Soc. Artif. Intern. Organs 28*, 43–48.

Chang, T. M. S., Shu, C. D., Yu, Y. T., and Grunwald, J. (1982c) in *Advances in the Treatment of Inborn Errors of Metabolism* (Crawford, M., Gibbs, D., and Watts, R. W. E., eds.), pp. 175–184, John Wiley, New York.

Chang, T. M. S., Yu, Y. T., and Grunwald, J. (1982d) *Enzyme Eng. 6*, 451–561.

Chavers, B. M., Kjellstrand, C. M., Wiegand, C., Ebben, J., and Mauer, S. M. (1980) *Kidney Int. 18*, 386.

Chirito, E., Reiter, B., Lister, C., and Chang, T. M. S. (1977) *Artif. Organs 1*(1), 76–83.

Cousineau, J., and Chang, T. M. S. (1977) *Biochem. Biophys. Res. Comm. 79*(1), 24–31.

Damon Corporation (1981) *Bulletin on Tissue Microencapsulation*, Needham Heights, Mass.

Ergan, F., Thomas, D., and Chang, T. M. S. (1984) *J. Appl. Biochem. Biotechnol. 10*, 61–71.

Gardner, D. L., Falb, R. D., Kim, B. C., and Emmerling, D. C. (1971) *Trans. Amer. Soc. Artif. Intern. Organs 17*, 239.

Gelfand, M. C., Winchester, J. F., Knepshield, J. H., Hansen, K. M., Cohan, S. L., Stranch, B. S., Kennedy, A. C., and Schreiner, G. E. (1977) *Trans. Amer. Soc. Artif. Intern. Organs 23*, 599–603.

Gelfand, M. C., Winchester, J. F., Knepshield, J. H., Cohan, S. L. and Schreiner, G. E. (1978) *Trans. Amer. Soc. Artif. Intern. Organs 24*, 239–242.

Gimson, A. E. S., Brande, S., Mellon, P. J., Canalese, J., and Williams, R. (1982) *Lancet 2*, 681–683.

Gordon, A., Greenbaum, M. A., Marantz, L. B., McArther, M. S., and Maxwell, M. D. (1969) *Trans. Amer. Soc. Artif. Intern. Organs 15*, 347–349.

Gregoriadis, G. (1979) *Drug Carriers in Biology and Medicine*, Academic Press, New York.

Gregoriadis, G., Leathwood, P. D., and Schnure, B. E. (1971) *FEBS Lett. 14*, 95.

Grunwald, J., and Chang, T. M. S. (1978) *Biochem. Biophys. Res. Comm. 81*(2), 565–570.

Grunwald, J., and Chang, T. M. S. (1979) *J. Appl. Biochem. 1*, 104–114.

Grunwald, J., and Chang, T. M. S. (1981) *J. Mol. Catal. 11*, 83–90.

Ihler, G. M., Glew, R. H., and Schnure, F. W. (1973) *Proc. Nat. Acad. Sci. U.S.A. 70*, 2663.

Jamieson, G. A., and Greenwalt, T. J. (1978) *Blood Substitutes and Plasma Expanders*, Alan R. Liss, New York.

Keipert, P. E., and Chang, T. M. S. (1983) *Trans. Amer. Soc. Artif. Intern. Organs 29*, 329–333.

Keipert, P. E., and Chang, T. M. S. (1984) *J. Applied Bioch. Biotech. 10*, 133–141.

Keipert, P. E., Minkowitz, J., and Chang, T. M. S. (1982) *Int. J. Artif. Organs 5*, 383–385.

Kitajima, M., and Kondo, A. (1971) *Bull. Chem. Soc. Jap. 44*, 3201.

Kjellstrand, C., Borges, H., Pru, C., Gardner, D., and Fink, D. (1981) *Trans. Amer. Soc. Artif. Intern. Organs 27*, 24–30.

Lim, F., and Sun, A. M. (1980) *Science 210*, 908.

Martin, A. M., Gibbins, T. K., Kimmit, T., and Rennie, F. (1979) *Dial. Transplant. 8*, 135.

May, S. W., and Li, N. N. (1972) *Biochem. Biophys. Res. Comm. 47*, 1179.

Mori, T., Tosa, T., and Chibata, I. (1973) *Biochim. Biophys. Acta 321*, 653.

Mueller, P., and Rudin, D. O. (1968) *J. Theoret. Biol. 18*, 222.

Odaka, M., Tabata, Y., Kobayashi, H., Nomura, Y., Soma, H., Hirasawa, H., and Sato, H. (1978) in *Artificial Kidney, Artificial Livers, and Artificial Cells* (Chang, T. M. S., ed.), pp. 79–88, Plenum Press, New York.

Odaka, M., Hirasawa, H., Kobayashi, H., Ohkawa, M., Soeda, K., Tabata, Y., Soma, M., and Sato, H. (1980) in *Hemoperfusion*, Part 1, *Kidney and Liver Support and Detoxification* (Sideman, S., and Chang, T. M. S. eds.), pp. 45–55, Hemisphere, Washington, D.C.

Østergaard, J. C. W., and Martiny, S. C. (1973) *Biotech. Bioeng. 15*, 561.

Papadopoulous, Z. L., and Novello, A. C. (1982) *Pediatric Clinics of North America 26*, 1039–1052.

Piskin, E., and Chang, T. M. S. (1982) *Hemoperfusion and Artificial Organs*, Artificial Organs Society, P.K. 716, Kizilay, Ankara, Turkey.

Poznansky, M., and Chang, T. M. S. (1974) *Biochim. Biophys. Acta 334*, 103.

Rosenthal, A. M., and Chang, T. M. S. (1980) *J. Membrane Sci. 6*(3), 329–338.

Sekiguchi, W., and Kondo, A. (1966) *J. Jap. Soc. Blood Transfusion 13*, 153–154.

Sessa, G., and Weissman, G. (1970) *J. Biol. Chem. 245*, 3295.

Shi, Z. Q., and Chang, T. M. S. (1982) *Trans. Amer. Soc. Artif. Intern. Organs 28*, 205–209.

Shiba, M., Tomioka, S., Koishi, M., and Kondo, T. (1970) *Chem. Pharm. Bull.* (Tokyo) *18*, 803.

Shu, C. D., and Chang, T. M. S. (1980) *Int. J. Artif. Organs 3*(5), 287–291.

Shu, C. D., and Chang, T. M. S. (1981) *Int. J. Artif. Organs 4*, 82–84.

Sideman, S., and Chang, T. M. S. (1980) *Hemoperfusion: Kidney and Liver Support and Detoxification*, Hemisphere, Washington, D.C.

Sideman, S., Mor, L., Fishler, L. S., Thaler, I., and Brandes, J. M. (1981) in *Artificial Liver Support* (Brunner, G., and Schmidt, F. W., eds.), pp. 103–109, Springer, Berlin, New York.

Siu Chong, E. D., and Chang, T. M. S. (1974) *Enzyme 18*, 218–239.

Sparks, R. E., Mason, N. S., Samuels, W. E., Litt, M. H., and Lindan, O. (1972) *Trans. Amer. Soc. Artif. Intern. Organs 18*, 458–464.

Stefoni, S., Coli, L., Feliciangeli, G., Baldrati, L., and Bonomini, V. (1980) *Int. J. Artif. Organs 3*, 348.

Tabata, Y., and Chang, T. M. S. (1980) *Trans. Amer. Soc. Artif. Intern. Organs 26*, 394–399.

Terman, D. S. (1980) in *Sorbents and Their Clinical Applications* (Giordano, C. ed.), pp. 470–490, Academic Press, New York.

Terman, D. S., Tavel, T., Petty, D., Racic, M. R., and Buffaloe, G. (1977) *Clin. Exp. Ummunol. 28*, 180.

Terman, D. S., Young, J. B., Shearer, W. T., and Daskal, Y. (1981) *New Engl. J. Med. 305*, 1195–1200.

Ton, H.-Y., Hughes, R. D., Silk, D. B. A., and Williams, R. (1979) *Artif. Organs 3*, 20.

Vale, J. A., Rees, A. J., Widdop, B., and Goulding, R. (1975) *Brit. Med. J. 1*, 5.

Williams, R., and Murray-Lyon, I. M. (1975) *Artificial Liver Support*, Pitman Press, London.

Winchester, J. F., Apiliga, M. T., MacKay, J. M., and Kennedy, A. C. (1976) *Kidney Int. 10*, S315.

Yu, Y. T., and Chang, T. M. S. (1981a) *FEBS Lett. 125*(1), 94–96.

Yu, Y. T., and Chang, T. M. S. (1981b) *Trans. Amer. Soc. Artif. Intern Organs 27*, 535–538.

Yu, Y. T., and Chang, T. M. S. (1982) *J. Microbial Enzyme Technol. 4*, 327–331.

Enzymatic Synthesis of Halohydrins and Their Conversion to Epoxides

John Geigert
Saul L. Neidleman

I. INTRODUCTION

As early as 1968 it was reported that haloperoxidase could convert an alkene to a halohydrin (Neidleman and Levine, 1968) and that halohydrin epoxidase could convert a halohydrin to an epoxide (Castro and Bartnicki, 1968). However, it was not until 1980 that the first coupling of these two enzymatic reactions was reported (Neidleman, 1980). As observed in Fig. 12.1, the overall reaction results in incorporation of an oxygen atom across the carbon–carbon double bond. Halide ion is conserved and suitable for recycle. Dilute H_2O_2 is the "replaceable cofactor." Potentially, this two-step enzymatic reaction presents advantages over the waste-intensive chemical chlorohydrin process (the $CaCl_2$ of Fig. 12.2a) and the cofactor-dependent one-step oxygenase systems (the NAD(P)H of Fig. 12.2b). This chapter reviews aspects of each of the two steps in the coupled enzymatic reaction: formation of halohydrins from alkenes followed by their conversion to epoxides.

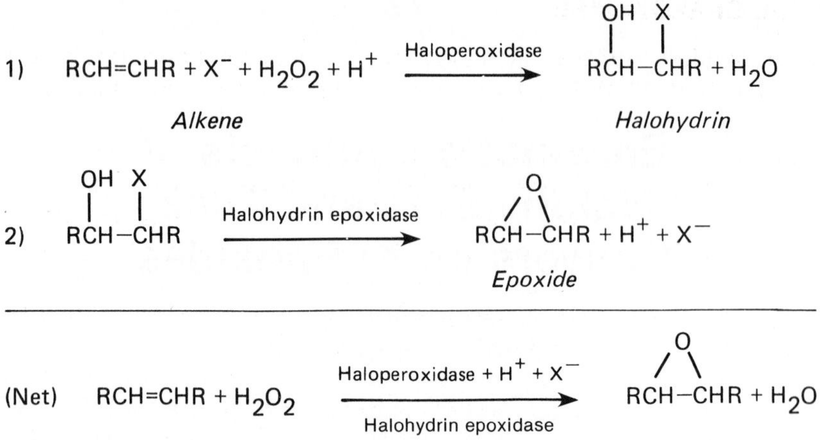

1) $RCH=CHR + X^- + H_2O_2 + H^+ \xrightarrow{\text{Haloperoxidase}} \overset{\displaystyle OH\ \ X}{\overset{\displaystyle |\ \ \ |}{RCH-CHR}} + H_2O$

      Alkene   *Halohydrin*

2) $\overset{\displaystyle OH\ \ X}{\overset{\displaystyle |\ \ \ |}{RCH-CHR}} \xrightarrow{\text{Halohydrin epoxidase}} \overset{\displaystyle O}{\overset{\displaystyle /\backslash}{RCH-CHR}} + H^+ + X^-$

                                                     Epoxide

(Net) $RCH=CHR + H_2O_2 \xrightarrow[\text{Halohydrin epoxidase}]{\text{Haloperoxidase} + H^+ + X^-} \overset{\displaystyle O}{\overset{\displaystyle /\backslash}{RCH-CHR}} + H_2O$

FIGURE 12.1   The two-step enzymatic process from alkene to epoxide via halohydrin.

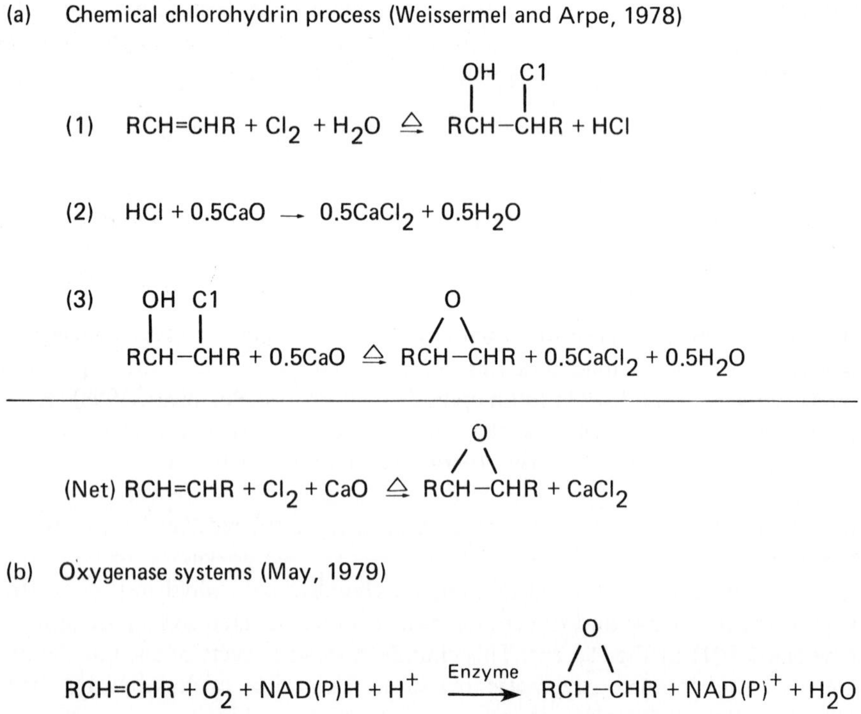

(a)   Chemical chlorohydrin process (Weissermel and Arpe, 1978)

(1) $RCH=CHR + Cl_2 + H_2O \;\triangleq\; \overset{\displaystyle OH\ \ Cl}{\overset{\displaystyle |\ \ \ |}{RCH-CHR}} + HCl$

(2) $HCl + 0.5CaO \longrightarrow 0.5CaCl_2 + 0.5H_2O$

(3) $\overset{\displaystyle OH\ \ Cl}{\overset{\displaystyle |\ \ \ |}{RCH-CHR}} + 0.5CaO \;\triangleq\; \overset{\displaystyle O}{\overset{\displaystyle /\backslash}{RCH-CHR}} + 0.5CaCl_2 + 0.5H_2O$

(Net) $RCH=CHR + Cl_2 + CaO \;\triangleq\; \overset{\displaystyle O}{\overset{\displaystyle /\backslash}{RCH-CHR}} + CaCl_2$

(b)   Oxygenase systems (May, 1979)

$RCH=CHR + O_2 + NAD(P)H + H^+ \xrightarrow{\text{Enzyme}} \overset{\displaystyle O}{\overset{\displaystyle /\backslash}{RCH-CHR}} + NAD(P)^+ + H_2O$

FIGURE 12.2   Comparative epoxidation processes.

II. HALOPEROXIDASE

Haloperoxidase producers are ubiquitous. Over 60 natural sources have been reported (Table 12.1). In addition, we have discovered an equal number of terrestrial microbial sources. These enzymes fall into three groups depending upon their ability to oxidize the halide series I^-, Br^-, and Cl^-: (1) those that can oxidize iodide, bromide, and chloride (referred to by the general name chloroperoxidase); (2) those that can oxidize iodide and bromide (referred to by the general name bromoperoxidase); and (3) those that can only oxidize iodide (referred to by the general name iodoperoxidase). Oxidation of F^- by these enzymes is not theoretically possible. In halogenating organic molecules, haloperoxidases consume H_2O_2.

As a class of enzymes, haloperoxidases have a broad range of important properties: (1) they are salt tolerant, being active in solutions of 3 M NaCl or KBr (Geigert et al., 1983a); (2) they have impressive turnover numbers; for example, a number of $3.5 \times 10^6 \, min^{-1}$ was reported for a marine bromoperoxidase (Baden and Corbett, 1980); and (3) they demonstrate broad substrate reactivity. A selection of halogenations catalyzed by these enzymes is shown in Fig. 12.3. This wide range of substrate reactivity is a mixed blessing, since close scrutiny of all components in a reaction mixture is required to identify and subsequently to eliminate competitive substrate reactions. Under certain conditions, even the enzyme can be a substrate. Haloperoxidases also act as classical peroxidases in the absence of halide ion; for example, they are capable of catalyzing phenol couplings and dealkylations.

TABLE 12.1 Known Haloperoxidases

Halide Oxidation	Enzyme Source	Common Enzyme Name
Cl^-, Br^-, I^-	Human/animal	
	Leucocytes	Myeloperoxidase (MPO)
	Fungal	
	Caldariomyces fumago	Chloroperoxidase (CPO)
Br^-, I^-	Human/animal	
	Milk, saliva, tears	Lactoperoxidase (LPO)
	Algal	
	Bonnemaisonica hamifera	
	Rhipocephalus phoenix	
	Penicillus capitatus	Bromoperoxidase (BPO)
	Rhodomela larix	
	> 50 others	
I^-	Human/animal	
	Thyroid	Thyroid peroxidase (TPO)
	Plant	
	Horseradish	Horseradish peroxidase (HRPO)

From Neidleman (1975) and Hewson and Hager (1980).

(1) Phenols, O-Alkyls, Anilines

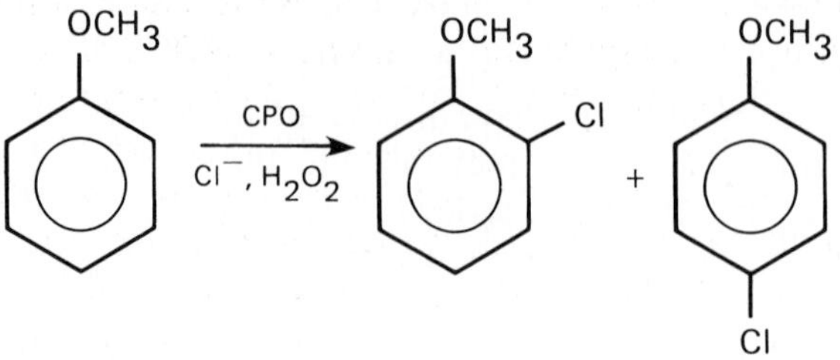

(2) β-Diketones

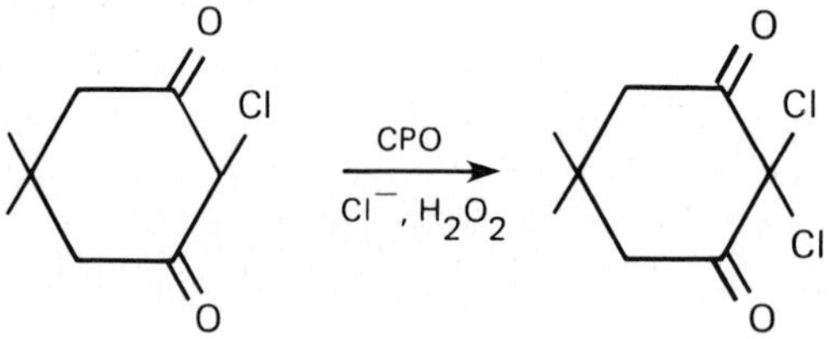

(3) β-Keto Acids

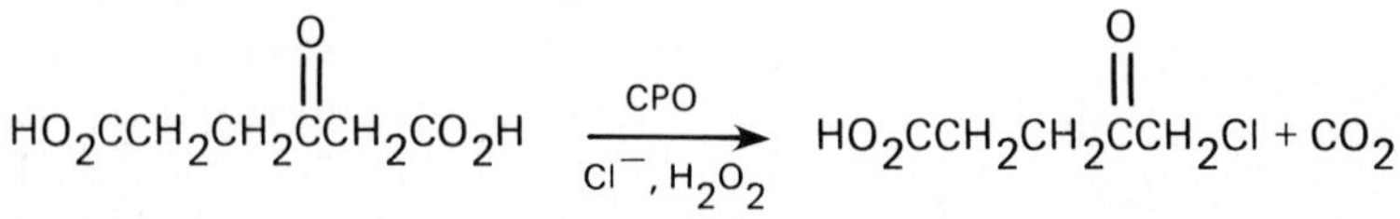

(4) Alkenes

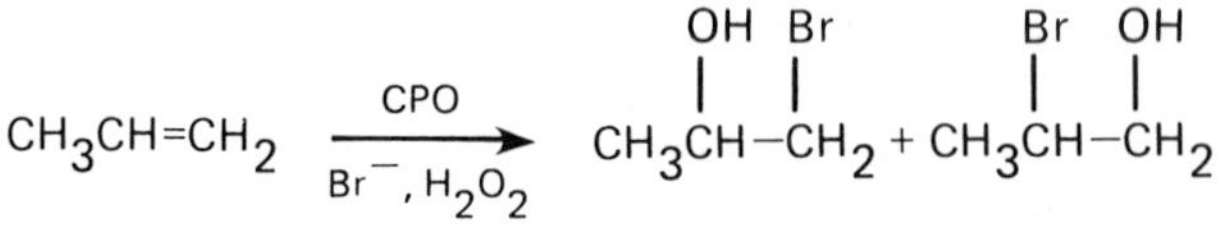

(5) Alkynes

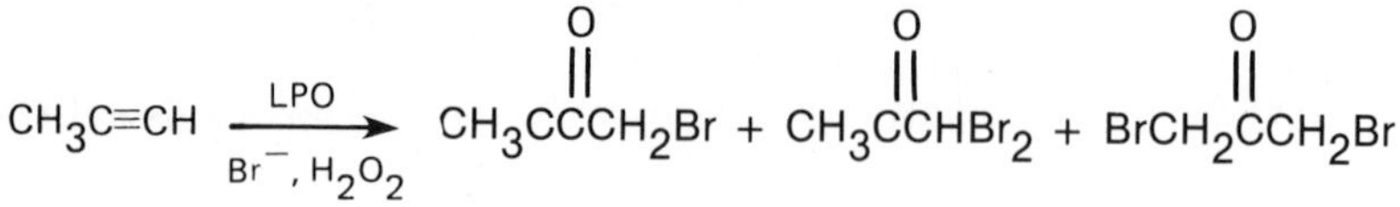

(6) Cyclopropanes

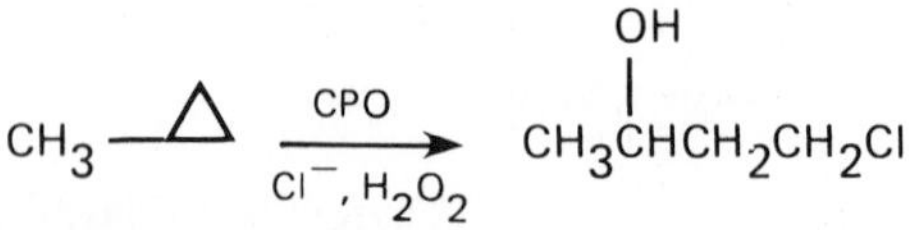

FIGURE 12.3 Selected halogenations catalyzed by haloperoxidases. (From Neidleman, 1975; Geigert et al., 1983b.)

286

A continuing area of experimental controversy is the nature of the active halogenating moiety. Is it free hypohalous acid, such as HOCl, or an enzyme-bound form of activated halogen (Fig. 12.4)? The primary argument for the formation of an enzyme-bound, electrophilic halogenating species is the observed differences in substrate specificities between enzymic and chemical halogenation reactions (Libby et al., 1982). However, the observed reaction rate differences could just as well be explained by the diffusion rate differences for the various substrates traveling to the enzyme active site where free hypohalous acid is being generated. The primary arguments for the active halogenating moiety being free hypohalous acid include: (1) the volatilization of radioactively labeled halide salt (Harrison and Schultz, 1976), (2) the absence of optically active halohydrin products from alkenes (Kollonitsch et al., 1970), and (3) the facile reaction on both alkenes and alkynes and many other substrates (Geigert et al., 1983b). Surprisingly, this mechanistic question has been raised only for chloroperoxidase (CPO) and never for myeloperoxidase (MPO), which always has been described as a free hypohalous acid generator (Zgliczynski, 1980).

III. ALKENE SPECIFICITY

A. Gaseous Alkenes

All of the gaseous alkenes have been examined (Geigert et al., 1983c). Ethylene, the simplest alkene, yields a single α,β-halohydrin product with each halide ion (Fig. 12.5a). Propylene, an asymmetric alkene, yields two positional isomers for

(a)   Active intermediate is enzyme-bound (EOX)

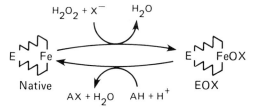

(b)   Active intermediate is free hypohalous acid (HOX)

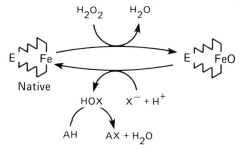

FIGURE 12.4   Proposed haloperoxidase reaction mechanisms.

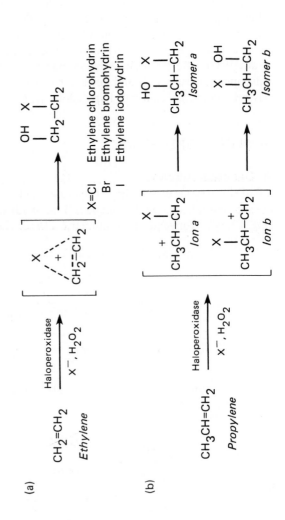

(a)

$CH_2=CH_2$
Ethylene

$\xrightarrow[\text{X}^-, \text{H}_2\text{O}_2]{\text{Haloperoxidase}}$

$\left[\begin{array}{c} \text{X} \\ \overset{+}{\underset{\text{CH}_2\text{---}\text{CH}_2}{\diagup\diagdown}} \end{array} \right]$

$\longrightarrow$

$\begin{array}{cc} \text{OH} & \text{X} \\ | & | \\ \text{CH}_2\text{---}\text{CH}_2 \end{array}$

X=Cl Ethylene chlorohydrin
Br Ethylene bromohydrin
I Ethylene iodohydrin

(b)

$CH_3CH=CH_2$
Propylene

$\xrightarrow[\text{X}^-, \text{H}_2\text{O}_2]{\text{Haloperoxidase}}$

$\left[\begin{array}{c} \text{X} \\ | \\ \overset{+}{\text{CH}_3\text{CH}\text{---}\text{CH}_2} \\ \textit{Ion a} \\[1em] \text{X} \\ | \\ \text{CH}_3\overset{+}{\text{CH}}\text{---}\text{CH}_2 \\ \textit{Ion b} \end{array} \right]$

$\longrightarrow$
$\begin{array}{cc} \text{HO} & \text{X} \\ | & | \\ \text{CH}_3\text{CH}\text{---}\text{CH}_2 \\ \textit{Isomer a} \end{array}$

$\longrightarrow$
$\begin{array}{cc} \text{X} & \text{OH} \\ | & | \\ \text{CH}_3\text{CH}\text{---}\text{CH}_2 \\ \textit{Isomer b} \end{array}$

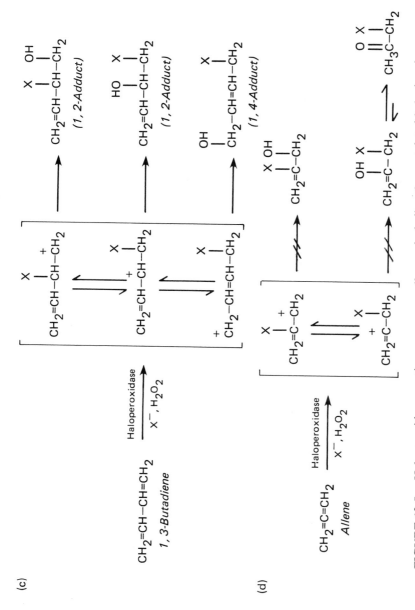

(c)

(d)

FIGURE 12.5 Haloperoxidase reaction on gaseous alkenes, including the postulated halonium ion.

each halohydrin product (Fig. 12.5b). The ratio formed (~9:1, isomer *a*:isomer *b*) correlates with the stability of the postulated halonium ion intermediates: ion *a* being more stable than ion *b* owing to the electron-donating ability of the methyl group. This ratio matches that obtained by chemical addition of hypohalous acid to propylene (Weissermel and Arpe, 1978). Butadiene, a conjugated alkene, yields two positional isomers for each halohydrin product, although it could have yielded a third (the 1,4-adduct) owing to the resonance of the intermediate allylic halonium ion (Fig. 12.5c). Exclusive formation of only the 1,2-adducts is also observed in the chemical addition of hypohalous acid to butadiene (Dalton and Davis, 1972). Allene, a cumulative alkene, yields a single isomer for each halohydrin product (Fig. 12.5d). None of the positional isomer that could tautomerize to the α-haloketone is detected in the reaction. Again, this follows the reaction observed in the chemical addition of hypohalous acid to allenic hydrocarbons (Bianchini and Cocordano, 1970).

B. Steroids

Numerous steroids containing the 9(11)-double bond and a few containing the 5(6)-double bond yield bromohydrins with chloroperoxidase (Neidleman and Levine, 1970; Neidleman, 1975) (Fig. 12.6). To circumvent the insolubility of

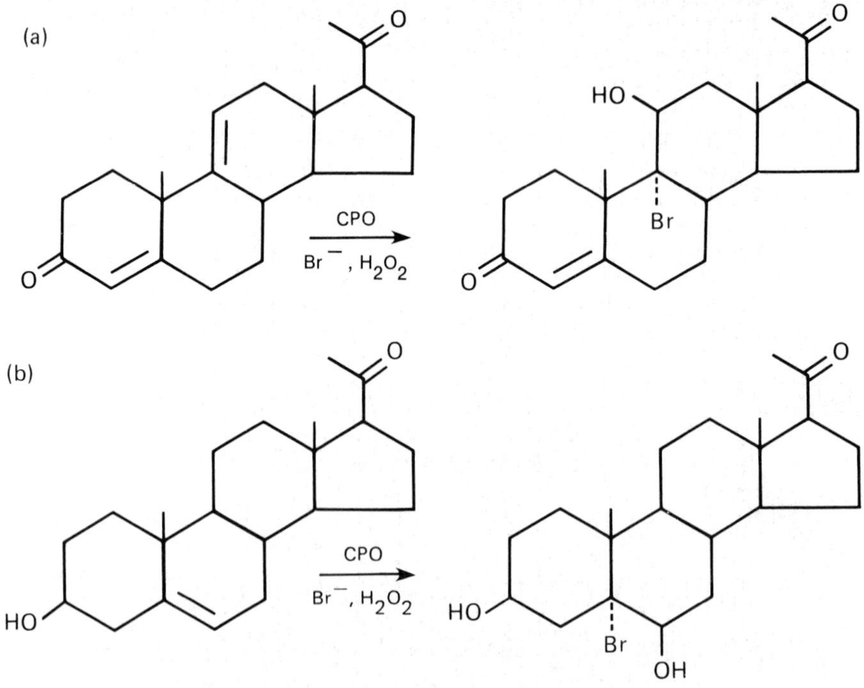

FIGURE 12.6 Reaction of chloroperoxidase (CPO) with unsaturated steroids. (a) Δ9(11)-Dehydroprogesterone, (b) pregnenolone.

these alkenes, the steroids are first dissolved in DMSO to yield a fine dispersion of the substrate upon addition to an aqueous reaction mixture.

C. Lipids

Two polyunsaturated fatty acids have been studied: arachidonic acid ($20:4\omega6$) and docosahexaenoic acid ($22:6\omega3$). In the presence of iodide, hydrogen peroxide, and lactoperoxidase these unsaturated lipids are converted into several iodinated compounds (Figure 12.7). For arachidonic acid the major product is the iodo-δ-lactone (Boeynaems et al., 1981a), whereas for docosahexaenoic acid the major product is the iodo-γ-lactone (Boeynaems et al., 1981b). These iodolactones, upon mild base treatment, can be converted to their respective iodohydrins. We have also observed the formation of these halolactones with much smaller acids; for example, 3-butenoic acid ($CH_2 = CHCH_2CO_2H$).

D. Functionalized Alkenes

One of the very first alkenes that was reacted with a haloperoxidase was propenylphosphonic acid (Fig. 12.8). The absence of optical activity in the chlorohydrin products surprised the authors, and they described the reaction as "an enzymatic synthesis of a chiral molecule in its racemic form from an achiral substrate" (Kollonitsch et al., 1970). With today's increased understanding, we would expect the formation of such racemic mixtures.

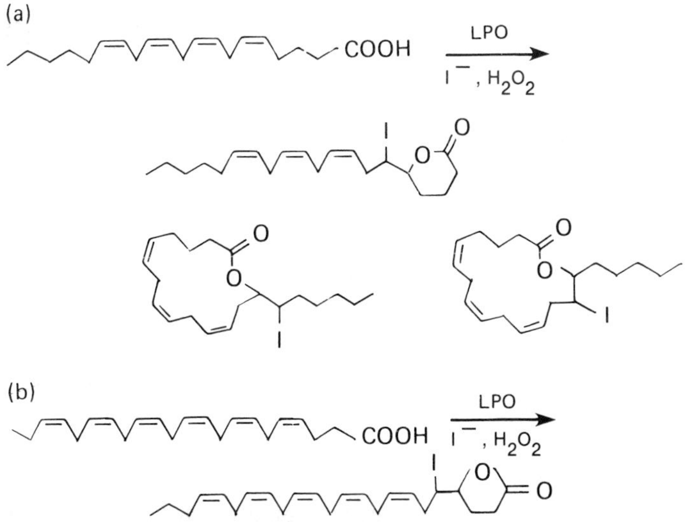

FIGURE 12.7   Reaction of lactoperoxidase (LPO) with polyunsaturated fatty acids. (a) Arachidonic acid, (b) docosahexaenoic acid.

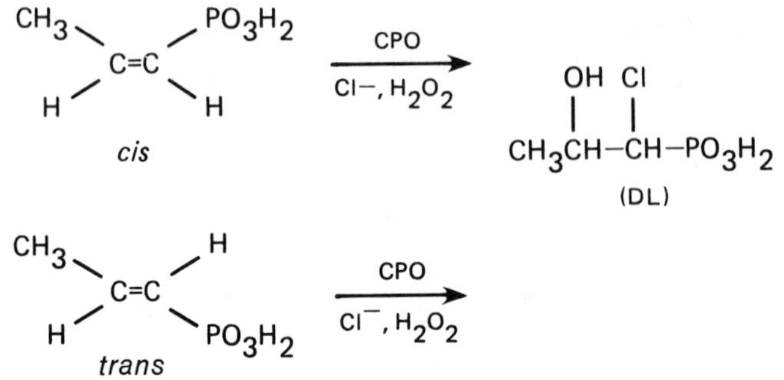

FIGURE 12.8 Reaction of chloroperoxidase with propenylphosphonic acids.

The presence of a functional group adjacent to the double bond opens the possibility of side reactions or, alternatively, additional final products. Alken-2-ols have been shown to yield both α,β-halohydrins and α,β-dihalo derivatives in the haloperoxidase reaction (Geigert et al., 1983a). Decreasing the halide concentration present favors formation of halohydrins (Fig. 12.9). Allyl halides have been shown to form three halohydrin positional isomers instead of the normal two (Lee et al., 1983). The third isomer arises from neighboring group participation of the original halogen (Fig. 12.10).

IV. HALOHYDRIN EPOXIDASE

To date, only a single halohydrin epoxidase producer has been reported: the bacterium *Flavobacterium* sp. (Castro and Bartnicki, 1968). This single reported occurrence is probably not due to the rarity of the enzyme class but most likely to the fact that no extensive screen has been conducted for pro-

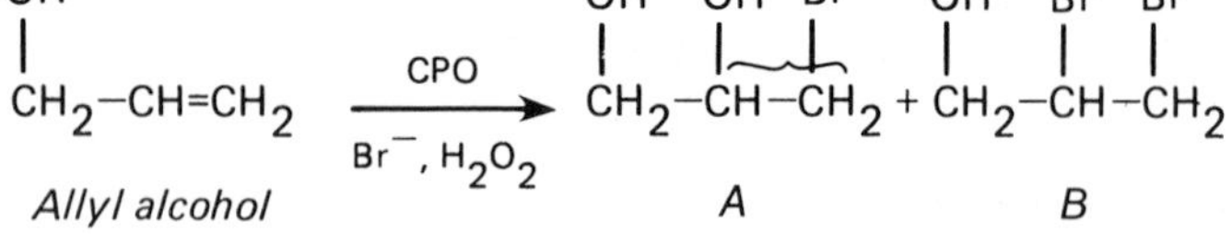

(Br^-)	A:B
2000 mM	1:15
200 mM	1:1
20 mM	8:1

FIGURE 12.9 Formation of bromohydrins and dibrominated products from an alkenol (allyl alcohol) and chloroperoxidase (CPO) at varying bromide ion concentrations.

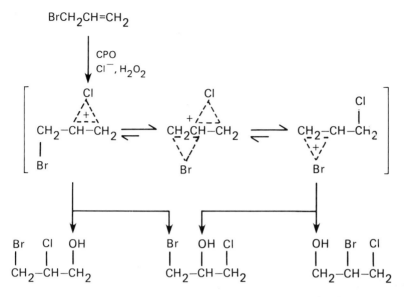

FIGURE 12.10 Neighboring group participation occurring in the enzymatic chlorination of allyl bromide with chloroperoxidase (CPO).

ducers of this enzyme. In fact, this single enzyme was discovered in a screen designed to look for reductive, dehalogenating microorganisms, not for microorganisms that catalyze an "internal nucleophilic substitution" reaction on halohydrins.

The reaction appears to be an equilibrium process, with halide ion concentrations of about 100 mM shifting the reaction back to the halohydrin. Because of this, if two different halide ions are present, transhalogenation (i.e., halogen exchange) can occur (Bartnicki and Castro, 1969). Product inhibition has been observed in the catalyzed reaction. However, for the volatile epoxide products (e.g., propylene oxide) we have circumvented the problem by removing the product as rapidly as it is formed.

V. HALOHYDRIN SPECIFICITY

α,β—Iodohydrins, bromohydrins, and chlorohydrins are suitable substrates for halohydrin epoxidase (Table 12.2). Comparing a halohydrin series (e.g., the 2-haloethanol series), the enzyme reacts more readily with the bromohydrin. Comparing halohydrin positional isomers (e.g., the two bromopropanol isomers), the enzyme catalyzes more effectively the conversion of the terminal halo-containing isomer. This same order of reactivity is observed with chemical base treatment of the halohydrins. α,γ-Halohydrins (e.g., $BrCH_2CH_2CH_2OH$) as well as α,β-fluorohydrins (e.g., FCH_2CH_2OH) are not reactive with this enzyme.

TABLE 12.2   Relative Activity of Whole Cells of *Flavobacterium* sp. on the Enzymatic Synthesis of Epoxides from Alpha, Beta-Halohydrins

$$\underset{\text{RCH--CHR'}}{\overset{\text{OH \ X}}{\underset{|\quad\ |}{}}} \xrightarrow[\text{epoxidase}]{\text{Halohydrin}} \underset{\text{RCH--CHR'}}{\overset{\text{O}}{\overset{/\backslash}{}}} + \ H^+ + X^-$$

Substrate	R	R'	X	Relative Activity[a]
2-Chloroethanol	H	H	Cl	18
2-Bromoethanol	H	H	Br	60
2-Iodoethanol	H	H	I	24
1-Chloro-2-propanol	CH_3	H	Cl	26
1-Bromo-2-propanol	CH_3	H	Br	100
2-Bromo-1-propanol	H	CH_3	Br	5
1-Iodo-2-propanol	CH_3	H	I	34
1-Bromo-3-butene-2-ol	CH_2=CH	H	Br	65

[a] Relative to activity on 1-bromo-2-propanol (taken as 100%) at 7 hours into the reaction. From Geigert et al. (1983d).

VI. THE COUPLED SYSTEM

In considering schemes for the potential utilization of a coupled enzyme system a number of operational and technological problems can be visualized, for example: (1) overcoming inherent enzyme instability, (2) obtaining adequate supplies of each biocatalyst, (3) matching the operating properties of the two enzymes, and (4) efficiently utilizing the reactants consumed in the reaction.

Enzyme immobilization is one approach to stabilizing an enzyme. For haloperoxidases, active research has given rise to a host of immobilization techniques. Lactoperoxidase covalently bound to Sepharose increased the temperature stability of this haloperoxidase (Tenovuo and Kurkijarvi, 1981). High operating temperatures would facilitate substrate solubility or product recovery. Lactoperoxidase covalently bound to Sepharose also exhibited increased stability to added H_2O_2 (Neidleman et al., 1981). Horseradish peroxidase covalently bound to an amino porous silica or entrapped in an acryloyl morpholine polymer demonstrated greater solvent tolerance than the soluble enzyme (Thibault et al., 1981; Epton et al., 1979). Solvent tolerance is desirable to increase the solubility of the substrates and products. For halohydrin epoxidase, one immobilization technique is whole-cell entrapment within cross-linked polyacrylamide gel (Neidleman et al., 1981). We have found that a column packed with this immobilized biocatalyst will continuously convert propylene bromohydrin to propylene oxide for at least three months (we used a flow rate of 1 ml per hour and a concentration of 1 mg/ml).

Ready availability of both biocatalysts with matched operating properties is essential. Techniques of molecular biology, screening for superior enzyme producers, as well as modern-day fermentation technology should be able to supply the relatively large amounts of haloperoxidase and halohydrin epoxidase required in an industrial process. A fledgling start in this area has already been reported for chloroperoxidase (Hager, 1982). One fallout from our application of screening techniques has been the discovery of new chloroperoxidase-producing microbes. Not only are these new sources amenable to advanced fermentation technology, but they also yield chloroperoxidases that operate at the same pH optimum as that for halohydrin epoxidase (pH 6–7). Neither of the two previously reported chloroperoxidases are both readily available and matched in pH with the second enzyme step: chloroperoxidase from the fungus *Caldariomyces fumago* is readily available but does not operate above pH 4, while myeloperoxidase from leukocytes operates between pH 5–8 but is not readily available. Our discovery eliminates the need for shifting the reaction pH between the two steps of the process.

Efficient utilization of H_2O_2 in the haloperoxidase reaction is required. H_2O_2 handling is complex, with possible undesirable reactions occurring with trace metals, free hypohalous acid, and the enzyme itself. Despite these difficulties, over 50% of added H_2O_2 is readily incorporated into the desired halohydrin products, and further optimization could improve this efficiency. In addition, the ability to use dilute and impure H_2O_2 helps in reducing substrate cost.

It is difficult to predict the commercial applications of the enzymatic production of halohydrins or epoxides. It would seem likely that if an industrial use of these reactions and the enzymatic production of other halogenated compounds have a future, it will be in the synthesis of useful specialty chemicals having valuable chemical or biological properties (Geigert et al., 1983a, b; Neidleman and Geigert, 1983). It is clear that much effort is needed to optimize the reactions to satisfy the requirements of an economic commercial process.

REFERENCES

Baden, D. G., and Corbett, M. D. (1980) *Biochem. J. 187*, 205–211.

Bartnicki, E. W., and Castro, C. E. (1969) *Biochem. 8*, 4677–4681.

Bianchini, J. P., and Cocordano, M. (1970) *Tetrahedron 26*, 3401–3411.

Boeynaems, J. M., Reagan, D., and Hubbard, W. C. (1981a) *Lipids 16*, 246–249.

Boeynaems, J. M., Watson, J. T., Oates, J. A., and Hubbard, W. C. (1981b) *Lipids 16*, 323–327.

Castro, C. E., and Bartnicki, E. W (1968) *Biochem. 7*, 3213–3218.

Dalton, D. R., and Davis, R. M. (1972) *Tetrahedron Lett. 13*, 1057–1060.

Epton, R., Hobson, M. E., and Marr, G. (1979) *Enzyme Microb. Technol. 1*, 37–40.

Geigert, J., Neidleman, S. L., Dalietos, D. J., and DeWitt, S. K. (1983a) *Appl. Environ. Microbiol. 45*, 1575–1581.

Geigert, J., Neidleman, S. L., and Dalietos, D. J. (1983b) *J. Biol. Chem. 258*, 2273–2277.

Geigert, J., Neidleman, S. L., Dalietos, D. J., and DeWitt, S. K. (1983c) *Appl. Environ. Microbiol. 45*, 366–374.

Geigert, J., Neidleman, S. L., Liu, T. E., DeWitt, S. K., Panschar, B. M., Dalietos, D. J., and Siegel, E. R. (1983d) *Appl. Environ. Microbiol. 45*, 1148–1149.

Hager, L. P. (1982) *Basic Life Sci. 19*, 415–429.

Harrison, J. E., and Schultz, J. (1976) *J. Biol. Chem. 251*, 1371–1374.

Hewson, W. D., and Hager, L. P. (1980) *J. Phycol. 16*, 340–345.

Kollonitsch, J., Marburg, S., and Perkins, L. M. (1970) *J. Amer. Chem. Soc. 92*, 4489–4490.

Lee, T. D., Geigert, J., Dalietos, D. J., and Hirano, D. S. (1983) *Biochem. Biophys. Res. Comm. 110*, 880–883.

Libby, R. D., Thomas, J. A., Kaiser, L. W., and Hager, L. P. (1982) *J. Biol. Chem. 257*, 5030–5037.

May, S. W. (1979) *Enzyme Microb. Technol. 1*, 15–22.

Neidleman, S. L. (1975) *Crit. Rev. Microbiol. 3*, 333–358.

Neidleman, S. L. (1980) *Hydrocarbon Proc. 59*, 135–138.

Neidleman, S. L., and Geigert, J. (1983) *Trends Biotechnol. 1*, 21–25.

Neidleman, S. L., and Levine, S. D. (1968) *Tetrahedron Lett. 9*, 4057–4059.

Neidleman, S. L., and Levine, S. D. (1970) U.S. Patent 3,528,886.

Neidleman, S. L., Amon, W. F., Jr., and Giegert, J. (1981) U.S. Patent 4,247,641.

Tenovuo, J., and Kurkijarvi, J. (1981) *Arch. Oral Biol. 26*, 309–314.

Thibault P., Monsan, P., and Jouret, C. (1981) *Sci. Aliments 1*, 55–66.

Weissermel, K., and Arpe, H. J. (1978) in *Industrial Organic Chemistry*, pp. 235–272, Verlag Chemie, New York.

Zgliczynski, J. M. (1980) *Reticuloendothel Syst. 2*, 255–278.

Synthetic Enzyme Analogs (Synzymes)

Garfield P. Royer

I. INTRODUCTION

Synthetic enzyme models have been of great importance to the bio-organic chemist in efforts to understand the mechanism of enzyme action. Could such artificial enzyme-like catalysts be constructed and used in industry? What advantages would the use of synzymes have over natural enzymes or whole cells as catalysts? Consider the points listed in Table 13.1. Although enzymes bring about selective rate enhancements at low temperature and pressure, they are usually expensive and often unstable. Starch and protein hydrolases account for more than 70% of the enzymes sold. In most cases these enzymes are extracellular; they are made by microorganisms and then secreted into the medium. Since they work in a relatively harsh environment, they have evolved in such a way that they are more stable than intracellular enzymes, which work within the protected environment of the cell. The low cost of these extracellular enzymes is related to ease of isolation; often, the culture filtrate can be dried and used directly as a crude enzyme preparation. Improvements in enzyme isolation techniques and recombinant DNA technology may result in a larger list for the selection of low-cost enzymes in the future. The length of the list and the definition of "future" are uncertain.

TABLE 13.1   Advantages and Limitations of Biocatalysts

Advantages

• Great rate enhancements under mild reaction conditions
• Specificity, especially regiospecificity and stereospecificity

Limitations

• Cost
• Instability
• Phase incompatibility
• Product inhibition
• Substrate inhibition

Industrial processes are often carried out at elevated temperatures. With the exception of enzymes from thermophilic organisms, enzymes usually denature at 60°. In many cases, one would prefer high temperature for chemical reactions in order to achieve favorable distribution of reactant and product (positive ΔH). For bulk chemical production, effective utilization of capital requires high reactor throughputs; gas phase reactions at elevated temperature and concentrations are the norm. In most reactions involving condensed phases, organic solvents are used. Enzymes do not work in the gas phase, nor do they generally work in organic solvents. It has been known for many years that proteins denature at oil/water interfaces. Moreover, many enzymes depend on hydrophobic bonding to provide the driving force for enzyme–substrate interaction. Recombinant DNA technology and novel enzyme reactors may help solve the instability problem. Another approach, of course, is to attempt to make an artificial enzyme that is stable under the conditions cited.

Enzyme reaction schemes are shown in Fig. 13.1. In the top scheme the back-reaction is ignored. In the cell the product may be rapidly consumed by a second enzyme-catalyzed reaction, which would justify neglecting of the back-reaction. Similarly, when kinetics studies are done in the lab under initial velocity conditions, the effect of the product can be discounted. Enzymes have evolved in such a way that K_m values approximate physiological concentrations

$$E + S \rightleftharpoons ES \longrightarrow E + P$$

$$E + S \rightleftharpoons ES \rightleftharpoons EP \rightleftharpoons E + P$$

$$V_m/K_m \longrightarrow 10^9 \ M^{-1}s^{-1}$$

$$K_m \longrightarrow [S]_{phys} = 10^{-3} - 10^{-4}M$$

FIGURE 13.1   Schemes depicting enzyme-catalyzed reactions.

of the substrates, namely, 10^{-3}–10^{-6} M. In other words, enzymes bind their substrates strongly (Table 13.2). When the substrate and product are structurally similar, the problem of product inhibition occurs. In industrial processing, complete conversion with a single pass over the catalyst is desired. Product inhibition may rule out this possibility. A synzyme with relatively poor binding ability would be more attractive than the natural enzyme with a very high affinity for reactant and product.

Substrate inhibition may be a problem when high feed concentrations are employed. The classical model to explain substrate inhibition is illustrated in Fig. 13.2. The substrate interacts with the enzyme binding site at two points. The dark wedge corresponds to the point of covalent reaction. When two molecules bind end-on, the active site is tied up in a nonproductive ternary complex. The quantitative aspect of substrate inhibition, which is worth mentioning, is that substrate concentrations five to ten times the K_m are sufficient in many cases to slow the rate substantially.

A discussion of attempts to mimic enzymes should logically begin with a statement of what properties we hope to have in the synthetic catalyst. Enzymes as catalysts have three remarkable characteristics: they bring about enormous rate enhancements (10^6–10^{14}) over uncatalyzed reactions; they are very selective (more on this later); and changes in catalytic activity are rapid

TABLE 13.2   Representative K_m Values for Enzyme-Catalyzed Reactions

Enzyme	Reactant	K_m
Chymotrypsin	Acetyl-L-tryptophanamide	5×10^{-3}M
β-Galactosidase	Lactose	4×10^{-3}M
Carbonic anhydrase	CO_2-	8×10^{-3}M
Citrate synthase	Oxaloacetate	2×10^{-5}M
Arginine-tRNA synthetase	Arginine	3×10^{-6}M
	tRNA	4×10^{-7}M
	ATP	3×10^{-4}M

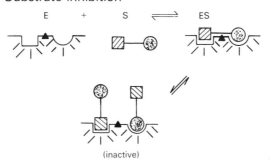

FIGURE 13.2   Schematic illustration of substrate inhibition.

and dramatic (covalent reaction and noncovalent interaction with small molecules or other proteins modulate enzyme activity). This last property is of less interest than the first two.

How do enzymes bring about such enormous increases in rate? What structural features of the enzyme are indispensable? Is the entire protein structure required, or would an active site replica function just as well as the complete protein? Enzyme active centers are composed of a binding site and a catalytic site. It is generally agreed that the enzyme–substrate interaction results in the destabilization of the substrate in relation to the transition state. The favorable free-energy change that accompanies binding is utilized in the promotion of reaction. Desolvation, dielectric effects, bond strain, and orbital overlap are factors that contribute to rate enhancement. Entropic factors include approximation and orientation of reactants in the active center. Enzyme active centers usually have multiple groups that participate in the positioning of the reactant and the induction of the bond-breaking steps. In the active center of carboxypeptidase the following groups play important roles: glutamic acid 270 (nucleophile), tyrosine 248 (general acid), arginine 145 (positioning), and a zinc atom (general acid and nucleophile as the hydroxide). The transition state of enzyme-catalyzed reactions can be comprised of many atoms. The positioning of the reactant amid the other contributors to the transition state is made possible by a favorable binding interaction of the enzyme with the reactant. The enzyme active center could be described as a transition state minus the reactant. Obviously, the spatial relationships among the contributing groups at the active center are crucial. To exactly duplicate the active centers of some enzymes would clearly be a formidable task. Replication of others may be less difficult. A much simpler approach would be the attachment of a coenzyme molecule to a polymer that provides the correct dielectric environment.

Another important question is how much rate enhancement is required? Must the synzyme be an exact duplicate? Turnover numbers of enzymes can be enormous—6×10^5 s^{-1} (Table 13.3). Clearly, all of the rate enhancement provided by many enzymes would not be needed. Sir John Cornforth has studied hydration of olefins with nonenzymic catalysts (Cornforth, 1978). He has calculated that one kilogram of an olefin hydration catalyst one thousandth as effective as fumarase would produce 5000 metric tons of product per year. In

TABLE 13.3   Turnover Numbers of Some Enzymes (s^{-1})

Enzyme	
Carbonic anhydrase	600,000
Acetylcholinesterase	25,000
Lactate dehydrogenase	1,000
Chymotrypsin	100
Tryptophan synthetase	2

other words, a near miss in the effort to mimic enzymes may result in a fabulous success.

Recent reviews of this subject have been organized with emphasis on the type of synthetic catalyst as the central theme, that is, synthetic polymers, cycoamyloses, macrocycles, and so on (Royer, 1980; Page, 1980). The goal of this chapter is not to comprehensively update past reviews, but rather to highlight some examples of efforts that have resulted in catalysts with enzyme-like properties; rate enhancement and specificity are discussed.

II. RATE ENHANCEMENT

A. Polymers

Most enzymes are globular proteins with compact structures. Mimicking enzymes with linear, extended synthetic vinyl polymers has met with some success, but experiments with branched polymers have yielded the most encouraging results. The differences in polymer shape and its relevance have been emphasized by Klotz; a schematic representation of a linear polymer and a globular protein is shown in Fig. 13.3. The approximation of binding groups, catalytic groups, and reactant would appear to be more likely when a globular or branched polymer is the starting point.

One branched polymer that has been studied extensively is poly(ethylene-imine) (PEI). The polymer is produced from aziridine; it has the branched structure shown schematically in Fig. 13.4. Substitution of the PEI nitrogens

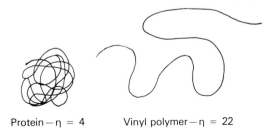

Protein — η = 4   Vinyl polymer — η = 22

FIGURE 13.3 Comparison of the shape of a protein (globular) and a vinyl polymer (linear). η refers to the intrinsic viscosity.

$$(-CH_2CH_2NH_x^-)_n$$

—distribution of amines: 25% primary, 50% secondary, and 25% tertiary

FIGURE 13.4 Polyethylenimine-branched structure and composition.

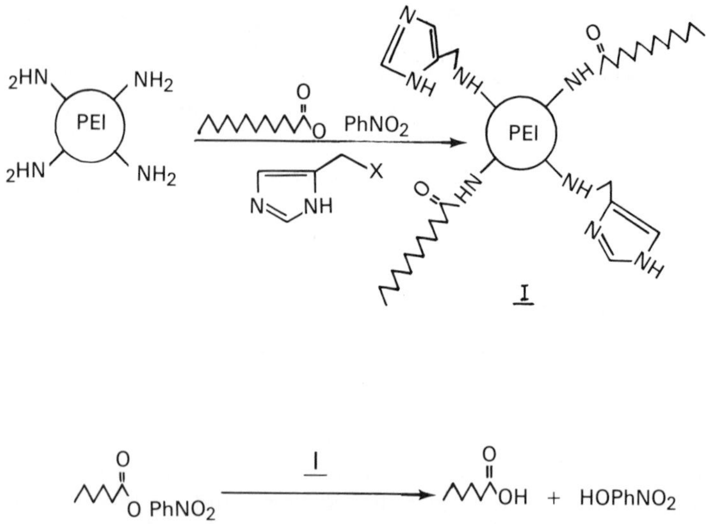

FIGURE 13.5   Polyethylenimine-substituted with lauroyl groups and imidazoylmethyl groups.

with lauroyl groups and imidazoylmethyl groups results in a catalyst that produces rate enhancements for the solvolysis of nitrophenylesters in the range 10^2–10^3 (Fig. 13.5) (Klotz et al., 1971). The rate enhancement is, at least in part, brought about as a result of the binding of the hydrophobic nitrophenyl ester in the vicinity of the imidazole groups on the surface of the polymer.

Lauroylated PEI derivatives substituted with 4-dialkyaminopyridines are good catalysts for the hydrolysis of nitrophenyl esters, presumably via the scheme shown in Fig. 13.6 (Hierl et al., 1979; Delaney et al., 1982). Acyl intermediates were detected spectrophotometrically, which means that the dialkylpyridine groups are acting as nucleophilic catalysts. The effect of the polymer is to bind the reactant in the vicinity of the nucleophilic groups. Also, a medium effect is probable; it is known that catalysis by free dialkylaminopyridine is enhanced by the use of apolar solvent (Hassner et al., 1978). Also, the polymer provides a microenvironment that is relatively rich in hydroxide ions in comparison with the bulk solution. The deacylation of the pyridinium intermediate is therefore favored.

Another reaction that is accelerated by the lauroylated PEI is the decarboxylation of 6-nitrobenzisoxazole-3-carboxylate:

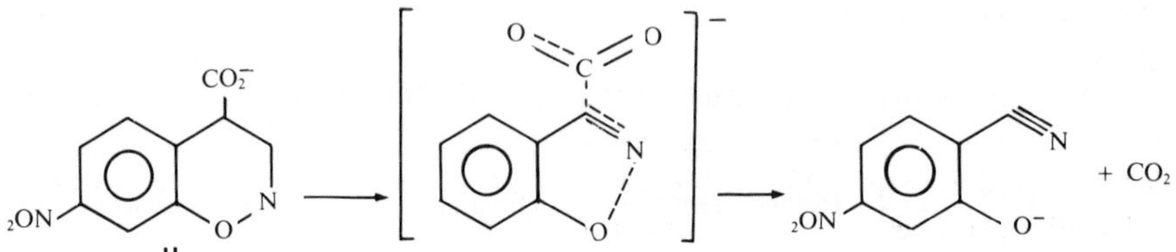

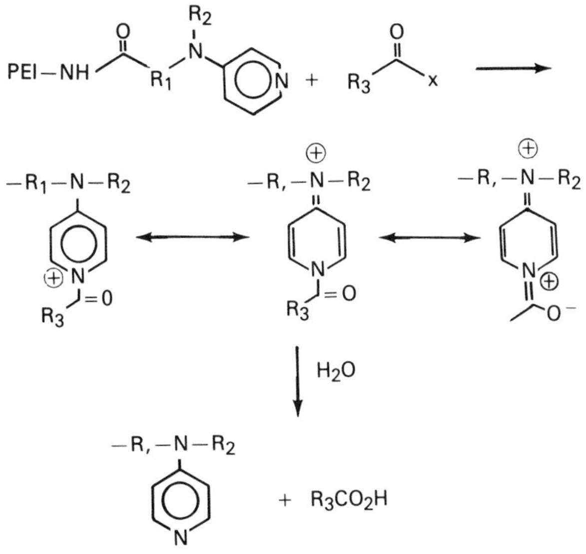

FIGURE 13.6   Catalysis by dialkylaminopyridine derivative of PEI.

Kemp and Paul (1970, 1975) have reported that the above reaction is relatively fast in aprotic media (dimethylsulfoxide, hexamethylphosphoramide) in comparison with water; the rate of enhancement is about 10^7. The apolar environment of the polymer resembles the organic medium in which the transition state is stabilized. In the presence of lauroylated PEI with quaternized nitrogens the rate of the decarboxylation of II is enhanced by a factor of 10^3.

B. Cyclodextrins

Breslow (1982) and others have studied the cyclodextrins as enzyme models. These compounds, also called cyclomylaoses, are naturally occurring cyclic oligosaccharides consisting of glucose residues in α-1,4 linkage. The molecules are toroidal in shape with apolar cavities. Three types of cyclodextrins have been studied actively: cyclohexaamylose (α-cyclodextrin), with a cavity 5 angstroms in diameter; cycloheptaamylose (β-cyclodextrin), with a 7-angstrom cavity; and cyclooctaamylose (α-cyclodextrin), with a 9-angstrom cavity. The cyclodextrins form strong inclusion complexes with apolar molecules. With appropriate molecules in the cavity, reaction can occur with one of the cyclodextrin hydroxyl groups. With β-cyclodextrin and the ferrocenyl derivative shown below a rate enhancement of 3.2×10^6 was observed (Trainor and Breslow, 1981); the authors point out that even this high rate is not fully optimized.

Research on cyclodextrins has also resulted in impressive demonstrations of selectivity (see below).

C. Synthetic Macrocycles (Cavitands)

Cram has defined the term "cavitand" as follows: "a synthetic organic compound that contains an enforced cavity of dimensions at least equal to those of the smaller ions, atoms, or molecules" (complexed by it). Additional subclasses have been defined (Cram, 1983). Cram and Katz (1983) have described the synthesis of the macrocyclic compound shown in Fig. 13.7, which is an interesting model for the serine proteases.

These enzymes (hydrolases) are characterized by the presence of a nucleophilic serine in the active center. In the first step of the hydrolysis of amides or esters in the serine group is acylated, and the amino or alcohol moeity of the substrate is displaced. Hydrolysis of the acyl serine intermediate completes the reaction. The complexation of the L-alanine nitrophenyl ester by III precisely aligns the carbonyl carbon for attachment by the phenolic hydroxyl, which represents the analog of the serine hydroxyl in the enzyme. Acylation of III is faster than the acylation of a reference compound (3-phenylbenzyl alcohol) by a factor of 10^{11}. Although the resulting ester intermediate does not turn over, this model system is quite interesting in that an enormous rate enhancement is observed in a rigid, well-defined system. Follow-up work on this catalyst will no doubt be interesting; it is hoped that turnover will be achieved.

III. SELECTIVITY

A. Geometric Specificity

Regiospecific chlorination of anisole has been demonstrated by Breslow and Campbell (1969). The chlorination of anisole (Fig. 13.8) normally results in the formation of both the ortho- and para-isomers.

When α-cyclodextrin is present, the para-isomer is formed exclusively. The explanation for the preferential formation of the para-isomer is that the inclusion of anisole in the cyclodextrin cavity blocks the ortho position but the para position is accessible through the end of the cavity (Fig. 13.8).

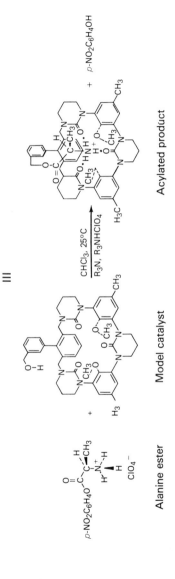

III

Alanine ester   Model catalyst   Acylated product

FIGURE 13.7 The "catalyst" of Cram and Katz (1983). The acylatioin of III is as fast as many enzyme catalyzed reactions. (Reprinted with permission—*C&EN* Feb. 14, 1983, p. 14. Copyright 1983 American Chemical Society.)

305

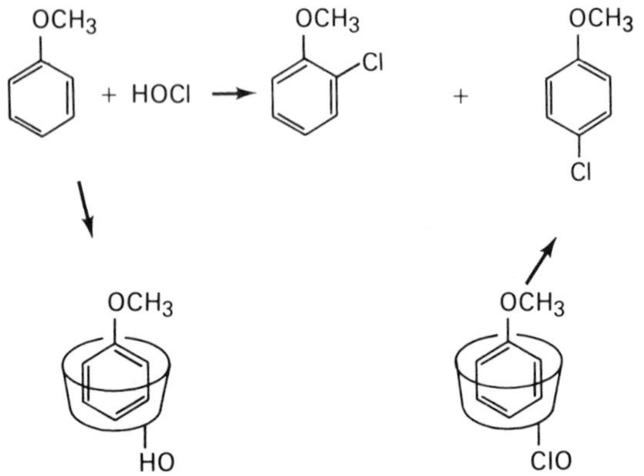

FIGURE 13.8 Chlorination of anisole in the presence of cyclodextrin. (From Breslow (1982). Reprinted with permission from *Science*, *218*, 533, 1982. Copyright 1982 by AAAS.)

Another impressive achievement of the Breslow group involves the coenzyme pyridoxal phosphate (Breslow, 1982). Enzymes that contain pryidoxal phosphate catalyze a variety of reactions: transamination, racemization, decarboxylation, deamination, and aldol cleavage. The transamination reaction is illustrated in Fig. 13.9. Using the sequence of reactions shown in Fig. 13.10, pyridoxamine was attached to β-cyclodextrin. The pyridoxamine adduct preferentially binds indolepyruvic acid in the presence of pyruvate. The amination reaction occurs to give tryptophan. In the early stage of the reaction the product is 97% tryptophan with equal starting concentrations of indolepyruvate and pyruvate. The basis for this selection is that the indole side chain of tryptophan has greater hydrophobic character than the side chain of alanine and is therefore bound strongly by the catalyst.

B. Stereochemical Specificity

1. Enantiomeric Selectivity.   Examples of the distinction between R,S pairs using synzymes have not been abundant. One report by Nango et al. (1980) describes the selection between D and L isomers of nitrophenyl esters of N-blocked amino acids in a hydrolysis. The catalyst was a PEI derivative that contained L-histidine as the catalytic group. Binding specificity was found to be greater than catalytic specificity. Neither was great in comparison with enzymatic selectivity. In a separate system, Lehn and Sirlin (1978) reported enantiomeric preference by a synthetic macrocycle in the thiolysis of dipeptide esters. In this case the selectivity was significant — a 50-fold enhancement of the thiolysis of glycyl-L-phenylalanine as compared to glycyl-D-phenylalanine. The structure of the catalyst is shown in Fig. 13.11.

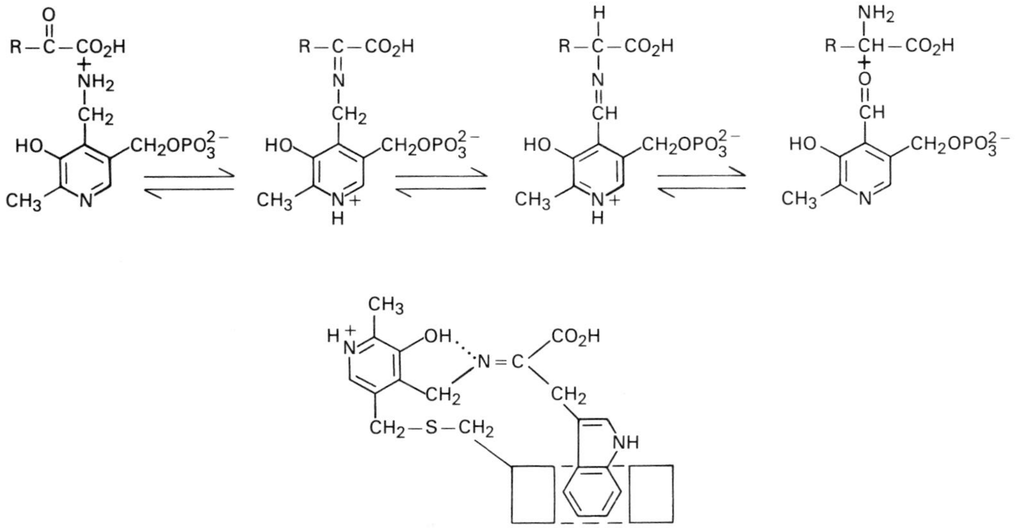

FIGURE 13.9 Transamination and pyridoxamine/cyclodextrin. (From Breslow (1982). Reprinted with permission from *Science*, *218*, 534, 1982. Copyright 1982 by AAAS.)

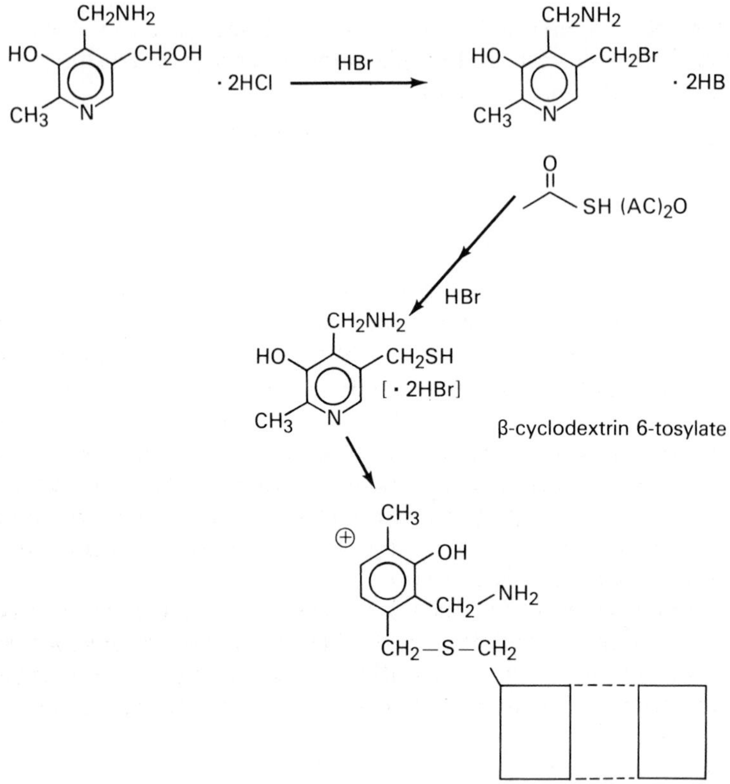

FIGURE 13.10 Preparation of pyridoxal/cyclodextrin.

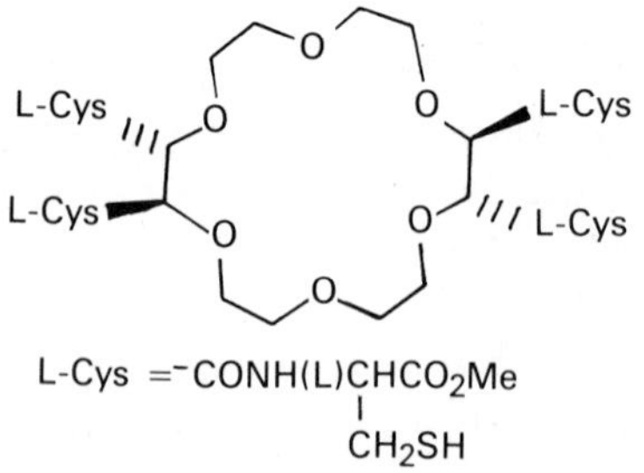

FIGURE 13.11 Crown ether catalyst of Lehn and Sirlin (1978).

2. Asymmetric Synthesis/Stereoheterotopic Selectivity. In 1948, Ogston published his classic note on the explanation of the asymmetric transformation of the symmetrical molecule citrate:

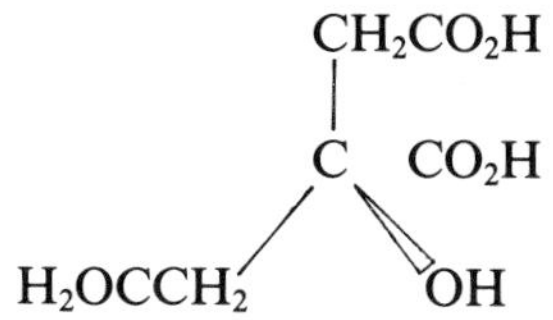

$$
\begin{array}{c}
CH_2CO_2H \\
| \\
C \quad CO_2H \\
\diagup \quad \diagdown \\
H_2OCCH_2 \qquad OH
\end{array}
$$

Since the appearance of Ogston's paper, much has been done on the stereoheterotopic selection at prochiral centers and *re, si* faces in enzyme-catalyzed reactions (see Cornforth (1976) and Bentley (1978) for overviews on this subject). There are a number of systems in which the asymmetric transformations catalyzed by enzymes have been worked out in detail. Enzymes are composed of optically active amino acids; they have complex three-dimensional structures that allow asymmetry at the active centers. A symmetrical molecule, such as citrate, is bound in the active center with multiple points of attachment in such a way that the asymmetry of the enzyme is expressed in the stereochemical course of the reaction.

Unlike the situation for enantiomeric selectivity, a number of very effective abiotic catalysts have been made that are capable of stereoheterotopic selectivity and hence asymmetric synthesis. One catalyst system is the basis for the commercial production of L-dopa by Monsanto (see Fig. 13.12).

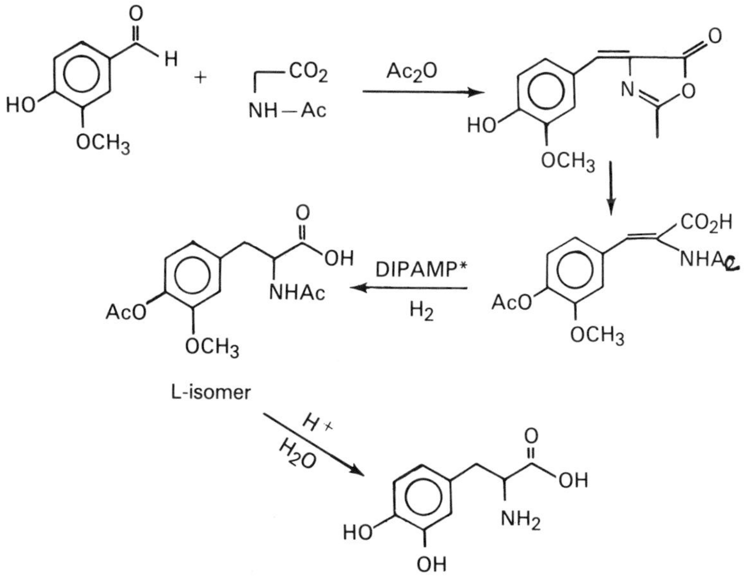

FIGURE 13.12   The synthesis of L-dopa by the Monsanto process, which is based on the chiral diphosphine–rhodium complex DIPAMP.

Mosher and Morrison have written an excellent article that surveys "The Current Status of Asymmetric Synthesis" (1983). In this article the authors point out that the field has experienced very rapid growth: from 1971 to 1976, *Chemical Abstracts* listed only 47 entries on "Asymmetric Synthesis"; in the period 1976–1981 there were 940 entries! Much of this expansion of interest can be attributed to some basic breakthroughs relating to the chiral phosphine–rhodium catalysts. Also, the system for asymmetric epoxidation of allylic alcohols to the epoxy carbinol has attracted much attention; Sharpless and Masamune with their co-workers have actually synthesized L-hexoses using these elegant reagents and procedures (see Ko et al. (1983), and references therein). Some examples of other reactions for which synthetic catalysts have been used for asymmetric synthesis follow: Michel addition (Cram and Sogar, 1981), borohydride reduction of ketones (Shida et al., 1979), and the hydroformylation of styrene (Stille and Parrinello, 1983).

Of the recent work on asymmetric synthesis the hydrogenation catalysts have attracted the most attention, probably because the work has led to a commerically viable process, namely, the synthetic process for production of L-dopa (Fig. 13.12) (Knowles, 1983). The asymmetry of the catalysts is provided by the chiral phosphine ligand. Two chiral ligands are shown in Fig. 13.13; DIPAMP has chirality at the phosphorus atoms, and DIOP has chiral carbon atoms. The rhodium diphosphines with the arrangement of aromatic rings shown in Fig. 13.13 seem to be the most effective in selective hydrogenation of compounds such as the acylamino acrylate (IV) shown in the reaction below.

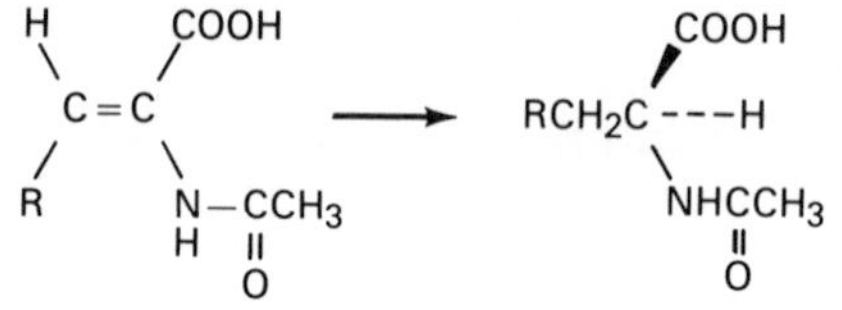

IV

The excess of the L-enantiomer over the D-enantiomer is 95% when rhodium diphosphine complexes are used as hydrogenation catalysts with polar organic solvents. In addition to impressive selectivity, the rates of reac-

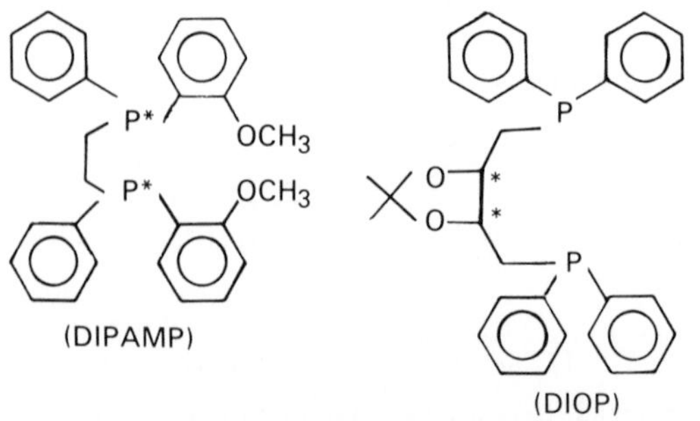

FIGURE 13.13   Chiral diphosphine ligands.

tion catalyzed by compounds of the type shown in Fig. 13.13 can be very fast. Chan and Halpern (1980) have estimated a turnover number of 100 s^{-1} for a diphosphine catalyst for the reaction shown above. To put this in perspective, the well-studied enzyme α-chymotrypsin also has a turnover number of 100 (Table 13.3). The mechanism of reaction has been studied by Halpern and co-workers with important and interesting results (see Halpern, 1982). The four-step process is illustrated in Fig. 13.14.

In the first step the reactant binds to the catalyst with the displacement of two molecules of solvent (methanol). Binding occurs via the carbonyl oxygen of the N-acetyl group and the olefin. In the second step, hydrogen reacts as shown, and the rhodium is oxidized—Rh (I) to Rh (III). The hydrogen migrates as indicated, and the process is completed by reductive elimination of the amino acid, which results in the regeneration of the catalyst with the rhodium as Rh (I). Using NMR spectroscopy, X-ray crystallography, and kinetic analysis, Halpern and co-workers were able to establish a plausible model for enantioselectivity. Since there are two possible binding orientations of the reaction (one is favored by a factor of 11 to 1), selectivity can be expressed in the binding step or the catalytic step in analogy to enzymes.

Generally, the most favored catalyst–reactant complex is expected to react at the fastest rate. However, if the various forms of the catalyst–reactant complex are rapidly interconvertible, it is not required that the most favored complex be the productive complex. In the case of the rhodium–diphosphine

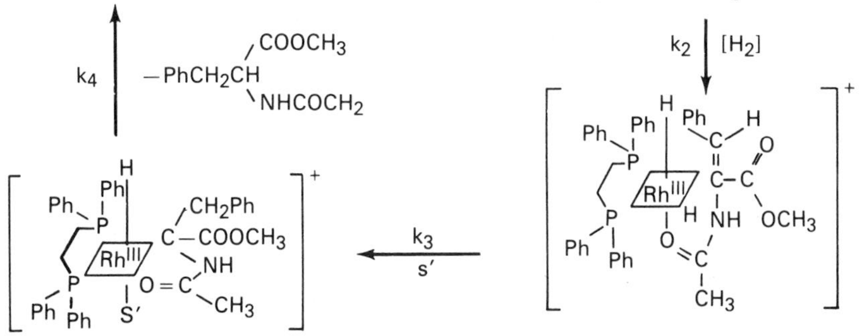

FIGURE 13.14 The mechanism of diphosphine–rhodium catalyzed hydrogenation of olefins. From Maugh (1983). (Reprinted with permission from *Science*, *221*, 352, 1983. Copyright 1983 by AAAS.)

catalysts this latter model appears to be applicable — the minor complex reacts much faster than the major complex.

IV. CONCLUSION

Enzyme engineering has a bright future. The specificity of enzymes is unique. Applications of enzymes that exploit specificity will continue to appear. However, as knowledge of the mechanism of enzyme action continues to advance along with methods in synthetic organic chemistry, enzymes will compete with synthetic catalysts even in instances in which stereochemical selectivity are involved. The progress of the enzymes/synzymes competition should be of great interest in the years to come.

REFERENCES

Bentley, R. (1978) *Nature 276*, 673.
Breslow, R. (1982) *Science 218*, 532.
Breslow, R., and Campbell, P. (1969) *J. Amer. Chem. Soc. 91*, 3085.
Chan, A. S. C., and Halpern, J. (1980) *J. Amer. Chem. Soc. 102*, 838.
Cornforth, J. W. (1976) *Science 191*, 121.
Cornforth, J. (1978) *Proc. Roy. Soc. London 203*(13), 101.
Cram, D. J. (1983) *Science 219*, 1177.
Cram, D. J., and Katz, H. E. (1983) *J. Amer. Chem. Soc. 105*, 135.
Cram, D. J., and Sogar, G. D. Y. (1981) *J.C.S. Chem. Comm.* 635.
Delaney, E. J., Wood, L. E., and Klotz, I. M. (1982) *J. Amer. Chem. Soc. 104*, 799.
Halpern, J. (1982) *Science 217*, 401.
Hassner, A., Krepski, L. R., and Alexanian, V. (1978) *Tetrahedron 34*, 2069.
Hierl, M. A., Gamsen, E. P., and Klotz, I. M. (1979) *J. Amer. Chem. Soc. 101*, 6020.
Kemp, A. S., and Paul, K. (1970) *J. Amer. Chem. Soc. 92*, 2553.
Kemp, A. S., and Paul, K. (1975) *J. Amer. Chem. Soc. 97*, 7305.
Klotz, I. M., Royer, G. P., and Scarpa, I. S. (1971) *Proc. Nat. Acad. Sci. U.S.A. 68*, 263.
Knowles, W. S. (1983) *Acc. Chem. Res. 16*, 106.
Ko, S. Y., Lee, A. W. M., Masamune, S., Reed, L. A., III, Sharpless, K. B., and Walker, F. J. (1983) *Science 222*, 949.
Lehn, J. M., and Sirlin, C. (1978) *J.C.S. Chem. Comm.* 949.
Maugh, T. H. (1983) *Science 221*, 352.
Mosher, H. S., and Morrison, J. D. (1983) *Science 221*, 1013.
Nango, M., Kozuka, H., Kimura, Y., Kuroki, N., Ihara, Y., and Klotz, I. M. (1980) *J. Polymer Sciences: Polymer Letters Ed. 18*, 647.
Ogston, A. G. (1948) *Nature 162*, 963.
Page, M. I. (1980) *Macromol. Chem. 1.* (London) 397.
Royer, G. P. (1980) *Adv. Catal. 29*, 197–224.
Shida, Y., Ando, N., Yamamoto, Y., Oda, J., and Inouye, Y. (1979) *Agr. Biol. Chem. 43*, 1797.
Stille, J. K., and Parrinello, G. (1983) *J. Mol. Catal. 21*, 203.
Trainor, G. L., and Breslow, R. (1981) *J. Amer. Chem. Soc. 103*, 154.